Mathematical Methods in Hydrodynamics and Integrability in Dynamical Systems
(La Jolla Institute 1981)

AIP Conference Proceedings
Series Editor: Hugh C. Wolfe
Number 88

Mathematical Methods in Hydrodynamics
and
Integrability in Dynamical Systems
(La Jolla Institute 1981)

Edited by
Michael Tabor and Yvain M. Treve

American Institute of Physics
New York
1982

Copying fees: The code at the bottom of the first page of each article in this volume gives the fee for each copy of the article made beyond the free copying permitted under the 1978 US Copyright Law. (See also the statement following "Copyright" below). This fee can be paid to the American Institute of Physics through the Copyright Clearance Center, Inc., Box 765, Schenectady, N.Y. 12301.

Copyright © 1982 American Institute of Physics

Individual readers of this volume and non-profit libraries, acting for them, are permitted to make fair use of the material in it, such as copying an article for use in teaching or research. Permission is granted to quote from this volume in scientific work with the customary acknowledgment of the source. To reprint a figure, table or other excerpt requires the consent of one of the original authors and notification to AIP. Republication or systematic or multiple reproduction of any material in this volume is permitted only under license from AIP. Address inquiries to Series Editor, AIP Conference Proceedings, AIP, 335 E. 45th St., New York, N. Y. 10017.

L.C. Catalog Card No. 82-072462
ISBN 0-88318-187-8
DOE CONF- 811225

Dedicated to
**Kenneth M. Watson
on the occasion of
his sixtieth birthday**

DEDICATION

I first met Ken Watson when he was passing through Chicago on his way to the University of Indiana in the fall of 1950. I had just come from MIT, and Ken had just left Berkeley. He was fresh from the triumph of demonstrating (with Brueckner and Serber) that the pi-meson was a pseudoscalar. He and Keith Brueckner had an extraordinarily productive collaboration at Indiana where Ken, among many other achievements, showed the importance of the role of isotopic spin in the interaction of mesons and nucleons.

We saw each other occasionally in that period, but first became friends during the summer of 1955 when Ken organized a small group of smart, but ignorant, particle physicists to work on the problem of controlled fusion at Los Alamos. He was already an expert in the field, and it was in that work that I first saw his unusual combination of incredible physical intuition, mathematical ability, and total lack of fear at attacking problems of enormous complexity. This was the beginning of a period of frequent interaction, both in physics and in issues of national security that we were both becoming involved with.

In the summer of 1959, again at Los Alamos, Ken and I and Keith Brueckner, along with Charlie Townes and Marvin Stern, invented JASON, and in December of 1959 brought it into being.

At some point in the early 1960's, I suggested to Ken that he spend a term at Princeton; and he agreed, on the condition that I would join him in writing a book on scattering theory. I didn't think he was really serious, so I said, "yes." But, the day he arrived at Princeton, he had an outline, and he already had assigned a substantial number of chapters to me with the responsibility of preparing a first draft. This ultimately lead to what we both refer to as our classic monograph, "Collision Theory."

Ken has a tremendous record of accomplishment in theoretical physics, ranging from nuclear physics to quantum field theory to plasma physics to statistical mechanics to his current involvement in nonlinear dynamics. In addition to these broad interests in physics, he has devoted a considerable fraction of his time to working on problems connected to national security. He is a man of absolute integrity and one of the giants in my generation of theoretical physicists.

> Marvin L. Goldberger, President
> California Institute of Technology
> Pasadena, California

PREFACE

A Workshop entitled "Mathematical Methods in Hydrodynamics and Integrability in Related Dynamical Systems" was held on the campus of the Scripps Institution of Oceanography from December 7th - 9th, 1981. It was organized by Drs. M. Tabor and Y. M. Treve under the auspices of the Center for Studies of Nonlinear Dynamics of the La Jolla Institute and held in honor of Professor Kenneth M. Watson. Professor Watson was the founding director of the Center (starting in the summer of 1979) while on leave of absence from the Physics Department at Berkeley and celebrated his sixtieth birthday just a few months prior to the workshop. He is now Director of the Marine Physical Laboratory of the Scripps Institution of Oceanography.

The slightly unwieldy title of this volume was required to embrace the wide ranging yet interrelated topics discussed at the Workshop. Almost half of the invited talks were devoted to Hamiltonian formulations of Hydrodynamics. These ranged from purely mathematical aspects to practical applications in Oceanography and Plasma physics. These topics then led on to a variety of talks on the integrability of partial differential equations, the Inverse Scattering Transform method and the connections between integrability and analytic structure. Also included in this volume are a number of contributed papers from workshop attendees and colleagues, past and present, of Professor Watson. The talks presented by Professors J. W. Miles and F. B. Estabrook are not included in these Proceedings, since they are being published elsewhere.

We would like to thank all the authors for their contributions and assistance with proofreading. We would particularly like to thank Donna McMain for her assistance with the workshop.

The Workshop and these Proceedings were financed by the Office of Naval Research and by independent research funds of the La Jolla Institute.

Michael Tabor
Yvain M. Treve

Center for Studies of Nonlinear Dynamics
La Jolla Institute
La Jolla, California

TABLE OF CONTENTS

GAUGE GROUPS AND POISSON BRACKETS FOR INTERACTING PARTICLES
AND FIELDS
    A. Weinstein 1

POISSON BRACKETS FOR FLUIDS AND PLASMAS
    P. J. Morrison 13

SINGULAR POISSON TENSORS
    R. G. Littlejohn 47

DYNAMICAL SYSTEMS DEFINED ON INFINITE DIMENSIONAL LIE ALGEBRAS OF
THE "CURRENT ALGEBRA" OR "KAC-MOODY" TYPE
    R. Hermann 67

GYROSCOPIC ANALOG FOR MAGNETOHYDRODYNAMICS
    D. D. Holm 73

GAUGE GROUPS AND NOETHER'S THEOREM FOR CONTINUUM MECHANICS
    F. S. Henyey 85

NONCANONICAL HAMILTONIAN MECHANICS
    J. M. Greene 91

PERFECT INVISICID FLUIDS AND GAUGE THEORY
    E. Detyna 99

THE LIE ALGEBRAIC INTERPRETATION OF THE COMPLETE INTEGRABILITY OF
THE ROSOCHATIUS SYSTEM
    T. Ratiu 109

HAMILTONIAN KINETIC THEORY OF PLASMA PONDEROMOTIVE PROCESSES
    S. W. McDonald and A. N. Kaufman 117

HAMILTONIAN STRUCTURE OF MULTI-SPECIES FLUID ELECTRODYNAMICS
    R. G. Spencer 121

HAMILTON'S PRINCIPLE AND ERTEL'S THEOREM
    R. Salmon 127

ADVANTAGES OF HAMILTONIAN FORMULATIONS IN COMPUTER SIMULATIONS
    O. Buneman 137

THE SIMPLE PARTICLE AND THE PERFECT FLUID
    E. A. Spiegel 145

DUAL HAMILTONIAN FORMULATIONS AND COMPLETELY INTEGRABLE SYSTEMS
    K. M. Case 163

ON COMMON ORIGINS OF INTEGRABLE HAMILTONIAN SYSTEMS
    H. Flaschka 189

THE METHOD OF ESTABROOK AND WAHLQUIST
    D. J. Kaup 193

COMMENTS ON INVERSE SCATTERING FOR THE KADOMTSEV-PETVIASHVILI EQUATION
    H. Segur 211

A DIRECT LINEARIZATION ASSOCIATED WITH THE BENJAMIN-ONO EQUATION
    M. J. Ablowitz and A. S. Fokas 229

DIRECT LINEARIZATIONS OF THE KORTEWEG-deVRIES EQUATIONS
    A. S. Fokas and M. J. Ablowitz 237

ANALYTIC STRUCTURE OF THE HENON-HEILES SYSTEM
    J. Weiss 243

LOGARITHMIC SINGULARITIES AND CHAOTIC BEHAVIOR IN HAMILTONIAN SYSTEMS
    T. Bountis and H. Segur 279

INTEGRALS OF THE TEST WAVE HAMILTONIAN: A SPECIAL CASE
    J. D. Meiss 293

ON THE STRUCTURE OF THE HYDRODYNAMICAL EQUATIONS FOR TWO-DIMENSIONAL FLOWS OF AN INCOMPRESSIBLE FLUID: THE ROLE OF INTEGRAL INVARIANCE
    P. D. Thompson 301

A GALERKIN METHOD TO STRONGLY NONLINEAR KdV EQUATIONS AND SCHRÖDINGER EQUATIONS
    F. Chen 319

MULTITIME ASYMPTOTIC METHODS FOR SLOWLY EVOLVING OSCILLATING SYSTEMS
    O. Regev and J. R. Buchler 327

FLUCTUATION-DISSIPATION RELATIONS FOR SYSTEMS WITH INTERNAL MULTIPLICATIVE NOISE
    K. Lindenberg and V. Seshadri 333

ON THE SYMPLECTIC STRUCTURE OF INTEGRABLE SYSTEMS
    A. M. Roos 339

NONOGAUSSIAN WATER-WAVE FIELDS
    B. J. West 347

GAUGE GROUPS AND POISSON BRACKETS FOR INTERACTING PARTICLES AND FIELDS

Alan Weinstein*
Department of Mathematics
University of California, Berkeley, CA 94720

ABSTRACT

A construction by Marsden and Weinstein is described for a Poisson bracket structure on the space of physical variables for the Maxwell-Vlasov equations. The coupling of the plasma and electromagnetic fields, which appears in the Poisson structure rather than the hamiltonian, is produced by the action of the gauge group on the uncoupled phase spaces of matter and fields. Other examples of coupling by the gauge group are given, including an application to the Maxwell-Schrödinger equation.

INTRODUCTION

A hamiltonian formulation of the Maxwell-Vlasov equations for a collisionless plasma in an electromagnetic field was given recently by Morrison[1] in terms of a non-canonical Poisson bracket structure on functionals of the dynamical variables $f(\underline{x},\underline{v})$ (density in phase space), $\underline{B}(\underline{x})$, and $\underline{E}(\underline{x})$. Morrison's structure did not satisfy the Jacobi identity[2], but a corrected version of the structure was found by Marsden and the author[3] which did satisfy that identity. Subsequently, a relativistic version was found by Bialynicki-Birula and Hubbard,[4] who also pointed out several antecedent papers[5,6,7] in which closely related Poisson brackets had been presented.

A feature which seems to be unique to the approach in Ref. 3 is the use of the gauge group to produce the terms in the Poisson structure which give the coupling between the matter and fields. It is this aspect of our work which I wish to emphasize in the present paper.

---

*Research partially supported by NSF grant MCS 80 23356.

POISSON MANIFOLDS

The basic setup is as follows. We begin with a manifold P carrying a Poisson structure $\{\,,\,\}$ on its space $C^\infty(P)$ of smooth real-valued functions.[8,9] A group G, with Lie algebra $\mathfrak{g}$, is given together with a hamiltonian action on P; that is, for each $X \in \mathfrak{g}$ there is a function $\alpha(X) \in C^\infty(P)$ such that the action on P of the 1-parameter group $\exp(tX) \in G$ is given by the evolution equation

$$\frac{dz}{dt} = \{z, \alpha(X)\}.$$

The map $\alpha : \mathfrak{g} \to C^\infty(P)$ is assumed to be a Lie algebra homomorphism. The individual functions $\alpha(X)$ can be assembled into a mapping J from P to the dual space $\mathfrak{g}^*$ of linear functionals on $\mathfrak{g}$, defined by

$$\langle J(z), X \rangle = [\alpha(X)](z)$$

where $\langle \mu, X \rangle$ denotes the value of $\mu \in \mathfrak{g}^*$ on $X \in \mathfrak{g}$. J is called the <u>momentum mapping</u> of the hamiltonian action, since it corresponds to the (vector-valued) linear or angular momentum when G is a group of translations or rotations.

The group G acts on $\mathfrak{g}^*$ by the <u>coadjoint representation</u>, and the momentum mapping is automatically equivariant on the identity component G^0 of G; i.e. $J(a \cdot z) = a \cdot J(z)$ for all $a \in G^0$, where $a\cdot$ denotes the appropriate action of a in each case. (If G is not connected, the equivariance of J is usually taken to be part of the definition of a hamiltonian action.)

The G-invariant functions on P form a subalgebra $C^\infty(P)^G$ of the Lie algebra $C^\infty(P)$, and they may be considered as the smooth functions on the orbit space P/G. If P/G is a manifold, the identification of $C^\infty(P/G)$ with $C^\infty(P)^G$ gives P/G its own Poisson structure. P/G is then called the <u>reduced Poisson manifold</u> of P by the hamiltonian action of G.

For instance, if P is the symplectic manifold T^*G with its usual "canonical" Poisson structure, and G acts by the lifts of left translations,[10] then P/G can be identified with $\mathfrak{g}^*$, and the reduced Poisson structure is given by the formula[11]

$$\{F, G\}(\mu) = \langle \mu, [DF(\mu), DG(\mu)] \rangle.$$

RIGID BODIES

A special case of this construction is $G = SO(3)$, where $\mathfrak{g}$ can be identified with $\mathbb{R}^3$ and the Lie algebra bracket $[\,,\,]$ with the negative of the cross product. Then the Poisson bracket may be written as

$$\{F,G\}(\underline{\mu}) = -\underline{\mu} \cdot (\nabla F(\underline{\mu}) \times \nabla G(\underline{\mu})),$$

and the equation of motion[12] of a system with hamiltonian H is

$$\underline{\dot{\mu}} = -\underline{\mu} \times \nabla H(\underline{\mu}).$$

If H is linear in $\underline{\mu}$, this is the equation of motion of an elementary spin system with spin vector $\underline{\mu}$ (in a constant magnetic field). If H is quadratic in $\underline{\mu}$, we have the Euler equations for the motion of a rigid body (where $\underline{\mu}$ is the angular momentum in body coordinates).

It is perhaps interesting to note that the 3-dimensional Poisson manifold $(\mathbb{R}^3, \times)$ can also be obtained by reduction from a symplectic manifold which is smaller than the 6-dimensional $T^*SO(3)$. We may begin with $\mathbb{C}^2 = \mathbb{R}^4$ with its canonical symplectic structure and the obvious hamiltonian action of the group $U(2)$. The subgroups $SU(2)$ (determinant = 1) and $U(1)$ (multiples of the identity) commute with one another, so the momentum mapping $\mathbb{C}^2 \to \mathfrak{su}(2)^*$ is invariant under the action of $U(1)$, and we can identify $\mathfrak{su}(2)^*$ as a Poisson manifold with $\mathbb{C}^2/U(1)$. But $\mathfrak{su}(2)^*$, $\mathfrak{so}(3)^*$, and $\mathbb{R}^{3*}$ are all isomorphic Poisson manifolds. This realization of the rigid body space by $\mathbb{C}^2$ is related to the Cayley-Klein parameters.[13]

It should be interesting to analyze this example of abelian ($U(1)$) and nonabelian ($SO(3)$) gauge groups for the rigid body in the context of Henyey's report.[14]

POISSON-VLASOV PHASE SPACE

An infinite-dimensional analogue to the rigid body example is given by taking $\mathfrak{g}$ to be the space $C_0^\infty(\mathbb{R}^6)$ of compactly supported smooth functions on the usual phase space of q's and p's, with the Lie algebra structure given by the usual Poisson bracket. (In fact, one could take $C^\infty(P)$ for any Poisson manifold P.) The dual space $\mathfrak{g}^*$ can be identified with the space $\mathcal{D}'(\mathbb{R}^6)$ of distribution densities f on $\mathbb{R}^6$, and the Poisson bracket $\{\!\{\,,\,\}\!\}$ on $\mathcal{D}'(\mathbb{R}^6)$ is

$$\{\{F,G\}\}(f) = \iint d^3q\, d^3p\; f\{\frac{\delta F}{\delta f}, \frac{\delta G}{\delta f}\}$$

where $\frac{\delta}{\delta f}$ is the functional derivative and $\{\,,\,\}$ is the ordinary Poisson bracket in $\underline{q}$ and $\underline{p}$. As Gibbons[15] has already noted, this is the natural phase space for the Poisson-Vlasov equation, in which the hamiltonian is

$$\mathcal{H}(f) = \frac{1}{2}\iint dq\, dp\left(\frac{p^2}{m} + V(\underline{q})\right) f(\underline{x},\underline{p})$$

$$+ \iiiint dq\,dq'\,dp\,dp'\, G(q,q') f(q,p) f(q',p')\; ,$$

$G(q,q')$ being the Green's function for two-particle interactions. This first term in $\mathcal{H}$ is analogous to a spin-system hamiltonian on $\mathbb{R}^3$, while the second term is like that for a rigid body.

COADJOINT ORBITS AND CASIMIR FUNCTIONS

Deriving the Poisson structure on $\mathfrak{g}^*$ from the symplectic structure on T^*G gives useful information which is not so easily seen by looking at $\mathfrak{g}^*$ alone. In general, we define the <u>Casimir functions</u>[16] on a Poisson manifold P to be those functions f for which $\{f,k\} = 0$ for all functions k.

On regions where the tensor defining the Poisson structure has constant rank, the maximal manifolds on which all Casimir functions are constant form a foliation of the Poisson manifold by symplectic manifolds.[8,16] We call these the <u>symplectic leaves</u> of the Poisson manifold; the Poisson bracket of functions on P may be reconstructed by putting together the ordinary Poisson brackets on the symplectic leaves.

In the case of a dual Lie algebra $\mathfrak{g}^*$, a decomposition into symplectic manifolds exists even though the rank of the Poisson structure varies. The Casimir functions on $\mathfrak{g}^*$ are the images of left <u>and right</u> invariant functions on T^*G, so they are just the functions which are invariant under the coadjoint representation. (The name Casimir functions comes from this example, since these functions correspond to bi-invariant operators on the group.) The Casimir functions are obviously constant on the orbits of the coadjoint representation, and those <u>coadjoint orbits</u> are known to be symplectic manifolds[11] whose individual Poisson brackets combine to give the Poisson structure on $\mathfrak{g}^*$. It would be interesting to know if the general (analytic) Poisson manifold has such a symplectic decomposition.

It was already noticed by Arnol'd[12] that the equilibrium states of a rigid body or a fluid are given by points where the hamiltonian on $\mathfrak{g}^*$ has a critical point when restricted to a coadjoint orbit. One may compare the original proof with that in Appendix 2 of Ref. 10 to see the utility of the symplectic and Poisson structures.

In the Poisson-Vlasov case of densities on phase space, each coadjoint orbit consists of all the densities which can be obtained from a given one by applying a canonical transformation. It would be interesting to study the stability of equilibrium solutions of the Poisson-Vlasov equation by using the methods of Refs. 10 and 12.

A class of Casimir functions is given by integrals of the form $\iint dqdp F(f(\underline{q},\underline{p}))$ for functions $F: \mathbb{R} \to \mathbb{R}$ (at least for sufficiently well-behaved densities f). In some sense, these may generate all the Casimir functions at generic points of $\mathfrak{g}^*$. This is roughly equivalent to saying that, for a generic hamiltonian system on $\mathbb{R}^6$, the only smooth constants of the motion are functions of the hamiltonian. The latter statement has never been proven, to my knowledge.

Two special classes of the Casimir functions are the power integrals $C_n(f) = \iint dqdp\, f(\underline{q},\underline{p})^n$ and the "Fourier" integrals $D_\lambda(f) = \iint dqdp\, e^{i\lambda f(\underline{q},\underline{p})}$. By studying the behavior of the latter as $\lambda \to \infty$, one can recover information about the critical points of f (stationary phase method).

To end this topic, we give a description of the Casimir functions on a reduced Poisson manifold P/G. The Casimir functions on P are invariant under any hamiltonian action, in particular that of G, so they pass to Casimir functions on P/G. On the other hand, Casimir functions on $\mathfrak{g}^*$ pulled back by the momentum mapping $J: P \to \mathfrak{g}^*$ are also G invariant. Their projections to P/G are again Casimir functions. Now one can show, at least near "regular" points, that all the Casimir functions on P/G are functions of the ones just described. This follows from the theory of symplectic reduction,[10,17] applied to the symplectic leaves of P and $\mathfrak{g}^*$.

ELECTROMAGNETIC GAUGE GROUP

From now on, the principal group of interest will be the additive "gauge" group G of C^∞ functions $\sigma(\underline{x})$ on $\mathbb{R}^3$ with values in $\mathbb{R}$ (actually in a 1-dimensional vector space carrying the physical dimension of ACTION·VELOCITY/CHARGE = FORCE·DISTANCE^2/CHARGE). The Lie algebra of G may be identified with G itself, and $\mathfrak{h}^*$ consists of densities on $\mathbb{R}^3$ (with values in CHARGE/ACTION·VELOCITY space).

The rest of this paper will be devoted to looking at various hamiltonian actions of G.

MAXWELL EQUATIONS

Let A be the configuration space of vector potentials $\underline{A}$ on $\mathbb{R}^3$ and $T^*A = A \times A^*$ the corresponding phase space. Elements of A^*, denoted by $\underline{Y}$, are vector field densities[18] on $\mathbb{R}^3$. The Poisson structure on A is the canonical one defined by

$$\{F,G\} = \int dx \left(\frac{\delta F}{\delta \underline{A}} \frac{\delta F}{\delta \underline{Y}} - \frac{\delta F}{\delta \underline{Y}} \frac{\delta F}{\delta \underline{A}} \right)$$

The hamiltonian $H(A,Y) = \frac{1}{2} \int dx (\varepsilon_0 \| \text{curl } \underline{A} \|^2 + \frac{c^2}{\varepsilon_0} \| \underline{Y} \|^2)$ leads to Maxwell's evolution equations for $\underline{B} = \text{curl } \underline{A}$ and $\underline{E} = -\frac{c}{\varepsilon_0} \underline{Y}$.

The gauge group acts on T^*A by $\sigma \cdot (\underline{A}, \underline{Y}) = (\underline{A} + \text{grad } \sigma, \underline{Y})$ and has momentum mapping $J(\underline{A}, \underline{Y}) = -\text{div } \underline{Y}$. The reduced Poisson manifold $\text{Max} = T^*A/G$ has coordinates $(\underline{B}, \underline{E})$, where $\underline{B} = \text{curl } \underline{A}$ is restricted by the condition $\text{div } \underline{B} = 0$, and $\underline{E} = -\frac{c}{\varepsilon_0} \underline{Y}$ is free. Since there were no non-constant Casimir functions on T^*A, the Casimir functions on Max are just the functions of $J(\underline{A},\underline{Y}) = -\text{div } \underline{Y}$. (Since G is abelian, all functions on $\mathfrak{h}^*$ are Casimirs.) The typical symplectic leaf in Max is described by an equation $\text{div } \underline{E} = \rho/\varepsilon_0$, where ρ is a given charge density. The gauge invariant[19] states are those with $J(A,Y) = 0$ and so correspond to Maxwell's equations in free space.

The reduced Poisson structure on Max is

$$\{F,G\} = \frac{1}{\varepsilon_0 c} \int dx \left(\frac{\delta F}{\delta \underline{E}} \cdot \text{curl } \frac{\delta G}{\delta \underline{B}} - \frac{\delta G}{\delta \underline{E}} \cdot \text{curl } \frac{\delta F}{\delta \underline{B}} \right)$$

and the hamiltonian for Maxwell's equations is $\frac{\varepsilon_0}{2} \int dx (\|\underline{B}\|^2 + \|\underline{E}\|^2)$.

We learned from Ref. 4 that Maxwell's equations were already written down in just this form by Born and Infeld.[6]

CHARGED PARTICLES

Consider $\mathbb{R}^6_e = T^*\mathbb{R}^3 = \{(q,p)\}$ as the phase space for a particle with charge e. The gauge group acts by $\sigma \cdot (\underline{q},\underline{p}) = (\underline{q}, \underline{p} + \frac{e}{c}(\text{grad }\sigma)(\underline{q}))$ and has momentum map $J(\underline{q},\underline{p}) = -\sum \frac{e}{c} \delta(\underline{x} - \underline{q})$. The reduced Poisson manifold is just the space of $\underline{q}$'s, with all Poisson brackets zero, so it does not carry any hamiltonian systems. Furthermore, if $e \neq 0$, $J^{-1}(0)$ is empty, so there are no gauge-invariant states. These observations are the classical analogue of the frequently noted fact that in quantum mechanics there are no significant gauge-invariant lagrangians until the so-called "gauge fields" are introduced.[20]

The particle picture becomes more interesting if we allow several particles with charges $e_1, \ldots, e_n$. The phase space is now $\mathbb{R}^{6n} = \{\underline{q}_1, \ldots, \underline{q}_n, \underline{p}_1, \ldots, \underline{p}_n\}$, $\sigma \cdot (\underline{q}_i, \underline{p}_i) = (\underline{q}_i, \underline{p}_i + \frac{e_i}{c}(\text{grad }\sigma)(\underline{q}_i))$, and $J(\underline{q},\underline{p}) = -\sum \frac{e_i}{c}\delta(\underline{x} - \underline{q}_i)$. Now $J^{-1}(0)$ consists of all configurations in which the particles are grouped in neutral "atoms". If the particles are just in pairs with opposite charges, the part of the reduced manifold with $J = 0$ is just the phase space for these "hydrogen atoms" considered as indivisible particles. Although there are no interesting gauge invariant hamiltonians on all of $\mathbb{R}^{6n}$, there are such on $J^{-1}(0)$.

When an atom contains three or more particles, the momentum mapping becomes degenerate in such a way that the atoms seem to have some internal structure which might be able to "spin". This phenomenon seems to be worth further study.

PLASMAS

If a Vlasov plasma consists of particles with charge e, the gauge group acts by

$$(\sigma \cdot f)(\underline{q},\underline{p}) = f(\underline{q}, \underline{p} - \frac{e}{c}(\text{grad }\sigma)(\underline{q}))$$

with

$$(Jf)(\underline{x}) = \frac{e}{c} \int dp\, f(\underline{x},\underline{p})$$

If $e \neq 0$ and the f's are non-negative, then $J^{-1}(0)$ is trivial. The case of two species with opposite charges leads to a non-trivial $J^{-1}(0)$, but it is not clear whether there are any reasonable gauge-invariant hamiltonians. They would have to preserve "neutrality" of the plasma.

We also note that the discrete particle phase spaces discussed above can be embedded in the Vlasov phase space as special coadjoint orbits consisting of delta-distribution densities on $\mathbb{R}^6$.

THE MAXWELL-VLASOV SYSTEM

To couple particles with fields, we simply take the product of the various phase spaces and apply the simultaneous action of the gauge group. This construction produces a coupling by linking all the phase spaces to the same physical space $\mathbb{R}^3$.

For instance, for the Maxwell-Vlasov system we start with the space $\mathcal{D}'(\mathbb{R}^6) \times T^*A$ of triples $(f,\underline{A},\underline{Y})$. (For simplicity, we use just one species.) The gauge group action is

$$\sigma \cdot (f,\underline{A},\underline{Y}) = (f(\underline{q},\underline{p} - \frac{e}{c}(\text{grad } \sigma)(\underline{q})), \underline{A} + \text{grad } \sigma, \underline{Y})),$$

and

$$J(f,\underline{A},\underline{Y}) = \frac{1}{c}\left[\varepsilon_0 \text{ div } \underline{E} - e \int dp\, f(\underline{x},\underline{p})\right],$$

where $\underline{E} = -\frac{c}{\varepsilon_0}\underline{Y}$ as before. Now the equation $J(f,\underline{A},\underline{Y}) = 0$ corresponds to the Maxwell equation div $\underline{E} = \rho/\varepsilon_0$, where ρ is the charge density of the plasma. Natural coordinates on the reduced manifold are given by the variables $\underline{E}$, $\underline{B} = \text{curl } \underline{A}$, and

$$f(\underline{q},\underline{p}_{kin}) = f(\underline{q}, \underline{p}_{kin} + \frac{e}{c}\underline{A}(\underline{q})),$$

where $\underline{p}_{kin}$ is the "kinetic momentum" $m\underline{v}$ as opposed to the canonical momentum $\underline{p}$.

The Poisson structure and hamiltonian structures on the reduced manifold, given in Ref. 3, lead to the Maxwell-Vlasov equations for the interacting plasma and field. As we noted in the introduction, this structure is a modification of one given in Ref. 1. A relativistic version is given in Ref. 4.

We could carry out the same procedure for interacting discrete particles and electromagnetic fields, but difficulties arise in the hamiltonian because of the infinite energy in the particle-produced fields.

MAXWELL-SCHRÖDINGER EQUATION

The phase space $L^2(\mathbb{R}^3)$ of complex-valued square integrable wave functions is a symplectic (hence Poisson) manifold with the bracket on functionals $F(\psi)$

$$\{F,G\} = \frac{1}{\hbar} \operatorname{Im} \langle \frac{\delta F}{\delta \psi}, \frac{\delta F}{\delta \psi} \rangle .$$

The real and imaginary parts of the wave function can be taken as canonical variables. If H is a self-adjoint operator, then the hamiltonian system generated by $\mathcal{H}(\psi) = \langle H\psi, \psi \rangle$ is just the Schrödinger equation if $\frac{\partial \psi}{\partial t} = H\psi$. The Poisson bracket on $L^2(\mathbb{R}^3)$ of two such hamiltonians corresponds to the commutator bracket of the operators, so the famouns Dirac[21] quantization problem becomes one of matching classical and quantum Poisson brackets.

The gauge group acts by $\sigma \cdot \psi = \exp\left(\frac{ie\sigma}{\hbar c}\right) \psi$, with $J(\psi) = \frac{e}{c} |\psi|^2$. When this system is coupled with the Maxwell system and reduced by the gauge group, one obtains a hamiltonian formulation of the <u>Maxwell-Schrödinger</u> equations in the variables $(\psi, \underline{B}, \underline{E})$. The current term in the Maxwell equation for $\partial \underline{B}/\partial t$ is proportional to the probability current $\operatorname{Re}\left[\overline{\psi}\left(\frac{\hbar}{i} \underline{\nabla} - \frac{e}{c} \underline{A}\right)\psi\right]$ associated with the wave function ψ. This is no surprise, but it is interesting to see how the probability current comes up automatically from the action of the gauge group.

* OTHER SYSTEMS

The methods described in this paper can also be applied to fluids with or without an electromagnetic field,[22,23] and should be applicable to the coupling of classical particles and Schrödinger fields as in certain approximate formulations of molecular physics, as well as to the relativistic Maxwell-Dirac equations.

POSSIBLE APPLICATIONS

There are many possible applications of the hamiltonian formalism for interacting matter and fields, but few have been carried out so far. Some possibilities are:

___ variational formulations;

___ quantization;

___ perturbation theory by Lie transform and other methods;

___ numerical methods²⁴

___ qualitative studies of periodic orbits, invariant tori, and homoclinic orbits by extensions (yet to be done) of finite dimensional theory;

___ search for conserved quantities and, for special equations on special coadjoint orbits, complete integrability.

REFERENCES

1. P. J. Morrison, Phys. Lett. 80A, 383(1980).

2. A. Weinstein and P. J. Morrison, Phys. Lett. 86A, 235(1981).

3. J. E. Marsden and A. Weinstein, Physica D (to appear).

4. I. Bialynicki-Birula and J. C. Hubbard, Gauge-independent canonical formulation of relativistic plasma theory, preprint (1981).

5. W. Pauli, General Principles of Quantum Mechanics, Springer-Verlag (1981). This is a translation of the 1933 German edition.

6. M. Born and L. Infeld, Proc. Roy. Soc 150A, 141 (1935).

7. I. Bialynicki-Birula and Z. Iwinski, Rep. Math. Phys. 4, 139(1973).

8. A. Lichnerowicz, J. Diff. Geom. 12, 253(1977).

9. V. Guillemin and S. Sternberg, Ann. of Phys. 127, 220(1980).

10. V. I. Arnold, Mathematical Methods of Classical Mechanics, Springer-Verlag (1978).

11. Strictly speaking, $Df(\mu)$ and $Dg(\mu)$ lie in the double dual $\mathfrak{g}^{**}$, but in most cases of interest we can identity this space with $\mathfrak{g}$ itself. The formula for the Poisson bracket on the dual of a Lie algebra appears on p. 100 of F. A. Berezin, Funct. Anal. Appl. 1, 91(1967). The symplectic structure on the coadjoint orbits, from which the Poisson structure can be built, was already discovered by A.A. Kirillov, Russian Math Surveys 17, 4,53(1962); its geometric content was explicated by

B. Kostant, Lect. Notes Math. $\underline{170}$, 87(1970) and J.-M. Souriau, Structure des Systemes Dynamiques, Dunod (1970).

12. V. I. Arnold, Ann. Inst. Fourier $\underline{16}$, 319(1966). Also see reference 10.

13. H. Goldstein, Classical Mechanics, 2nd ed., Addision-Welsey (1980).

14. F. S. Henyey, Gauge groups and Noether's theorem for continuum mechanics, these proceedings.

15. J. Gibbons, Physica $\underline{3D}$, 503(1981).

16. R. Littlejohn, Singular Poisson structures, these proceedings. Casimir functions are called invariants in reference 9.

17. J. Marsden and A. Weinstein, Rep. Math. Phys. $\underline{5}$, 121(1974). Also see reference 10.

18. For units to work out correctly, the components of the various quantities must have the following dimensions:

   σ ACTION·VELOCITY/CHARGE
   A   ENERGY/CHARGE
   Y   CHARGE/VELOCITY·DISTANCE^2
   ε_0 CHARGE^2/ACTION VELOCITY
   c   VELOCITY

   The Poisson bracket of two real-valued functions has the dimensions of 1/ACTION.

19. This identification is motivated by quantum mechanics. For instance, the translation-invariant wave functions of a system of particles in $\mathbb{R}^3$ are those which are annihilated by the total linear momentum operator.

20. See, for instance, C. N. Yang and R. L. Mills, Phys. Rev. $\underline{96}$, 191(1954).

21. P.A.M. Dirac, The Principles of Quantum Mechanics, Oxford: Clarendon Press (1947).

22. P. J. Morrison and J. M. Greene, Phys. Rev. Lett. $\underline{45}$, 790(1980).

23. R. G. Spencer and A. N. Kaufman, Hamiltonian structure of two-fluid plasma dynamics, preprint (1981).

24. O. Buneman, these proceedings.

POISSON BRACKETS FOR FLUIDS AND PLASMAS

Philip J. Morrison

Institute for Fusion Studies
Department of Physics
The University of Texas at Austin
Austin, Texas 78712
U. S. A.

1. INTRODUCTION

1.1 Overview

The traditional method for obtaining a Hamiltonian system is by way of a Lagrangian, that is obtained by physical considerations. The system is then Legendre transformed (if possible) to obtain Hamilton's equations in canonical form, a form that is conveniently representable in terms of the Poisson bracket. Canonical transformations preserve the form of the Poisson bracket; the idea of canonical conjugacy is maintained. An arbitrary coordinate transformation does not preserve the form of the Poisson bracket and consequently the canonical form of Hamilton is obscured. Conjugate variables cannot be discerned and the Poisson bracket may depend explicity on the dynamical variables. In spite of the obscured form, certain algebraic properties of the Poisson bracket are maintained: bilinearity, antisymmetry, and the Jacobi condition (c.f., below). This motivates an alternate definition of Hamiltonian: A system is Hamiltonian if one can find a Poisson bracket, with these algebraic properties, and a Hamiltonian, such that together they generate the time evolution of the system. For the case of even-(nondegenerate) finite-dimensional systems, the theorem of Darboux[1,2] provides an algorithm for locally constructing canonical variables. Also, there exists an extension of Darboux's theorem[3] for the case of infinite dimensional systems. (The situation here is subtle -- gauge conditions may be necessary for a canonical description.)

In this paper we present noncanonical yet Hamiltonian descriptions of many of the non-dissipative field equations that govern fluids and plasmas. The dynamical variables here are the usually encountered physical variables. These descriptions have the advantage that gauge conditions are absent, but at the expense of introducing peculiar Poisson brackets. Clebsch-like potential descriptions that reverse this situation are also introduced.

In the remainder of Sec. 1 the ideas sketched above are considered. The presentation here is admittedly non-rigorous. The reader who is interested in a more rigorous formulation of some of these ideas is directed to Refs. 4 - 11. Section 2 deals with the ideal three-dimensional compressible fluid. The noncanonical Poisson bracket for ideal magnetohydrodynamics[12] is presented. Various fluid descriptions are seen to be represented by portions of this bracket. The plasma equations of Chew, Goldberger and Low[13] are considered. The constants of motion for MHD are discussed and the bracket is shown to generate the infinitesimal transformations of the ten-parameter Galilean group. This section is concluded by presenting a canonical formalism. Various potential decompositions of the fluid velocity and the magnetic field are discussed. Section 3 deals with the Hamiltonization of the equations of two-dimensional vortex fluids and guiding center plasmas.[14] The sole noncanonical dynamical variable in this case is the scalar vorticity. The canonical description is given. Section 4 is concerned with the equations that govern fully nonlinear ion-acoustic waves in plasmas. This is the system from which the Korteweg-de Vries equation is obtained by approximation. Section 5 covers the Maxwell-Vlasov[15-18] equations. The noncanonical Poisson bracket is presented. The way to "canonize" this form[19] is indicated at the end of Sec. 6. The body of Sec. 6 deals with the Vlasov-Poisson equations.[15] It is observed that these equations possess the same noncanonical Poisson bracket as that for two-dimensional vortex fluids.[19] A Clebsch-like potential decomposition is seen to yield a canonical Hamiltonian description.[19]

1.2 Generalized Hamiltonian Field Theory

Consider the following system of autonomous evolution equations:

$$u^i_t(t,\vec{x}) = F^i(\vec{u},\vec{x}) \qquad i = 1,2,\ldots,m \ . \qquad (1.1)$$

Here, each u^i is a function of time t and $\vec{x}$, where $\vec{x} \in V \subset R^n$ for some integer n. The F^i are general nonlinear partial differential or integral operators on $\vec{u}$. Specifically the F^i may be any functions (with a finite number of arguments) of the following:

i) $\vec{u}$ and $\vec{x}$

ii) $u^i_k \equiv \dfrac{\partial^k u^i}{\partial x_1^{k_1} \partial x_2^{k_2} \ldots \partial x_n^{k_n}}$

where the X_i's are the components of $\vec{X}$,

$$k = |\vec{k}| = \sum_{i=1}^{n} k_i$$

and $\vec{k}$ has components k_i which are positive integers.

iii) $\int_V K(\vec{x}|\vec{x}')f(\vec{u}_{\vec{k}})d\tau$ where

f is some function and the kernel K is independent of $\vec{u}$.

We denote this class of operators by $\mathscr{L}$. (I.e., $F^i \in \mathscr{L}$.)

We are not concerned with specific auxiliary conditions necessary for existence and uniqueness of solutions, but suppose solutions do exist and are elements of a vector space ω (over R) that is equipped with the inner product

$$<f|g> = \int_V fg\, d\tau \quad , \tag{1.2}$$

where $d\tau$ is the volume element for $V \subset R^n$.

Customarily in field theory certain integrals or functionals arise. For example, the integral of the Hamiltonian density is that particular functional that generates the evolution. Here the evolution will be generated via generalized Poisson brackets that operate on functionals. To this end we define a vector space Ω (over R) of differentiable functionals that have the form

$$G[\vec{u}] = \int_V G(\vec{u},\vec{x})\, d\tau \tag{1.3}$$

where $G \in \mathscr{L}$ is an operator on ω. We define differentiation of functionals in the usual way.

$$\left.\frac{d}{d\epsilon} F[u^i + \epsilon w]\right|_{\epsilon=0} = \left<\frac{\delta F}{\delta u^i}\bigg| w\right>, \tag{1.4}$$

where the variation is taken with respect to functions w that vanish at the boundary of V. Equation (1.4) defines the functional derivative $\delta F/\delta u^i$, which is in general a nonlinear operator of the class $\mathscr{L}$ that operates on ω.

Before proceeding, consider the following examples of functional differentiation:

i) Suppose $F[u] = \int_0^{2\pi} F(x, u, u_x, u_{xx}, \ldots) dx$

where the function u is defined on $(0, 2\pi)$ and F is C^∞ in all its (finite number) of arguments. By Eq. (1.4) we observe

$$\frac{\delta F}{\delta u} = \frac{\partial F}{\partial u} - \frac{d}{dx}\frac{\partial F}{\partial u_x} + \frac{d^2}{dx^2}\frac{\partial F}{\partial u_{xx}} - \ldots$$

ii) Suppose $F[\vec{u}] = u^i(\vec{x}')$, i.e., the functional composed of functions u^i evaluated at the point $\vec{x}'$. Using the Dirac delta function $\delta(\vec{x})$, we can represent this in the form of Eq. (1.3) as

$$F[\vec{u}] = \int_V u^i(\vec{x}) \delta(\vec{x} - \vec{x}') d\tau ,$$

then from Eq. (1.4) we obtain

$$\frac{\delta u^i(\vec{x}')}{\delta u^j(\vec{x})} = \delta_{ij} \delta(\vec{x} - \vec{x}') ,$$

where

$$\delta_{ij} = \begin{cases} 0 & i \neq j \\ 1 & i = j \end{cases} .$$

Continuing now, we recall that the usual Poisson brackets[20] of field theory have the form

$$[F, G] = \sum_k \int_V \left[\frac{\delta F}{\delta \eta_k}\frac{\delta G}{\delta \pi_k} - \frac{\delta G}{\delta \eta_k}\frac{\delta F}{\delta \pi_k} \right] d\tau \quad (1.5)$$

where the dynamical equations generated by some Hamiltonian, H, are

$$\frac{\partial \eta_k}{\partial t} = [\eta_k, H] \quad , \quad \frac{\partial \pi_k}{\partial t} = [\pi_k, H] \quad .$$

Clearly the operator of Eq. (1.5) is antisymmetric and it is well known that it satisfies the Jacobi condition (cf. below). We generalize this form by defining the following generic bilinear product on Ω :

$$[F,G] = \left\langle \frac{\delta F}{\delta u^i} \middle| O^{ij} \frac{\delta G}{\delta u^j} \right\rangle , \qquad (1.6)$$

where repeated index notation is used and $O^{ij} \in \mathscr{L}$. We desire our form, Eq. (1.6), to possess the same algebraic properties as Eq. (1.5), i.e.,

i) $[F,G] = -[G,F]$ for $F, G \in \Omega$

ii) the Jacobi condition

$$[E,[F,G]] + [F,[G,E]] + [G,[E,F]] = 0$$

for every $E, F, G \in \Omega$.

The first condition requires that the operator O^{ij} be anti-self-adjoint with respect to the inner product on ω. The second condition is more stringent and will be discussed in the next subsection. We note that a bracket of the form of Eq. (1.6), with properties i) and ii), defined on Ω defines a Lie Algebra of functionals. We now define what we mean by Hamiltonian.

Definition. A system of equations of the form (1.1) is Hamiltonian if there exists an operator $O^{ij} \in \mathscr{L}$ and a functional H such that Eq. (1.1) can be cast into the form

$$\frac{\partial u^i}{\partial t} = [u^i, H]$$

where [,] makes Ω a Lie Algebra.

1.3 The Jacobi Condition

We now pinpoint what is required of an anti-self-adjoint O^{ij} in order for the Jacobi condition to be satisfied. Since the Jacobi condition involves nested Poisson brackets, we require the functional derivative of a Poisson bracket. To this end, we first obtain a property of second variation. We conclude this subsection by considering two general classes of O^{ij}: those that are independent of $\vec{u}$ and those that are linear in $\vec{u}$ in a particular way.

Recall Eq. (1.4)

$$\frac{d}{d\epsilon} F\left[u^i + \epsilon w\right]\bigg|_{\epsilon=0} = \left\langle \frac{\delta F}{\delta u^i} \bigg| w \right\rangle \equiv G .$$

G as defined here is clearly an element of Ω. Differentiating again, we obtain

$$\frac{d}{d\eta} G\left[u^j + \eta z\right]\bigg|_{\eta=0} = \left\langle \frac{\delta^2 F}{\delta u^j \delta u^i} z \bigg| w \right\rangle . \qquad (1.7)$$

Equation (1.7) defines the operator $\delta^2 F/\delta u^j \delta u^i \in \mathscr{L}$ that operates on $\vec{u}$ as well as operating linearly on z. Since the order of differentiation is immaterial, we must have the following:

$$\left\langle \frac{\delta^2 F}{\delta u^i \delta u^j} w \bigg| z \right\rangle = \left\langle w \bigg| \frac{\delta^2 F}{\delta u^j \delta u^i} z \right\rangle . \qquad (1.8)$$

Since the Poisson bracket of any two functionals is also a functional, formally we have

$$\frac{d}{d\epsilon} [F,G]\left[u^k + \epsilon w\right]\bigg|_{\epsilon=0} = \left\langle \frac{\delta [F,G]}{\delta u^k} \bigg| w \right\rangle . \qquad (1.9)$$

By Eq. (1.6) we also have

$$\left.\frac{d}{d\epsilon}[F,G]\left[u^k + \epsilon w\right]\right|_{\epsilon=0} = \left.\frac{d}{d\epsilon}\left\langle \frac{\delta F}{\delta u^i} \middle| O^{ij} \frac{\delta G}{\delta u^i}\right\rangle \left[u^k + \epsilon w\right]\right|_{\epsilon=0}$$

$$= \left\langle \frac{\delta^2 F}{\delta u^k \delta u^i} w \middle| O^{ij} \frac{\delta G}{\delta u^j}\right\rangle + \left\langle \frac{\delta F}{\delta u^i} \middle| O^{ij} \frac{\delta^2 G}{\delta u^k \delta u^j} w\right\rangle$$

$$+ \left\langle \frac{\delta F}{\delta u^i} \middle| \frac{\delta O^{ij}}{\delta u^k}(w) \frac{\delta G}{\delta u^j}\right\rangle . \qquad (1.10)$$

The first two terms of Eq. (1.10) come from the $d/d\epsilon$ acting on $\delta F/\delta u^i$ and $\delta G/\delta u^j$ respectively. The last term arises from the dependence of the operator O^{ij} on $\vec{u}$. This term is complicated in that the symbol $\delta O^{ij}(w)/\delta u^k \in \mathscr{L}$ is used for an operator that acts on $\vec{u}$, linearly on w, and also on $\delta G/\delta u_j$ to its right. We require that this term be written as follows:

$$\left\langle K_k^{ij}\left(\frac{\delta F}{\delta u^i}, \frac{\delta G}{\delta u^j}\right) \middle| w \right\rangle \qquad (1.11)$$

where w is now isolated from the operator. (For the case when O^{ij} only involves partial differentiation, this is obtained by integration by parts.) Using Eqs. (1.8), (1.9), (1.10), and (1.11), the anti-self-adjointness of O^{ij}, and the fact that these relations hold for arbitrary w within a wide class, we strip away the integration to obtain

$$\frac{\delta[F,G]}{\delta u^k} = \frac{\delta^2 F}{\delta u^i \delta u^k} O^{ij} \frac{\delta G}{\delta u^j} - \frac{\delta^2 G}{\delta u^j \delta u^k} O^{ij} \frac{\delta F}{\delta u^i}$$

$$+ K_k^{ij}\left(\frac{\delta F}{\delta u^i}, \frac{\delta G}{\delta u^j}\right) . \qquad (1.12)$$

Using Eq. (1.12) in the Jacobi condition yields

$$[E,[F,G]] + \text{cyc} = \left\langle \frac{\delta E}{\delta u^m} \middle| 0^{m\ell} K^{ij}_\ell \left(\frac{\delta F}{\delta u^i}, \frac{\delta G}{\delta u^j} \right) \right\rangle$$

$$+ \text{cyc} . \qquad (1.13)$$

Here cyc means cyclic permutation and we observe that the only surviving terms are those that involve the K^{ij}_ℓ. The terms that involve the second variation cancel by virtue of the anti-self-adjointness of 0^{ij} and Eq. (1.8). The following theorem is apparent.

Theorem 1. If 0^{ij} is independent of $\vec{u}$ (including any operator of class $\mathscr{L}$ on $\vec{u}$), then anti-self-adjointness is sufficient for the Jacobi condition.

Now we consider a special case where 0^{ij} depends linearly on $\vec{u}$. We suppose 0^{ij} has the manifestly anti-self-adjoint form

$$0^{ij} = \sum_{k,r} \left[a^{ij,k}_r u_k \partial_r + a^{ji,k}_r \partial_r u_k \right] \qquad (1.14)$$

where $k = 1,2,\ldots,m$; $r = 1,2,\ldots,n$; $\partial_r = \partial/\partial x_r$; and $a^{ij,k}_r \in R$ for all $i, j, k,$ and r. With this form, the quantity $\delta 0^{ij}(w)/\delta u^k$ of Eq. (1.10) is seen to be

$$\frac{\delta 0^{ij}}{\delta u^k}(w) = \sum_r \left(a^{ij,k}_r w \partial_r + a^{ji,k}_r \partial_r w \right) .$$

From this we obtain the quantity K_k^{ij} by integration by parts

$$K_k^{ij}\left(\frac{\delta F}{\delta u^i}, \frac{\delta G}{\delta u^j}\right) =$$

$$\sum_r \left(\frac{\delta F}{\delta u^i} a_r^{ij,k} \partial_r \frac{\delta G}{\delta u^j} - \frac{\delta G}{\delta u^j} a_r^{ji,k} \partial_r \frac{\delta F}{\delta u^i}\right).$$

(1.15)

Inserting Eq. (1.15) into Eq. (1.13) yields a complicated expression that vanishes if the coefficients $a_r^{ij,k}$ satisfy certain properties.

<u>Theorem 2.</u> Poisson brackets made up of operators O^{ij} of the form of Eq. (1.14) satisfy the Jacobi condition if

i) $\quad \sum_k \left(a_r^{\ell k,m} a_t^{ij,k} - a_r^{ik,m} a_t^{\ell j,k}\right) = 0$

and

ii) $\quad \sum_k \left(a_r^{\ell k,m} a_t^{ij,k} + a_r^{ki,m} a_t^{\ell j,k} - a_t^{\ell k,m} a_r^{ji,k}\right.$

$\left. - a_t^{kj,m} a_r^{\ell i,k}\right) = 0$

for all $r, t, \ell, m, i,$ and j.

We conclude this section by noting that many of the Poisson brackets presented in this paper are of the forms of Theorems 1 and 2. The Jacobi condition for the others can similarly be established by the procedures discussed here.

2. The Ideal Fluid and Magnetohydrodynamics (Double Adiabatic Equations)

The equations of ideal magnetohydrodynamics are

$$\vec{v}_t = -\nabla\left(\frac{v^2}{2}\right) + \vec{v} \times (\nabla \times \vec{v}) - \rho^{-1}\nabla(\rho^2 U_\rho) - \rho^{-1}\nabla \cdot \overleftrightarrow{T}_B \tag{2.1}$$

$$\rho_t = -\nabla \cdot (\rho\vec{v}) \tag{2.2}$$

$$s_t = -\vec{v} \cdot \nabla s \tag{2.3}$$

$$\vec{B}_t = -\vec{B}\nabla \cdot \vec{v} + \vec{B} \cdot \nabla\vec{v} - \vec{v} \cdot \nabla\vec{B} . \tag{2.4}$$

The variables of Eqs. (2.1) - (2.4), ρ, $\vec{v}$, $\vec{B}$ and s, are functions of three spatial coordinates and one time coordinate. Equation (2.1) is the equation of motion for a fluid with density ρ and velocity $\vec{v}$. The magnetic body force term is represented in terms of a symmetric stress tensor $\overleftrightarrow{T}_B = (B^2/2)\overleftrightarrow{I} - \vec{B}\vec{B}$ where $\vec{B}$ is the magnetic field. The symmetry of $\overleftrightarrow{T}_B$ precludes the existence of internal torque densities; the equation obtains a symmetry without the use of the initial condition $\nabla \cdot \vec{B} = 0$. Also in Eq. (2.1) the internal energy per unit mass, $U(\rho,s)$, is a prescribed function of and the entropy per unit mass, s. The intensive variable, pressure p and temperature T, are obtained from this function $p = \rho^2 U_\rho$, $T = U_s$. Equation (2.2) is mass conservation and Eq. (2.3) expresses entropy advection. Equation (2.4) is Faraday's law assuming $\vec{E} + \vec{v} \times \vec{B} = 0$. It is written in a form which is manifestly Galilean invariant. Below we obtain the Poisson brackets for specific subsets of Eqs. (2.1) - (2.4). The equations of Chew, Goldberger and Low are also expressed in Poisson bracket form.

2.1 Noncanonical Poisson Brackets

The MHD equations are known to possess several conservation laws. In addition to ρ, the momentum density $\rho\vec{v}$ and the energy density $\frac{1}{2}\rho v^2 + \rho U + (B^2/2)$ are densities of conservation laws. The symmetry of $\overleftrightarrow{T}_B$ assures that the angular momentum density $(\vec{x} \times \vec{v})\rho$, is conserved and also one can show that $\rho(\vec{x} - \vec{v}t)$ is conserved.[21] Similarly, the entropy per unit volume $\sigma \equiv s\rho$ is conserved (more generally) $\rho f(s)$ for arbitrary function f). Also, $\vec{B}$, $\vec{A}\cdot\vec{B}$ (where $\vec{B} = \nabla \times \vec{A}$) and $\vec{v}\cdot\vec{B}$ are conserved densities if $\nabla\cdot\vec{B} = 0$. Below we will discuss these constants in the context of our Poisson structure.

The natural choice for the Hamiltonian is the energy functional

$$H = \int_V \left(\frac{1}{2}\rho v^2 + \rho U(\rho,s) + \frac{B^2}{2} \right) d\tau . \qquad (2.5)$$

With this as Hamiltonian, the following Poisson bracket[12] produces the Eqs. (2.1) - (2.4):

$[F,G] =$ \hfill (2.6)

$$-\int_V \left\{ \left[\frac{\delta F}{\delta \rho} \nabla \cdot \frac{\delta G}{\delta \vec{v}} + \frac{\delta F}{\delta \vec{v}} \cdot \nabla \frac{\delta G}{\delta \rho} \right] \right. \qquad (2.6.1)$$

$$+ \left[\rho^{-1}(\nabla \times \vec{v}) \cdot \left(\frac{\delta G}{\delta \vec{v}} \times \frac{\delta F}{\delta \vec{v}} \right) \right] \qquad (2.6.2)$$

$$+ \left[\rho^{-1}\nabla s \cdot \left(\frac{\delta F}{\delta s}\frac{\delta G}{\delta \vec{v}} - \frac{\delta G}{\delta s}\frac{\delta F}{\delta \vec{v}} \right) \right] \qquad (2.6.3)$$

$$+ \vec{B}\cdot\left[\frac{1}{\rho}\frac{\delta F}{\delta \vec{v}} \cdot \nabla \frac{\delta G}{\delta \vec{B}} - \frac{1}{\rho}\frac{\delta G}{\delta \vec{v}} \cdot \nabla \frac{\delta F}{\delta \vec{B}} \right]$$

$$\left. + \vec{B}\cdot\left[\left(\nabla\frac{1}{\rho}\frac{\delta F}{\delta \vec{v}}\right)\cdot\frac{\delta G}{\delta \vec{B}} - \left(\nabla\frac{1}{\rho}\frac{\delta G}{\delta \vec{v}}\right)\cdot\frac{\delta F}{\delta \vec{B}} \right] \right\} d\tau$$
$$(2.6.4)$$

The first term, Eq. (2.6.1), is a natural extension to higher dimension of the K-dV bracket obtained by Gardner.[4] Considered as a binary operation of functionals of ρ and $\vec{v}$, Eq. (2.6.1) satisfies the Jacobi condition. If Eq. (2.5) with $\vec{B}$ set to zero and the s dependence of U suppressed, is used as Hamiltonian, one obtains the ideal fluid equation of motion with $\nabla \times \vec{v} = 0$, and the continuity equation (2.2). The inclusion of Eq. (2.6.2) with the same Hamiltonian produces Eq. (2.1) with the $\nabla\times\vec{v}$ term. The

The sum of these terms satisfies the Jacobi condition. If Eq. (2.6.3) is added to the previous two, then the resulting bracket considered as an operator on functionals of ρ, $\vec{v}$, and s can be shown to satisfy the Jacobi condition. If Eq. (2.5), with the s dependence of U included and $\vec{B} = 0$, is used as Hamiltonian then Eqs. (2.1) (with $\vec{B} = 0$), (2.2) and (2.3) are produced. The remaining term, Eq. (2.6.4), accounts for the magnetic field. The last two terms here are doubly contracted dyads, i.e.,

$$B \cdot \left(\nabla \frac{1}{\rho} \frac{\delta F}{\delta \vec{v}}\right) \cdot \frac{\delta G}{\delta \vec{B}} = \left(\nabla \frac{1}{\rho} \frac{\delta F}{\delta \vec{v}}\right) : \left(\frac{\delta G}{\delta \vec{B}} \vec{B}\right)$$

$$= \sum_{i,j=1}^{3} B_i \frac{\partial G}{\partial B_j} \partial_i \left(\frac{1}{\rho} \frac{\partial F}{\partial v_j}\right)$$

If one considers a bracket composed of this term added to Eqs. (2.6.1) and (2.6.2), then Eqs. (2.1), (2.2) and (2.4) are produced with $H = (\rho v^2/2 + \rho U(\rho) + B^2/2) d\tau$. It can be shown that this satisfies the Jacobi requirements. (We note that the Jacobi condition in no way depends upon the initial condition $\nabla \cdot \vec{B} = 0$). Finally, the entire bracket, Eq. (2.6) satisfies the Jacobi requirement. If Eq. (2.5) is used as Hamiltonian then as noted Eqs. (2.1) - (2.4) are obtained. We summarize the above paragraph in Table (1).

Let us now return to the constants. We divide them into three groups, the first we call generators

$$H = \int_V \left(\frac{1}{2} \rho v^2 + \rho U(\rho,s) + \frac{B^2}{2}\right) d\tau \qquad (2.7)$$

$$\vec{P} = \int_V \rho \vec{v} \, d\tau \qquad (2.8)$$

$$\vec{l} = \int_V \vec{x} \times \rho \vec{v} \, d\tau \qquad (2.9)$$

$$\vec{G} = \int_V \rho (\vec{x} - \vec{v}t) d\tau \qquad (2.10)$$

These constants together with the Poisson bracket defined by Eq. (2.6) generate the infinitesimal transformations of the ten parameter Galilean group. H, of course, generates time translation, while $\vec{P}$ and $\vec{\mathscr{L}}$ generate infinitesimal changes due to space translations and rotations respectively. For example, using the kth component of $\vec{P}$ we obtain

$$[\rho, P_k] = -\partial_k \rho$$

$$[v_\ell, P_k] = -\partial_k v_\ell$$

$$[s, P_k] = -\partial_k s$$

$$[B_\ell, P_k] = -\partial_k B_\ell \quad .$$

The remaining constant, $\vec{G}$, physically corresponds to uniform motion of the center of mass of the fluid, i.e., $\vec{x}_{cm} = \vec{v}_{cm} t + \text{const}$. It can be interpreted as an embodyment of Newton's third law; all internal forces occur in action-reaction pairs. The only forces that can be imparted to the center of mass occur through surface terms that here are assumed to vanish. This constant generates changes due to Galilean transformation. We obtain

$$[\rho, G_k] = t\partial_k \rho$$

$$[v_\ell, G_k] = t\partial_k v_\ell - \delta_{\ell k}$$

$$[s, G_k] = t\partial_k s$$

$$[B_\ell, G_k] = t\partial_k B_\ell \quad .$$

The Kronecker delta term of $[v_\ell, G_k]$ allows for the boost in velocity.

The second group of constants commute with any functional of the dynamical variables. That is, for

$$M = \int_V \rho \, d\tau \qquad (2.11)$$

$$S = \int_V \rho s \, d\tau \qquad (2.12)$$

we have

$$[\chi, M] = [\chi, S] = 0$$

for arbitrary χ.

The third group of constants is composed of the magnetic constants

$$\vec{\mathcal{B}} = \int_V \vec{B} \, d\tau \qquad (2.13)$$

$$\mathcal{T} = \int_V \vec{A} \cdot \vec{B} \qquad (2.14)$$

$$\mathcal{W} = \int_V \vec{v} \cdot \vec{B} \, d\tau \qquad (2.15)$$

These functionals commute with the Hamiltonian [Eq. (2.7)] only for the initial condition $\nabla \cdot \vec{B} = 0$. The constant $\mathcal{W}$ also requires constant entropy per unit mass.

The double adiabatic equations of Chew, Goldberger and Low can also be produced from the bracket Eq. (2.6). These equations account for the presence of a strong magnetic field through an anisotropic pressure tensor. The pressure parallel to the direction of the magnetic field $p_\parallel$ differs from that perpendicular, $p_\perp$.

If the internal energy depends on B, the magnitude of the magnetic field, in addition to ρ and s, then our bracket produces the double adiabatic equations, if we make the following identifications:

$$p_\parallel = \rho^2 \frac{\partial U}{\partial \rho}$$

and

$$p_\perp = \rho^2 \frac{\partial U}{\partial \rho} + \rho B \frac{\partial U}{\partial B}.$$

To conclude this subsection we present an alternate, more symmetric form of the bracket defined in Eq. (2.6). If we transform to the set of dynamical variables $\{\rho,\sigma,\vec{M},\vec{B}\}$, where $\sigma = \rho s$ is the entropy per unit volume and $\vec{M} = \rho\vec{v}$ is the momentum density, then Eqs. (2.1) - (2.4) become eight conservation equations (if one adjoins $\nabla\cdot B = 0$). The pressure is now determined by $p = \rho^2(\widetilde{U}_\rho + \sigma\rho^{-1}\widetilde{U}_\sigma)$ where $\widetilde{U}(\rho,\sigma) = U(\rho,s)$. As a result of the transformation

$$\left.\frac{\delta}{\delta\rho}\right|_{\vec{v},s} = \left.\frac{\delta}{\delta\rho}\right|_{\vec{M},\sigma} + \rho^{-1}\vec{M}\cdot\frac{\delta}{\delta\vec{M}} + \sigma\rho^{-1}\frac{\delta}{\delta\sigma} ,$$

together with similar transformations for the other variables, Eq. (2.6) becomes

$$[F,G] = -\int_V \left\{\rho\left[\frac{\delta F}{\delta\vec{M}}\cdot\nabla\frac{\delta G}{\delta\rho} - \frac{\delta G}{\delta\vec{M}}\cdot\nabla\frac{\delta F}{\delta\rho}\right]\right.$$

$$+ \vec{M}\cdot\left[\frac{\delta F}{\delta\vec{M}}\cdot\nabla\frac{\delta G}{\delta\vec{M}} - \frac{\delta G}{\delta\vec{M}}\cdot\nabla\frac{\delta F}{\delta\vec{M}}\right]$$

$$+ \sigma\left[\frac{\delta F}{\delta\vec{M}}\cdot\nabla\frac{\delta G}{\delta\sigma} - \frac{\delta G}{\delta\vec{M}}\cdot\nabla\frac{\delta F}{\delta\sigma}\right] + \vec{B}\cdot\left[\frac{\delta F}{\delta\vec{M}}\cdot\nabla\frac{\delta G}{\delta\vec{B}} - \frac{\delta G}{\delta\vec{M}}\cdot\nabla\frac{\delta F}{\delta\vec{B}}\right]$$

$$+ \vec{B}\cdot\left[\left(\nabla\frac{\delta F}{\delta\vec{M}}\right)\cdot\frac{\delta G}{\delta\vec{B}} - \left(\nabla\frac{\delta G}{\delta\vec{M}}\right)\cdot\frac{\delta F}{\delta\vec{B}}\right]\right\} d\tau . \qquad (2.16)$$

Notice that each term of Eq. (2.16) is linear in one Eulerian variable and there are no terms, like those of Eq. (2.6), with the density ρ in the denominator. This feature facilitates evaluating the bracket when polynomial or Fourier representations are used for the dynamical variables. Also we observe that Eq. (2.16) is of the form discussed in Theorem 2 of 1.3.

2.2 Potential Representations

The use of potentials to represent vector fields has a history that transcends the familiar potential decomposition of electricity and magnetism. In this subsection, we discuss potential representations that pertain to our Poisson bracket [Eq. (2.6)]. (We note that the historical account presented here should not be taken as complete. Such a task is hampered by a great deal of rediscovery in this area. The interested reader is directed to Refs. 22 - 29). In particular, our main goal is to represent the fluid velocity field in a form that facilitates a canonical Hamiltonian description, and to show how this form transforms to Eq. (2.6). Various forms of potential representations "canonize" the subsets of the MHD equations discussed in 2.1. The magnetic field, of course, can also be subjected to potential decomposition. We conclude this subsection with a highly symmetric description where this decomposition, in addition to that for the velocity field, is done.

Euler (1769) in his investigation of fluids, represented the solenoidal vector field $\vec{v}$, where $\nabla \cdot \vec{v} = 0$, in the form

$$\vec{v} = \nabla F \times \nabla G \qquad (2.17)$$

This decomposition in terms of the "Euler potentials" F and G can be shown to be locally general. This contravariant representation manifestly assures $\nabla \cdot \vec{v} = 0$. Locally, F and G must define independent surfaces. The intersection of these surfaces defines flow lines. (In plasma physics it is common, as we do below, to represent the magnetic field in this form; the intersection of these surfaces in this case defines field lines.) This representation is clearly not unique, since any function of G may be added to F (and vice versa) without changing $\vec{v}$. More generally any two functions $\alpha(F,G)$ and $\beta(F,G)$ can replace F and G provided the Jacobian $\partial(\alpha,\beta)/\partial(F,G) = 1$. [Note, one can add the gradient of an arbitrary harmonic function, ϕ, to Eq. (2.17) without destroying the solenoidal property. In the case where $\nabla \cdot \vec{v} \neq 0$ and ϕ is not harmonic, we have a form, in the same vein as the Helmholtz representation, which was presented by Monge (1784).]

We now present (as a stepping stone) a representation due to Clebsch (1859), which yields a variational description of the incompressible Eulerian fluid equations. If

$$\vec{v} = \alpha \nabla \beta + \nabla \phi \, , \qquad (2.18)$$

where ϕ is chosen such that $\nabla \cdot \vec{v} = 0$, then Euler's equations can be represented in Hamiltonian form. The potential α is seen to be canonically conjugate, in the usual sense, to the potential β.

A generalization of Eq. (2.18) that includes density variation is the following;

$$\rho \vec{v} = \lambda \nabla \mu + \rho \nabla \phi \quad . \tag{2.19}$$

This decomposition allows (at the expense of obtaining gauge conditions) a Hamiltonian description for a compressible fluid. The density ρ is seen to be conjugate to the potential ϕ and similarly, λ and μ are canonically conjugate. The Poisson bracket in terms of these potentials is

$$[F, G] = \int \left[\left(\frac{\delta F}{\delta \rho} \frac{\delta G}{\delta \phi} - \frac{\delta G}{\delta \rho} \frac{\delta F}{\delta \phi} \right) + \left(\frac{\delta F}{\delta \lambda} \frac{\delta G}{\delta \mu} - \frac{\delta G}{\delta \lambda} \frac{\delta F}{\delta \mu} \right) \right] d\tau , \tag{2.20}$$

where F and G are functionals of ρ, ϕ, λ and μ. If the Hamiltonian $H = \int [\tfrac{1}{2}\rho v^2 + \rho U(\rho)] d\tau$ is represented in terms of these variables by making use of Eq. (2.19), then the equations of motion are obtained in the usual manner (e.g., $\phi_t = [\phi, H]$). Now suppose

$$F[\rho, \phi, \lambda, \mu] = \tilde{F}[\rho, \vec{v}] ,$$

then the chain rule for functional differentiation yields

$$\frac{\delta F}{\delta \phi} = -\nabla \cdot \frac{\delta \tilde{F}}{\delta \vec{v}} , \qquad \frac{\delta F}{\delta \rho} = \frac{\delta \tilde{F}}{\delta \rho} - \frac{\lambda}{\rho^2} \nabla \mu \cdot \frac{\delta \tilde{F}}{\delta \vec{v}} \tag{2.21}$$

and similar expressions for λ and μ. Substitution of these expressions into Eq. (2.20) yields a portion of our Poisson bracket, Eq. (2.6.1) plus Eq. (2.6.2). [Note by Eq. (2.21), exclusion of λ and μ yields the irrotational portion of the bracket Eq. (2.6.1).]

Similarly, entropy advection is alloted for by the inclusion of an additional potential. Consider the following covariant form:

$$\rho \vec{v} = \lambda \nabla \mu + \rho \nabla \phi + \sigma \nabla \psi \quad . \tag{2.22}$$

Here ψ, the additional potential, is canonically conjugate to σ the entropy per unit volume. As above, the chain rule for Eq. (2.22) yields the Poisson bracket that is the sum of Eqs. (2.6.1), (2.6.2), and (2.6.3).

Consider now a form that includes the magnetic field

$$\rho\vec{v} = \vec{B} \times (\nabla \times \vec{T}) + \nabla\phi . \tag{2.23}$$

Zakharov and Kuznetsov[29] (1971) presented a Hamiltonian description for MHD (with constant entropy/mass), where the vector potential $\vec{T}$ of Eq. (2.23) is seen to be conjugate to $\vec{B}$ in addition to maintaining the ρ,ϕ conjugacy. We emphasize that this form cannot be transformed into our bracket. The appropriate form, which respects the distinction between the initial condition $\nabla \cdot \vec{B} = 0$ and the dynamical symmetries of invariance under Galilean transformation and rotation, is

$$\rho\vec{v} = (\nabla\vec{T}) \cdot \vec{B} - \vec{B} \cdot \nabla\vec{T} - \vec{T}\nabla \cdot \vec{B} + \rho\nabla\phi . \tag{2.24}$$

The following Poisson bracket:

$$[F,G] = \int_V \left\{ \left(\frac{\delta G}{\delta\phi} \frac{\delta F}{\delta\rho} - \frac{\delta F}{\delta\phi} \frac{\delta G}{\delta\rho} \right) + \left(\frac{\delta F}{\delta\vec{B}} \cdot \frac{\delta G}{\delta\vec{T}} - \frac{\delta G}{\delta\vec{B}} \cdot \frac{\delta F}{\delta\vec{T}} \right) \right\} d\tau$$

yields with Eq. (2.24) and the chain rule, the Poisson bracket Eq. (2.6) with the exception of the entropy term [Eq. (2.6.3)]. The entire bracket is obtained by adding $\sigma\nabla\psi$ to Eq. (2.24) and considering the canonical structure that includes σ conjugate to ψ.

To conclude this subsection, we present a formulation that entails a decomposition of $\vec{B}$ as well as $\vec{v}$. If we expand $\vec{B}$ in terms of Euler potentials as in Eq. (2.17)

$$\vec{B} = \nabla\alpha \times \nabla\beta ,$$

then the appropriate expression for $\vec{v}$ is

$$\rho\vec{v} = a\nabla\alpha + b\nabla\beta + \rho\nabla\phi . \tag{2.25}$$

In this representation the advected field labels α and β are seen to be conjugate to the potentials a and b. The initial condition $\nabla \cdot \vec{B} = 0$ is now inherent to the dynamics. The connection to the formulation of Eq. (2.23) is easily seen to be made through the following:

$$a = -\nabla\beta \cdot (\nabla \times \vec{T}) , \quad b = \nabla\alpha \cdot (\nabla \times \vec{T}) .$$

We note that the entire canonical formuation is obtained by appending $\sigma\nabla\psi$ to Eq. (2.25). These results are summarized in Table 2.

3. Two-Dimensional Vortex Fluids and Guiding Center Plasmas

The equations for vortex advection in two spatial dimensions are used to model large scale motions that occur in atmospheres and oceans. The same equations have arisen in the study of plasma transport perpendicular to a uniform magnetic field. If we assume the usual euclidean coordinate system with uniformity in the $\hat{z}$ direction then the scalar vorticity is $\omega(\vec{x},t) = \hat{z}\cdot\nabla\times\vec{v}(\vec{x},t)$, where $\vec{v}$ is the fluid velocity such that $\vec{v}\cdot\hat{z} = 0$. For the guiding center plasma, ω corresponds to the charge density and $\vec{v}$ to the $\vec{E}\times\vec{B}$ drift velocity. The equations under consideration are the following:

$$\omega_t = -\vec{v}\cdot\nabla\omega \qquad (3.1)$$

$$\nabla\cdot\vec{v} = 0 \quad . \qquad (3.2)$$

For an unbounded fluid $\vec{v}$ can be eliminated from Eq. (3.1) by

$$\vec{v} = \int \omega(\vec{x}') \, \vec{M}(\vec{x}|\vec{x}') \, d\tau' \quad , \qquad (3.3)$$

where we display only the arguments necessary to avoid confusion. Here $\vec{M} = \hat{z}\times\nabla k(\vec{x}|\vec{x}')$ and $k(\vec{x}|\vec{x}')$ is the Green function for Laplace's equation in two dimensions.

$$k(\vec{x}|\vec{x}') = \frac{1}{4\pi} \ln[(x-x')^2 + (y-y')^2] \quad .$$

The integration in Eq. (3.3) is over the entire x-y plane; $d\tau \equiv dxdy$. Observe Eq. (3.2) is manifestly satisfied by Eq. (3.3). Eq. (3.1) becomes

$$\omega_t = -\int \omega(\vec{x}') \, \vec{M}(\vec{x}|\vec{x}') d\tau' \cdot \nabla\omega(\vec{x}) \quad . \qquad (3.4)$$

Equations (3.1) and (3.2) are known to possess conserved densities, e.g. any function of ω is conserved. In addition, the kinetic energy, which is the natural choice for the Hamiltonian, is conserved. With the density set to unity we have

$$H[\omega] = \int \frac{v^2}{2} d\tau = -\frac{1}{2} \int k(\vec{x}|\vec{x}')\, \omega(\vec{x}')\, \omega(\vec{x})\, d\tau\, d\tau'. \tag{3.5}$$

The functional derivative of Eq. (3.5) is

$$\frac{\delta H}{\delta \omega} = -\int k(\vec{x}|\vec{x}')\, \omega(\vec{x}')\, d\tau' \ .$$

The Poisson bracket[14] that produces Eq. (3.4) is the following:

$$[F,G] = \int \omega(\vec{x}) \left\{ \frac{\delta F}{\delta \omega}, \frac{\delta G}{\delta \omega} \right\} d\tau \tag{3.6}$$

where $\{f,g\} = (\partial F/\partial x)(\partial g/\partial y) - (\partial f/\partial y)(\partial g/\partial x)$. We note that the bracket defined by Eq. (3.6) is precisely that for the one-dimensional Vlasov-Poisson equations[15,19] (see Sec. 6) if one replaces the vorticity by the phase space density and the phase space (x,y) by (x,v). Also observe that any two functionals composed of functions of ω alone are in involution with respect to Eq. (3.6).

We conclude this section by transforming Eq. (3.6) to canonical form. The discussion of potentials in 2.2 indicates the following representation of the vorticity

$$\omega = \{\alpha,\beta\} \ . \tag{3.7}$$

The chain rule for functional differentiation yields

$$\frac{\delta F}{\delta \alpha} = \left\{\beta, \frac{\delta F}{\delta \omega}\right\}, \quad \frac{\delta F}{\delta \beta} = \left\{\frac{\delta F}{\delta \omega}, \alpha\right\} \tag{3.8}$$

where on the left F is now regarded as a functional of α and β. The canonical Poisson bracket for α and β is

$$[F,G] = \int \left(\frac{\delta F}{\delta \alpha}\frac{\delta G}{\delta \beta} - \frac{\delta F}{\delta \beta}\frac{\delta G}{\delta \alpha}\right) d\tau \tag{3.9}$$

which upon substitution of Eqs. (3.8) yields the bracket Eq. (3.6). (This is easily accomplished by making use of the relation $\int f\{g,h\} \, d\tau = \int g\{h,f\} \, d\tau$ and the Jacobi requirement).

4. Fully Nonlinear Ion-Acoustic Waves

In this seciton we present the Poisson bracket for a particular approximation of the two-fluid equations of plasma physics that models nonlinear ion-acoustic waves.[30] In the limit that the electron temperature greatly exceeds the ion temperature, the ion dynamics are governed by the cold fluid momentum transport and continuity equations,

$$v_t = -vv_x - \phi_x \qquad (4.1)$$

$$N_t = -(Nv)_x \quad . \qquad (4.2)$$

Here, v is the ion fluid velocity, normalized to the ion sound speed $c_s = \sqrt{T_e/m_i}$ where T_e is the electron temperature and m_i the ion mass, N is the ion density that is normalized to n_0 the quasi-neutral electron or ion density, x and t are expressed in units of the electron Debye length $\lambda_D = \sqrt{T_e/4\pi n_0 e^2}$ and ion plasma frequency $\omega_{pi} = \sqrt{4\pi n_0 e^2/m_i}$ respectively. The electrostatic potential ϕ couples the ion dynamics to the electrons through Poisson's equation

$$\phi_{xx} = n(\phi) - N \quad . \qquad (4.3)$$

Here, ϕ is normalized to e/T_e and the electron density, $n(\phi)$, is assumed to be a function of ϕ. Typically, since the electron mass is greatly exceeded by the ion mass, electron inertial terms are neglected and the approximation of isothermal electrons is justifiable. In this case $n(\phi) = e^\phi$. The structure that we present makes no restrictions on n except that it be a function of ϕ.

In the case $n(\phi) = e^\phi$, since $\phi_x = n_x/n$, it is customary to envision the electrons as supplying the ion pressure. Alternatively with $n(\phi)$ specified the constraint equation (4.3) can be interpreted as supplying a non-local equation of state for the ion pressure. It is through this non-local equation of state that dispersion is introduced into the dynamics. It is well-known that in addition to shock wave solutions these equations possess solitary wave solutions. Equations (4.1) - (4.3) are the starting point for the reductive perturbation procedure which yields the K-dV equation for ion-acoustic solutions.[31]

The three known integral constants for Eqs. (4.1) - (4.3) are

$$N = \int_R N\, dx \tag{4.4}$$

$$P = \int_R Nv\, dx \tag{4.5}$$

$$H = \int_R \left(\frac{Nv^2}{2} + \mathcal{L}N\right) dx \tag{4.6}$$

where in Eq. (4.6) $\mathcal{L}$ is a nonlocal operator determined by Eq. (4.3) such that

$$\mathcal{L}N = \frac{\phi_x^2}{2} + \int^\phi \phi' \frac{\partial n(\phi')}{\partial \phi'} d\phi' \quad . \tag{4.7}$$

Equation (4.7) represents a nonlocal internal energy function. The obvious choice for the Hamiltonian is, of course, the energy, Eq. (4.7). (We note that in terms of the Poisson bracket presented below Eq. (4.9), P and N are in involution.)

Observe

$$\frac{\delta H}{\delta v} = Nv$$

and subsequently we will show

$$\frac{\delta H}{\delta N} = \frac{v^2}{2} + \phi \quad . \qquad (4.8)$$

The following bracket, which is the one-dimensional restriction of the first term of Eq. (2.6), yields the equations of motion:

$$[F,G] = \int_R \left(\frac{\delta G}{\delta v} \partial \frac{\delta F}{\delta N} - \frac{\delta F}{\delta v} \partial \frac{\delta G}{\delta N} \right) dx \qquad (4.9)$$

where $\partial \equiv d/dx$. Clearly

$$N_t = [N,H] = -(Nv)_x$$

and assuming Eq. (4.8),

$$v_t = [v,H] = -(v^2/2 + \phi)_x \quad .$$

To justify Eq. (4.8), suppose $P[\phi]$ is some functional of ϕ, i.e.,

$$P[\phi] = \int_R P(\phi) dx \quad .$$

Varying this we obtain

$$\delta P(\phi;\delta\phi) = \int_R \frac{\delta P}{\delta \phi} \delta\phi \, dx \quad . \qquad (4.10)$$

To see how a variation in ϕ is related to a variation in N we linearize Eq. (4.3) and obtain

$$\left(\partial^2 - \frac{\partial n}{\partial \phi}(\phi)\right) \delta\phi = -\delta N \quad,$$

which upon formally inverting yields

$$\delta\phi(x) = -\int_R K(\phi,x,x') \, \delta N(x') \, dx' \qquad (4.11)$$

where K satisfies

$$\left(\partial^2 - \frac{\partial n}{\partial \phi}\right) K = \delta(x-x') \quad. \qquad (4.12)$$

Here, $\delta(x)$ is the Dirac delta function and we seek solutions with asymptotic charge neutrality and vanishing electric field. Substituting Eq. (4.11) into Eq. (4.10) yields

$$\int_R \left[\frac{\delta P}{\delta N} + \int_R \frac{\delta P}{\delta \phi} K \, dx'\right] \delta N \, dx = 0 \quad.$$

For our special case where

$$P[\phi] = \int_R \left[\frac{\phi_x^2}{2} + \int^\phi \phi' \frac{\partial n}{\partial \phi'} d\phi'\right] dx \quad,$$

we obtain

$$\frac{\delta P}{\delta N} = \int_R \left(\partial'^2 \phi + \phi(x') \frac{\partial n}{\partial \phi}\right) K \, dx' \quad,$$

which with Eq. (4.12) implies

$$\frac{\delta P}{\delta N} = \phi \quad.$$

To conclude this section we obtain a canonical form for the bracket Eq. (4.9). With the substitution

$$v = \psi_x ,$$

where ψ now replaces v as a dynamical variable, and the chain rule for functional differentation

$$\frac{\delta F}{\delta \psi} = - \partial \frac{\delta F}{\delta v} ,$$

Eq. (4.9) becomes

$$[F,G] = \int_R \left(\frac{\delta F}{\delta N} \frac{\delta G}{\delta \psi} - \frac{\delta F}{\delta \psi} \frac{\delta G}{\delta N} \right) dx .$$

(Observe that the substitution $N = \psi_x$ will also achieve the same end). Clearly the substitution (4.13) makes Eqs. (4.1) - (4.3) variational in the sense that we can construct the action

$$J = \int_R \int_T N_t \psi \, dx \, dt - \int_T H[N,\psi] \, dt$$

which upon variation with respect to N and ψ produces the dynamical equations.

5. The Vlasov-Maxwell Equations

If a plasma is sufficiently hot and tenuous, then collisions become unimportant. When this is the case, fast time scale plasma phenomena is described by the following set of equations:

$$f_{\alpha t}(\vec{x},\vec{v},t) = -\vec{v} \cdot \frac{\partial f_\alpha}{\partial \vec{x}} - \frac{e_\alpha}{m_\alpha} [\vec{E}+\vec{v}\times\vec{B}] \cdot \frac{\partial f_\alpha}{\partial \vec{v}} \qquad (5.1)$$

$$\vec{B}_t(\vec{x},t) = -\nabla \times \vec{E} \qquad (5.2)$$

$$\vec{E}_t(\vec{x},t) = \nabla \times \vec{B} - \sum_\alpha e_\alpha \int_R \vec{v} f_\alpha \, d^3v . \qquad (5.3)$$

Equation (5.1) is the evolution equation for the single particle distribution function, f_α, which is a function of the six phase-space coordinates together with time. Here α designates species and e_α and m_α are the signed charge and mass respectively. Equation (5.2) is Faraday's law relating the magnetic field $\vec{B}$ and the electric field $\vec{E}$. Equation (5.3) is Ampere's law with the inclusion of the displacement current. (We use rationalized Gaussian units with the speed of light set to unity.)

It is well known that this system, Eqs. (5.1) - (5.3), conserves energy. The natural choice for the Hamiltonian functional is the following:

$$H[f_\alpha, \vec{E}, \vec{B}] = \sum_\alpha \int_D \frac{1}{2} m_\alpha v^2 f_\alpha \, d^3x \, d^3v + \frac{1}{2} \int_R (E^2 + B^2) \, d^3x \ . \quad (5.4)$$

For this Hamiltonian observe

$$\frac{\delta H}{\delta f_\alpha} = \frac{1}{2} m_\alpha v^2 \ , \quad \frac{\delta H}{\delta \vec{E}} = \vec{E} \ , \quad \frac{\delta H}{\delta \vec{B}} = \vec{B} \ .$$

With Eq. (5.4) as Hamiltonian it is not difficult to show that the following Poisson bracket[15-17] produces Eqs. (5.1) - (5.3):

$$[F,G] = \quad (5.5)$$

$$\sum_\alpha \int_D \frac{f_\alpha}{m_\alpha} \left\{ \frac{\delta F}{\delta f_\alpha}, \frac{\delta G}{\delta f_\alpha} \right\} d^3x \, d^3v \quad (5.5.1)$$

$$+ \int_R \left(\frac{\delta F}{\delta \vec{E}} \cdot \nabla \times \frac{\delta G}{\delta \vec{B}} - \frac{\delta G}{\delta \vec{B}} \cdot \nabla \times \frac{\delta F}{\delta \vec{E}} \right) d^3x \quad (5.5.2)$$

$$+ \sum_\alpha \frac{e_\alpha}{m_\alpha} \int_D \frac{\partial f_\alpha}{\partial \vec{v}} \cdot \left[\frac{\delta F}{\delta \vec{E}} \frac{\delta G}{\delta f_\alpha} - \frac{\delta G}{\delta \vec{E}} \frac{\delta F}{\delta f_\alpha} \right] d^3x \, d^3v \quad (5.5.3)$$

$$+ \sum_\alpha \frac{e_\alpha}{m_\alpha^2} \int_D f_\alpha \vec{B} \cdot \left[\frac{\partial}{\partial \vec{v}} \frac{\delta F}{\delta f_\alpha} \times \frac{\partial}{\partial \vec{v}} \frac{\delta G}{\delta f_\alpha} \right] d^3x \, d^3v \quad (5.5.4)$$

In the first term, Eq. (5.5.1), the curly brackets are used to indicate the usual particle Poisson bracket of two phase functions $\{g,h\} = \partial g/\partial \vec{x} \cdot \partial h/\partial \vec{v} - \partial g/\partial \vec{v} \cdot \partial h/\partial \vec{x}$. This term with Eq. (5.4) produces Eq. (5.1) without the terms which couple in the electric and magnetic fields. It can be shown to satisfy the Jacobi requirement. (In Section 6 we present a construction where this bracket is the entire bracket for the Vlasov-Poisson system.) The second term, Eq. (5.5.2), produced Maxwell's equations in vacuum. This term was apparently first written down by Born and Infeld.[32] It satisfies the Jacobi condition. The next two terms, Eqs. (5.5.3) and (5.5.4) supply the coupling between Eqs. (5.1) and (5.3). Observe the e_α/m_α multiplying each. The first of these yields the electric field coupling term. The last term, Eq. (5.5.4), completes the coupling. This term is due to J. Marsden and A. Weinstein[16], who obtained it through consideration of the underlying Lie group. The Jacobi condition is satisfied for this term only if the space of functionals, on which the bracket acts, is restricted to vector fields $\vec{B}$ that satisfy $\nabla \cdot \vec{B} = 0$. For arbitrary functionals E, F and G we obtain

$$[E,[F,G]] + \text{cyc} = \int f \, \nabla \cdot \vec{B} \, \epsilon_{s\ell t} \frac{\delta E}{\delta v_\ell} \frac{\delta F}{\delta v_s} \frac{\delta G}{\delta v_t} d^3v \, d^3x \, ,$$

where $\epsilon_{s\ell t}$ is the Levi-Civita tensor. If initially $\nabla \cdot \vec{B} = 0$ then the Jacobi condition is satisfied for all time.

We conclude this section by pointing out a recent motivation[33] of the bracket, Eq. (5.5), (see Ref. 18). Here, the relativistic generalization is made and the generators of the full Poincaré group are pointed out. Table 3 summarizes the above.

6. The Vlasov-Poisson Equations

In this section we write the Vlasov-Poisson equations in a form[15,19] very similar to that of the two-dimensional vortex fluid equations of Sec. 3. We will observe that these sets of equations possess the same noncanonical and canonical formulations. The Vlasov-Poisson equations are

$$\frac{\partial f_\alpha}{\partial t}(\vec{x},\vec{v},t) = -\vec{v} \cdot \frac{\partial f_\alpha}{\partial \vec{x}} + \frac{e_\alpha}{m_\alpha} \frac{\partial \phi}{\partial \vec{x}} \cdot \frac{\partial f_\alpha}{\partial \vec{v}} \qquad (6.1)$$

$$\Delta \phi(\vec{x},t) = -\sum_\alpha e_\alpha \int f_\alpha d^3v \, . \qquad (6.2)$$

Here the only symbol not defined in Sec. 5 is ϕ, the electrostatic potential. If we seek solutions where ϕ is defined on R, and if we assume asymptotic charge neutrality and vanishing electric field, then the Laplacian operator Δ can be inverted Equations (6.1) and (6.2) can be written compactly as follows:

$$\frac{\partial f_\alpha}{\partial t} = -\vec{w}_\alpha \cdot \nabla_p f_\alpha \quad . \tag{6.3}$$

Here ∇_p is the six-dimensional phase-space gradient $(\partial/\partial\vec{x}, \partial/\partial\vec{v})$ and $\vec{w}_\alpha$ is defined by

$$\vec{w}_\alpha = \left(v, \frac{e_\alpha}{m_\alpha} \frac{\partial}{\partial \vec{x}} \sum_\beta e_\beta \int K(\vec{x}|\vec{x}') f_\beta d^3 x' \right) . \tag{6.4}$$

Observe $\nabla_p \cdot \vec{w}_\alpha = 0$. In Eq. (6.4) $K(\vec{x}|\vec{x}')$ is the kernel for the inverse Laplacian; e.g., in one-dimension $K(\vec{x}|\vec{x}') = \frac{1}{2}|\vec{x} - \vec{x}'|$. The Hamiltonian for this system is

$$H = \sum_\alpha \frac{1}{2} m_\alpha \int v^2 f_\alpha d^3 z - \frac{1}{2} \sum_{\alpha\beta} e_\alpha e_\beta \int K(\vec{x}|\vec{x}') f_\alpha(z) f_\beta(z') d^3 z d^3 z' . \tag{6.5}$$

Here we have used $z \equiv (\vec{x},\vec{v})$. The Poisson bracket is the first term of Eq. (5.5)

$$[F,G] = \sum_\alpha \int \frac{f_\alpha(z)}{m_\alpha} \left\{ \frac{\delta F}{\delta f_\alpha}, \frac{\delta G}{\delta f_\alpha} \right\} d^3 z , \tag{6.6}$$

where the braces are as defined in Sec. 5. It is not difficult to see that

$$\frac{\partial f_\alpha}{\partial t} = [f_\alpha, H] = -\vec{w}_\alpha \cdot \nabla_p f_\alpha .$$

To obtain canonical form we consider the three-dimensional generalization of the potential representation of Sec. 3,

$$f_\alpha = \frac{1}{m_\alpha} \{\psi_\alpha, \chi_\alpha\} . \tag{6.7}$$

With this substitution, ψ_α and χ_α become canonically conjugate variables. We note, in conclusion, that the entire bracket of Sec. 5, Eq. (5.5), can be put into canonical form by the substitution of Eq. (6.8) together with the usual canonical description of the fields in terms of the vector potential $\vec{A}$ and its conjugate $\vec{E}$.

Acknowledgements

This work began with the consideration of ideal MHD. My collaborator in this effort was John M. Greene. Our principle motivation was the work of Robert Littlejohn on the use of non-canonical variables for perturbation theory (Ref. 2) and the work of C.S. Gardner on the KdV equation (Ref. 4). It is with pleasure that I thank Allan Kaufman for his help and encouragement. Section 3 was motivated by several conversations with Harvey Segur. I have learned much about 2-D turbulence from Segur, Guido Sandri, and Rick Salmon. The work of Sec. 4 was prompted by many discussions with Predhiman Kaw. I have also benefited from conversations and/or correspondences with the following: D. Barnes, F. Henyey, D. Holm, J. Hubbard, M. Kruskal, R. Kulsrud, B. Kupershmidt, J. Marsden, C. Oberman, J.B. Taylor, W.B. Thompson, and A. Weinstein.

References

1. V. I. Arnold, Mathematical Methods of Classical Mechanics, (Springer-Verlag, New York, 1978).

2. R. G. Littlejohn, J. Math. Phys. 20, 2445 (1979).

3. J. E. Marsden, Lectures on Geometric Methods in Mathematical Physics, (Siam, Philadelphia, 1981).

4. C. S. Gardner, J. Math. Phys. 12, 1548 (1971).

5. P. D. Lax, Comm. Pure Appl. Math. 28, 141 (1975).

6. D. R. Lebedev and Y. I. Manin, Funk. Anal. Appl. 13, 40 (1979).

7. I. M. Gel'fand and I. Y. Dorfman, Funk, Anal. Appl. 13, 248 (1979).

8. B. A. Kupershmidt, Rutgers 1981 Proceedings, Lecture Notes in Mathematics (to appear).

9. Y. I. Manin, J. Sov. Math. 11, 1 (1979).

10. P. R. Chernoff and J. E. Marsden, Properties of Infinite Dimensional Hamiltonian Systems, Lect. Notes in Math. 425, (Springer-Verlay, Berlin, 1974).

11. R. Abraham, J. E. Marsden, Foundations of Mechanics, (Benjamin-Cummings, London, 1978).

12. P. J. Morrison and J. M. Greene, Phys. Rev. Lett. 45, 790 (1980); Phys. Rev. Lett. 48, 569 (1982).

13. G. F. Chew, M. L. Goldberger, and F. E. Low, Proc. Roy. Soc. A236, 112 (1956).

14. P. J. Morrison, Princeton University Plasma Physics Laboratory Report, PPPL-1783 (1981).

15. P. J. Morrison, Phys. Lett. 80A, 383 (1980).

16. J. E. Marsden and A. Weinstein, Physica D. (to appear).

17. A. Weinstein and P. J. Morrison, Phys. Lett. 86A, 235 (1981).

18. I. Bialynicki-Birula and J. C. Hubbard, preprint.

19. P. J. Morrison, Princeton University Plasma Physics Laboratory Report, PPPL-1788 (1981).

20. H. Goldstein, Classical Mechanics, (Addison-Wesley, Reading 1950).

21. Constants of this general form for dynamical systems are treated by I. L. Caldas and H. Tasso, Lett. Nuovo Cimento 24, 500 (1979).

22. D. P. Stern, Am. J. Phys. 38, 494 (1970).

23. H. Lamb, Hydrodynamics, (Dover, New York, 1945).

24. J. Serrin, Handbuch der Physik 8, 169 (1959).

25. P. A. M. Dirac, Proc. Roy. Soc. A212, 330 (1952).

26. A. Clebsch, J. Reine Angeev, Math. 54, 293 (1857); 56, 1 (1859).

27. R. L. Seliger and G. B. Whitham, Proc. Roy. Soc. A305, 1 (1968).

28. M. G. Calkin, Can. J. Phys. 41, 2241 (1963).

29. V. E. Zakharov and E. A. Kuznetzov, Sov. Phys. Dokl. 15, 913 (1971).

30. N. A. Krall and A. W. Trivelpiece, Principles of Plasma Physics, (McGraw Hill, New York, 1973).

31. C. H. Su and C. S. Gardner, J. Math. Phys. $\underline{10}$, 536 (1969).

32. M. Born and L. Infeld, Proc. Roy. Soc. $\underline{150}$, 141 (1935).

33. This motiviation is based on the early work of I. Bialynicki-Birula. See I. Bialynicki-Birula and z. Bialynicka-Birula, Quantum Electrodynamics, (Pergamon, Oxford, 1975).

SATISFIES JACOBI	COMMENTS
Eq. (2.6.1)	Defined on functionals of ρ & $\vec{v}$. With $H = \int [\rho v^2/2 + \rho U(\rho)] d\tau$ produces Eq. (2,1) with $\nabla \times \vec{v} = 0$ and $\vec{B} = 0$, and Eq. (2.2).
Eq. (2.6.1) + Eq. (2.6.2)	Defined on functionals of ρ & $\vec{v}$. With $H = \int [\rho v^2/2 + \rho U(\rho)] d\tau$ produces Eq. (2.1) with $\vec{B} = 0$ and Eq. (2.2).
Eq. (2.6.1) + Eq. (2.6.2) + Eq. (2.6.3)	Defined on functionals of $\rho, \vec{v}$ and s. With $H = \int [\rho v^2/2 + \rho U(\rho,s)] d\tau$ produces Eq. (2.1) with $\vec{B} = 0$, Eq. (2.2) and Eq. (2.3).
Eq. (2.6.1) + Eq. (2.6.2) + Eq. (2.6.4)	Defined on functionals of $\rho, \vec{v}$ and $\vec{B}$. With $H = \int [\rho v^2/2 + \rho U(\rho) + B^2/2] d\tau$ produces Eq. (2.1), Eq. (2.2) and Eq. (2.4).
Eq. (2.6.1) + Eq. (2.6.2) + Eq. (2.6.3) + Eq. (2.6.4)	Defined on functionals of $\rho, \vec{v}, \vec{B}$ and s. With $H = \int [\rho v^2/2 + \rho U(\rho,s) + B^2/2] d\tau$ produces Eqs. (2.1) - (2.4).

TABLE 1

Kind of Fluid	Noncanonical Variables	Canonical Variables	Velocity Representation
Ideal, Irrotational, and Isentropic	$\vec{v}$ and ρ	ϕ and ρ	$\rho\vec{v} = \rho\nabla\phi$
Ideal and Isentropic	$\vec{v}$ and ρ	λ, μ, ϕ, and ρ	$\rho\vec{v} = \lambda\nabla\mu + \rho\nabla\phi$
Ideal	$\vec{v}$, ρ and s	λ, μ, ϕ, ψ, σ and ρ	$\rho\vec{v} = \lambda\nabla\mu + \rho\nabla\phi + \sigma\nabla\psi$
Ideal Isentropic MHD	$\vec{v}$, ρ and $\vec{B}$	$\vec{T}$, $\vec{B}$, ϕ and ρ	$\rho\vec{v} = (\nabla\vec{T})\cdot\vec{B} - \vec{B}\cdot\nabla\vec{T} - \vec{T}\nabla\cdot\vec{B} + \rho\nabla\phi$
Ideal Isentropic MHD, $\vec{B} = \nabla\alpha\times\nabla\beta$	$\vec{v}$, ρ and $\vec{B}$	α, β, a, b, ϕ and ρ	$\rho\vec{v} = a\nabla\alpha + b\nabla\beta + \rho\nabla\phi$
Ideal MHD	$\vec{v}$, ρ, $\vec{B}$ and s	$\vec{T}$, $\vec{B}$, σ, ψ, ϕ and ρ	$\rho\vec{v} = (\nabla\vec{T})\cdot\vec{B} - \vec{B}\cdot\nabla\vec{T} - \vec{T}\nabla\cdot\vec{B} + \rho\nabla\phi + \sigma\nabla\psi$
Ideal MHD, $\vec{B} = \nabla\alpha\times\nabla\beta$	$\vec{v}$, ρ, $\vec{B}$ and s	α, β, a, b, σ, ψ, ϕ and ρ	$\rho\vec{v} = a\nabla\alpha + b\nabla\beta + \rho\nabla\phi + \sigma\nabla\psi$

TABLE 2

SATIFIES JACOBI	COMMENTS
Eq. (5.5.1)	with $H = \sum_\alpha \int \frac{1}{2} m_\alpha v^2 f_\alpha d^3x d^3v$ produces Eq. (5.1) with $\vec{E} = \vec{B} = 0$
Eq. (5.5.2)	with $H = \int \frac{1}{2}(E^2+B^2)d^3x$ produces Maxwell's equations in vacuum.
Eq. (5.5.1) + Eq. (5.5.2) + Eq. (5.5.3) + Eq. (5.5.4)	With Eq. (5.4) as Hamiltonian produces the Vlasov-Maxwell equations. Requires the constraint $\nabla \cdot \vec{B} = 0$

TABLE 3

SINGULAR POISSON TENSORS

Robert G. Littlejohn
University of California, Los Angeles, Cal. 90024

ABSTRACT

The Hamiltonian structures discovered by Morrison and Greene for various fluid equations were obtained by guessing a Hamiltonian and a suitable Poisson bracket formula, expressed in terms of noncanonical (but physical) coordinates. In general, such a procedure for obtaining a Hamiltonian system does not produce a Hamiltonian phase space in the usual sense (a symplectic manifold), but rather a family of symplectic manifolds. To state the matter in terms of a system with a finite number of degrees of freedom, the family of symplectic manifolds is parametrized by a set of Casimir functions, which are characterized by having vanishing Poisson brackets with all other functions. The number of independent Casimir functions is the corank of the Poisson tensor J^{ij}, the components of which are the Poisson brackets of the coordinates among themselves. Thus, these Casimir functions exist only when the Poisson tensor is singular.

INTRODUCTION

Recently, Morrison and Greene[1-3] have discovered that several important systems of partial differential equations in common use in classical physics (inviscid hydrodynamics, MHD, and the Vlasov equation) can be represented as classical Hamiltonian fields. These discoveries have stimulated a remarkable amount of interest on the part of both physicists and mathematicians, expecially those concerned with the formal properties of dynamical systems. Evidently, a resonant chord has been struck with other problems of current interest in physics and mathematics, such as the difficult and vexing problems of quantizing nonlinear fields, and the rapidly expanding subject of solitons and integrable systems. Marsden and Weinstein,[4] in particular, have shown that Morrison's and Greene's Hamiltonian systems can be understood in terms of the modern theory of dynamical systems with symmetries, which incorporates a number of ideas from group theory and differential geometry.

However, attempts to extract practical (i.e. physical) information from these researches are hindered by the abstract language and recondite notation of the mathematical medium in which they are expressed. So I think it is important not to lose sight of the physical point of view, or even the physical way of thinking, and it is important to approach these problems from a number of different perspectives. This is not simply a question of bridging the gap between physics and mathematics, but rather it is an appreciation of the fact that the physical point of view is a fertile source of new and interesting insights, both physical and mathematical, as it has been in the case of Morrison's and

Greene's discoveries.

Therefore I shall adopt the physical way of thinking, at least, in the following discussions. This was the spirit of the work of Morrison and Greene, and I shall continue it. To someone with the right mathematical perspective, much of what I have to say will seem obvious and overly laborious. But my intent is to promote further physical researches, and I shall even point out some new physical systems to which these ideas can be applied.

Although I do not indend to discuss Morrison's and Greene's field Hamiltonians themselves as much as to discuss some issues raised by them, nevertheless it is useful to write down one of these systems for future reference. I choose the Poisson-Vlasov system, because it has an especially simple Poisson bracket formula. For this system, the one and only dynamical (field) variable is the distribution function $f=f(\underset{\sim}{x},\underset{\sim}{p})$, i.e. the particle density on the $(\underset{\sim}{x},\underset{\sim}{p})$ phase space, where $\underset{\sim}{p}=m\underset{\sim}{v}$. The equation of motion is the Vlasov equation,

$$\frac{\partial f}{\partial t} + \frac{1}{m} \underset{\sim}{p} \cdot \frac{\partial f}{\partial \underset{\sim}{x}} + e\underset{\sim}{E} \cdot \frac{\partial f}{\partial \underset{\sim}{p}} = 0, \qquad (1)$$

where the electric field $\underset{\sim}{E}$ is a functional of f through Poisson's equation. One can introduce a neutralizing background or multiple species, if desired. The Hamiltonian functional for this system is the total energy:

$$H[f] = \int d^3\underset{\sim}{x}\, d^3\underset{\sim}{p}\, f(\underset{\sim}{x},\underset{\sim}{p}) \frac{p^2}{2m} + \int d^3\underset{\sim}{x}\, \frac{E^2(\underset{\sim}{x})}{8\pi}. \qquad (2)$$

Finally, the Poisson bracket which yields the correct equation of motion is

$$\{F,G\} = \int d^3\underset{\sim}{x}\, d^3\underset{\sim}{p}\, f(\underset{\sim}{x},\underset{\sim}{p})\, \{\frac{\delta F}{\delta f}, \frac{\delta G}{\delta f}\}_{xp}, \qquad (3)$$

where F and G are two functionals of f, and where the subscript xp on the bracket in the integrand indicates the usual Poisson bracket on the single particle $(\underset{\sim}{x},\underset{\sim}{p})$ phase space. (The bracket on the left hand side is the field bracket.)

One of the reasons that the Hamiltonian systems of Morrison and Greene took so long to be discovered is the fact that, for the systems they considered, canonically conjugate coordinates (the ϕ's and π's of ordinary field theory) are either hard to come by, or are nonphysical, or involve a number of constraints whose physical significance is obscure. This situation in regard to hydrodynamics is nicely summarized by Bretherton,[5] who reviews some of the older literature on Hamiltonian formulations of fluid equations. So one of the salient features of the new discoveries is the use of noncanonical variables, i.e. field quantities whose Poisson brackets with one another are not expressible in terms of simple delta functions. This immediately raises the following question, which will occupy our attention for the remainder of this talk: What exactly is needed to specify a Hamiltonian in the abstract, i.e. without reference to a given set of canonical variables? And, obviously: Do canonical variables necessarily exist?

POISSON BRACKETS AND THEIR PROPERTIES

Before proceeding with these questions, let me restrict the notation of my algebraic arguments to systems with a finite number of degrees of freedom. I do this for the sake of simplicity, and also because I will not specifically need to call on the Hamiltonian structures of Morrison and Greene. In any case, it will be obvious how to extend most of the following arguments to continuum systems.

So let us suppose that we have a physical system described by a finite set of D variables, $z^1, \ldots, z^D$ (as a vector, $\underset{\sim}{z}$), which are physically complete in the sense that every quantity of physical interest can be expressed as a function of the z's, and let us suppose that the time derivatives $\dot{z}^i$ can also be expressed as functions of the z's. That is, let us begin with

$$\dot{z}^i = F^i(z^1, \ldots, z^D), \quad i=1, \ldots, D. \tag{4}$$

We use superscripts to indicate contravariant indices. The z's are coordinates in a phase space of D dimensions, and part of our task here is to describe the relationship between this phase space and the familiar 2N-dimensional phase spaces of Hamiltonian systems. We do not assume that D is even. Note that the field equivalent of Eq. (4) is a system of field variables which satisfy equations which are first-order differential equations in the time.

Now, to the question of what defines a Hamiltonian system, Morrison and Greene supplied an answer of the following sort, expressed here in terms of the system (4). The system (4) is Hamiltonian if the equations of motion can be written in the form

$$\dot{z}^i = \{z^i, H\}, \tag{5}$$

where the curly brackets represent the Poisson bracket, and where $H = H(\underset{\sim}{z})$ is the Hamiltonian. The Poisson bracket is some prescription for taking two functions of $\underset{\sim}{z}$, say f and g, and turning them into a third, denoted by $\{f, g\}$. It cannot be defined in the

usual way, in terms of q's and p's, because as yet we do not know if
q's and p's exist. Instead, a formula must be given for evaluating
Poisson brackets, and this formula must cause the Poisson bracket
to satisfy some set of formal properties expected of Poisson
brackets.

So we see that this approach to finding a Hamiltonian structure
to Eqs. (4) involves guessing both a Hamiltonian and a formula for
the Poisson bracket. Sometimes the Hamiltonian is not hard to
guess, because it is the energy of the system, but the Poisson
bracket formula is more tricky, because of the formal properties it
must satisfy.

For the record, I should probably point out that it can be
shown that a Hamiltonian structure for Eqs. (4) always exists;
this is rather easy to show in terms of a set of D constants of
motion. But these constructions are seldom of any practical value,
because one does not have explicit formulas for the constants of
motion, nor are these constants useful in general, since typically
they are not isolating and cannot be defined in a continuous manner
over large regions of phase space. On the other hand, it is
usually true that if one can find a Hamiltonian structure in terms
of simple and well-behaved expressions, then this structure will
have a physical significance. These considerations are clarified
quite a bit by the researches of Marsden and Weinstein.[4]

In any case, the formal properties of the Poisson bracket
demanded by Morrison and Greene, explictly or implicitly, are
the following. The first is antisymmetry:

$$\{f,g\} = -\{g,f\}. \tag{6}$$

The second is the Jacobi identity:

$$\{f,\{g,h\}\} + \{g,\{h,f\}\} + \{h,\{f,g\}\} = 0. \tag{7}$$

And the third is the chain rule formula. If one of the operands
of the Poisson bracket, say g, is a function of a collection of
functions of $\underset{\sim}{z}$, say $h_1,\ldots,h_s$, then

$$\{f,g(h_1,\ldots,h_s)\} = \sum_{j=1}^{s} \{f,h_j\} \frac{\partial g}{\partial h_j}. \tag{8}$$

Similar formulas hold if f, or both f and g, depend on a collection
of other functions. The condition (8) is actually stronger than
the demands of deductive logic require, but it is convenient for
our purposes.

The Poisson brackets of Morrison and Greene satisfy these
identities (although the Jacobi identity is often difficult to
prove); notice that no use is made of coordinates (either q's and
p's or z's) in them. The rationale for demanding the properties

(6)-(8) is that the usual Poisson bracket, defined in terms of p's and q's, satisfies them. Another is that these properties are the classical analogs of well known relations involving commutators in quantum mechanics. Thus, if one is interested in studying a classical analog of a quantum system, such Hamiltonian structures as we are considereing here arise quite naturally.

As an example of this, consider a classical spin system, which forms a useful model for the rather general discussion at hand. Here we consider a single spin $\underline{S}$, with components S_x, S_y, S_z, which satisfy the Poisson bracket relation

$$\{S_x, S_y\} = S_z, \qquad (9)$$

and cyclic permutations thereof. A physically reasonable Hamiltonian might be

$$H = -\underline{\mu} \cdot \underline{B}, \qquad (10)$$

where $\underline{B}$ is a constant magnetic field, and μ is the magnetic moment. H becomes a function of $\underline{S}$ when we write $\underline{\mu} = \kappa \underline{S}$, for a constant κ. So this system can be regarded as a classical system, living on a phase space of three dimensions, with coordinates $(z^1, z^2, z^3) = (S_x, S_y, S_z)$. The equations of motion can be derived from Eq. (5), Hamilton's equations, and the chain rule formula for Poisson brackets:

$$\dot{\underline{S}} = -\kappa \underline{B} \times \underline{S}. \qquad (11)$$

The actual time evolution can be represented as a trajectory in $\underline{S}$-space; but notice that p's and q's have not been used.

To return to the general case, it is convenient to introduce a D×D matrix J^{ij} of functions of $\underline{z}$, defined by

$$J^{ij}(\underline{z}) = \{z^i, z^j\}. \qquad (12)$$

The importance of J^{ij} is that the Poisson bracket of any two functions f and g of $\underline{z}$ can be expressed in terms of J^{ij}, simply by using the chain rule formula. (To see this, simply set $h_1 = z^1$, etc., in Eq. (8)). One finds the relation

$$\{f, g\} = \sum_{i,j}^{D} \frac{\partial f}{\partial z^i} J^{ij} \frac{\partial g}{\partial z^j}. \qquad (13)$$

It is easy to see that J^{ij} represents a second rank, contravariant tensor on phase space, by considering a new coordinate system z', and writing $J'^{ij} = \{z'^i, z'^j\}$. Because of this, we will call J^{ij} "the Poisson tensor." So we see that guessing a Poisson bracket formula really involves guessing a tensor field. For future reference, we display here the Poisson tensor for the spin system:

$$J^{ij} = \begin{pmatrix} 0 & S_z & -S_y \\ -S_z & 0 & S_x \\ S_y & -S_x & 0 \end{pmatrix}. \tag{14}$$

The result of this is that if one is looking for a suitable Poisson bracket formula, then it must have the form of Eq. (13), for some tensor J^{ij}. But not just any tensor will do. To be consistent with the properties (6)-(7), the Poisson tensor must satisfy the following requirements. First, it must be antisymmetric:

$$J^{ij} = -J^{ji}. \tag{15}$$

And second, it must satisfy the Jacobi identity, in the form of a nonlinear, first-order differential equation:

$$\sum_{\ell=1}^{D} \left(J^{i\ell} J^{jk}{}_{,\ell} + J^{j\ell} J^{ki}{}_{,\ell} + J^{k\ell} J^{ij}{}_{,\ell} \right) = 0, \tag{16}$$

where commas represent derivatives with respect to the coordinates z. This comes directly from substituting Eq. (13) into Eq. (7). It is easy to show that these requirements are satisfied by the spin Poisson tensor of Eq. (14).

A SUFFICIENT SET OF PROPERTIES?

So far, we have collected a number of abstract properties to be satisfied by a Poisson tensor. And since many of the operations one would like to perform on a Hamiltonian system (such as finding the equations of motion) can be written in terms of Poisson brackets, it appears that perhaps one could dispense with canonical coordinates altogether. But are the requirements (6)-(8) really complete, i.e. do they give us everything we normally expect from a Hamiltonian system? The answer to this is no, as may be seen by considering the fixed points of a Hamiltonian system.

The behavior of Hamiltonian systems in the neighborhood of fixed points has been and continues to be an active area of research. In the usual canonical (q,p) coordinates, a fixed point is characterized as a point where $\dot{q} = \dot{p} = 0$, or, alternatively, where $\partial H/\partial q = \partial H/\partial p = 0$. The fixed points are the extremal points of H. The obvious generalization of this to an arbitrary system of coordinates z^i is to ask for $\dot{z}^i=0$ or $\partial H/\partial z^i=0$. For certainly, if the gradient of H is zero in any coordinate system, then it will be zero in all coordinate systems, including canonical ones.

The trouble with this is easily seen from Eqs. (10)-(11). The fixed points of the spin system, i.e. the points where $\dot{S}=0$, are given by $S=cB$, for some constant c. They lie on a line in S-space, passing through the origin and parallel to B. But there is no point where $\partial H/\partial S=0$; H has no extremal point. So in these Hamiltonian systems we are considering, the fixed points may not coincide with the extremal points of the Hamiltonian.

Completely analogous situations arise with the Poisson brackets of Morrison and Greene. For example, in the electrostatic Vlasov system, Eqs. (1)-(3), the Vlasov function $f=f(x,p)$ is the only field dynamical variable. That is, the analog of our D-dimensional phase space is a function space, a point of which is a particular function $f=f(x,p)$. Thus, a stationary solution to the Vlasov equation can be considered to be a fixed point in function space. But these fixed points are usually not extremal points of the Hamiltonian functional, $H=H[f]$. So we find that a Hamiltonian treatment of Landau damping, for example, which concerns the behavior of the system in a neighborhood of a fixed point, must necessarily take into account the nonstandard relationship between $\dot{z}^i$ and $\partial H/\partial z^i$.

The difficulty here is easily traced to the properties of the Poisson tensor J^{ij}. If we write Hamilton's equations, Eq. (5), in terms of J^{ij}, we have

$$\dot{z}^i = \sum_{j=1}^{D} J^{ij} \frac{\partial H}{\partial z^j} . \qquad (17)$$

In fact, if the coordinates z^i are canonical coordinates, i.e. if $(z^1,\ldots,z^D) = (q_1,\ldots,q_N,p_1,\ldots,p_N)$, with D=2N, and if the Poisson bracket is defined in the usual way, then J^{ij} can be represented by its partition into four N×N matrices:

$$J^{ij} = \begin{pmatrix} 0 & I \\ -I & 0 \end{pmatrix} . \qquad (18)$$

This is equivalent to Hamilton's equations in the usual sense, when substituted into Eq. (17).

Now, from Eq. (17) it is easy to see that $\dot{z}^i = 0$ implies $\partial H/\partial z^i = 0$ only if $\det(J^{ij}) \neq 0$. And, indeed, in the case of canonical coordinates, Eq. (18) gives us $\det(J^{ij}) = 1$. Even if we are not using canonical coordinates, but we know they exist, then we will still have $\det(J^{ij}) \neq 0$, because if $\det(J^{ij})$ is nonzero in one coordinate system, then it will be nonzero in all coordinate systems. On the other hand, the Poisson tensor of the spin system, Eq. (14), is a singular matrix, and the spin system will give $\det(J^{ij}) = 0$ in all coordinate systems. This means that canonical coordinates, at least in the usual sense, cannot exist for the spin system; this was obvious anyway, from the fact that the dimensionality of the spin phase space, namely three, is odd, and cannot be broken up into an equal number of q's and p's. In any case, we have arrived at the title of this talk, singular Poisson tensors.

It so happens that the rank of an antisymmetric matrix is always an even integer, so any Poisson tensor in a phase space of odd dimensionality, like the spin system, must be singular. But even-dimensional Poisson tensors can also be singular, so the oddness or evenness of the dimensionality of the phase space is not really the issue. For example, even if we had D=4, we could not necessarily conclude the existence of canonical variables (q_1, q_2, p_1, p_2).

But before considering singular Poisson tensors further, let us briefly describe the situation when J^{ij} is not singular (over some region of interest in phase space). In this case, D must be even, so we write D=2N. Then J^{ij} has an inverse at each point, denoted by ω_{ij}; this represents a "symplectic 2-form," which is responsible for the Poincaré invariants. It can be shown (conveniently starting with ω_{ij}) that in these circumstances, canonical coordinates $(q_1, \ldots, q_N, p_1, \ldots, p_N)$ always exist. This is Darboux's theorem, of which straightforward proofs have been given by Arnold[6] and myself.[7] These properties qualify the phase space as a "symplectic manifold," which can be thought of as a 2N-dimensional space possessing canonical coordinates and a Poisson bracket structure in the ordinary sense. Thus, Darboux's theorem shows that if $\det(J^{ij}) \neq 0$, then we really have an ordinary Hamiltonian phase space, although it may be disguised by noncanonical coordinates.

SINGULAR POISSON TENSORS

In the more general case that the Poisson tensor is singular, let us assume that the rank R of J^{ij} is constant over some region of interest. For example, in the case of the spin system, Eq. (14) gives R=2, except at the one point $\underset{\sim}{S}=0$, where R=0; so here we might want to exclude (or beware of) the origin. Since R is necessarily an even integer, we will write R=2N, and since we are assuming that J^{ij} is singular, we have $2N < D$; and we write K for the corank of J^{ij}, so that D=2N+K.

The fact that J^{ij} has corank K means that at each point of phase space, there exist K linearly independent vectors $\beta_{(k)}$, k=1,...,K, with components $\beta_{(k)i}$, i=1,...,D, such that

$$\sum_{j=1}^{D} J^{ij} \beta_{(k)j} = 0, \quad k=1,\ldots,K \qquad (19)$$

The $\beta_{(k)}$ are really covariant vectors, which explains the lower index, and they are eigenvectors of J^{ij} with eigenvalue zero. Since there is a K-fold degeneracy, the $\beta_{(k)}$ are not unique, but may be redefined according to

$$\beta'_{(\ell)} = \sum_{k=1}^{K} A_{\ell k} \beta_{(k)}, \qquad (20)$$

where the K×K matrix $A_{\ell k}$ is nonsingular. Relations of this sort hold at every point of phase space, so the $\beta_{(k)}$ determined at each point can be promoted into a set of K covariant vector fields, $\beta_{(k)i} = \beta_{(k)i}(\underset{\sim}{z})$. Also, the matrix $A_{\ell k}$ can be allowed to depend upon position, as long as $\det(A_{\ell k})$ is nowhere zero.

For example, in the case of the spin system, we have K=1, and the one $\beta = \beta_{(1)}$ is given by

$$\beta_i = S_i, \qquad (21)$$

for i=1,2,3 or x,y,z. This can be multiplied by any nonzero function of $\underset{\sim}{S}$ (the equivalent of the matrix $A_{\ell k}$).

CASIMIR FUNCTIONS

The existence of the $\beta_{(k)}$ has some interesting consequences. To begin, let us suppose that one of the $\beta_{(k)}$ can be represented in the form

$$\beta_{(k)i} = \frac{\partial C}{\partial z^i}, \qquad (22)$$

for some scalar function $C=C(z)$. It is not obvious that this is possible; but we must recognize that even if it were impossible for some original selection of $\beta_{(k)}$, it might still be possible under a redefinition of the form of Eq. (20). In any case, if such a function C exists, then it has some curious properties. In particular, its Poisson bracket with any function whatsoever is zero:

$$\{f,C\} = 0, \qquad (23)$$

as is easily seen from Eq. (13). We will call such a function a Casimir function, because it commutes with everything.

In the case of the spin system, a Casimir function is not hard to find. We simply look for a solution $C=C(\underline{S})$ to the equation

$$\frac{\partial C}{\partial \underline{S}} = A(\underline{S})\, \underline{S}, \qquad (24)$$

where $A(\underline{S})$ represents a possible redefinition of the one β, in accordance with Eq. (20). Inspection shows that a simple solution exists if we take $A=2$, whereupon we have

$$C(\underline{S}) = S^2. \qquad (25)$$

In view of the familiar properties of angular momentum, this result is not surprising, since it is equivalent to

$$\{\underline{S}, S^2\} = 0. \qquad (26)$$

In the case of Morrison's electrostatic Vlasov system, an example of a Casimir function (really a functional) is the $(\underline{x},\underline{p})$ phase space average of any function F of the Vlasov function f, with respect to the Liouville measure:

$$C[f(\underline{x},\underline{p})] = \int d^3\underline{x}\, d^3\underline{p}\; F(f(\underline{x},\underline{p})). \qquad (27)$$

A surprising consequence of the existence of a Casimir function is that the physical Hamiltonian is not unique. For example, if a presumably physical Hamiltonian in Eq. (5) is replaced by H', defined by

$$H' = H + \lambda C, \qquad (28)$$

for some constant λ, then the equations of motion are not altered, since

$$\{z^i, H'\} = \dot{z}^i + \lambda \{z^i, C\} = \dot{z}^i. \qquad (29)$$

But the extremal points of H may change under the substitution (28).

Note also that any Casimir function is necessarily a constant of the motion,

$$\dot{C} = \{C, H\} = 0, \qquad (30)$$

so that the physical motion is confined to a surface C=const. Thus, in the case of the spin system, the motion in $\underset{\sim}{S}$-space always takes place on the surfaces S^2=const. But a Casimir function is quite different from an ordinary constant of motion, as on a symplectic manifold, because a Casimir function is a constant of motion for any Hamiltonian. For example, in the spin system, the trajectories would lie on the surfaces S^2=const. even if another Hamiltonian, different from Eq. (10), were chosen. (One should not dismiss this fact, just because in physics there is only one "real" Hamiltonian. For example, in perturbation theory, one would like to break H up into two parts, $H=H_0+H_1$, and consider the unperturbed motion generated by H_0.) It is important to note that the Casimir functions are properties of the Poisson bracket, not the Hamiltonian.

So far we have not shown that Casimir functions even exist, except in special cases. We will now prove our central theorem, which says that if corank(J^{ij})=K=D-2N, and J^{ij} satisfies all the properties demanded above for a Poisson tensor, especially the Jacobi identity, Eq. (16), then there exist exactly K independent Casimir functions. Furthermore, we will show that the D-dimensional phase space can be decomposed or foliated into a K-parameter family of 2N-dimensional surfaces, each of which is a symplectic manifold, i.e. a Hamiltonian phase space in the ordinary sense, possessing canonical q's and p's. These symplectic manifolds are the intersections of the contour surfaces of the Casimir functions, i.e. they are given by $C_k(\underset{\sim}{z})=c_k$, k=1,...,K, for K constants c_k.

I will state here, and not belabor the point further, that the following constructions may be limited to a restricted region of phase space. If they are pushed too far, then one may expect to see the appearance of multivalued functions (such as for the Casimir functions), and other things that may be considered undesirable from a physical standpoint. Nevertheless, for the sake of simplicity I will continue to speak as if the constructions were valid everywhere in phase space, as indeed they sometimes are.

The proof of this central theorem is a simple application of Frobenius' theorem, which is a basic building block of differential geometry.[8] To begin, observe that the stated conclusion on the existence of the Casimir functions will follow if we can solve

$$\frac{\partial C_k}{\partial z^i} = \sum_{\ell=1}^{K} A_{k\ell} \, \beta_{(\ell)i}, \qquad (31)$$

where the $\beta_{(\ell)}$ are some given, linearly independent set of K solutions to Eq. (19), where $\det(A_{k\ell}) \neq 0$, and where everything depends on $\underset{\sim}{z}$. As a special case, if K=1, we will be expected to solve

$$\frac{\partial C}{\partial z^i} = A\beta_i . \qquad (32)$$

AN EXAMPLE OF THE FROBENIUS THEOREM

To obtain a geometrical picture of the situation, let us consider a problem in ordinary three-dimensional space which is familiar in plasma physics. Suppose we have a magnetic field $\underset{\sim}{B}=\underset{\sim}{B}(\underset{\sim}{x})$, with $\underset{\sim}{B} \neq 0$ in some region of interest, and suppose we wish to find a family of surfaces such that $\underset{\sim}{B}$ is everywhere orthogonal to these surfaces. It is often desirable to do this. The family of surfaces, if they exist, can be regarded as a set of contour surfaces of some function $C=C(\underset{\sim}{x})$, simply by assigning values of C to the members of the family. We will then have $\underset{\sim}{B}$ parallel to ∇C, or

$$\nabla C = A\underset{\sim}{B}, \qquad (33)$$

for some scalar function $A=A(\underset{\sim}{x})$. The ability to find a solution C to this equation, for given $\underset{\sim}{B}$ (and whatever A will work), is equivalent to finding the required family of surfaces. It is also identical in form to Eq. (32).

The solution to this problem is well known; a solution $C=C(\underset{\sim}{x})$ will exist if and only if we have $\underset{\sim}{B} \cdot (\nabla \times \underset{\sim}{B}) = 0$. It is easy to prove the necessity of this condition; one simply takes the curl of Eq. (33) and dots with $\underset{\sim}{B}$. For sufficiency, we will provide a geometrical argument.

Let us consider two vector fields, $\underset{\sim}{X}$ and $\underset{\sim}{Y}$, which are linearly

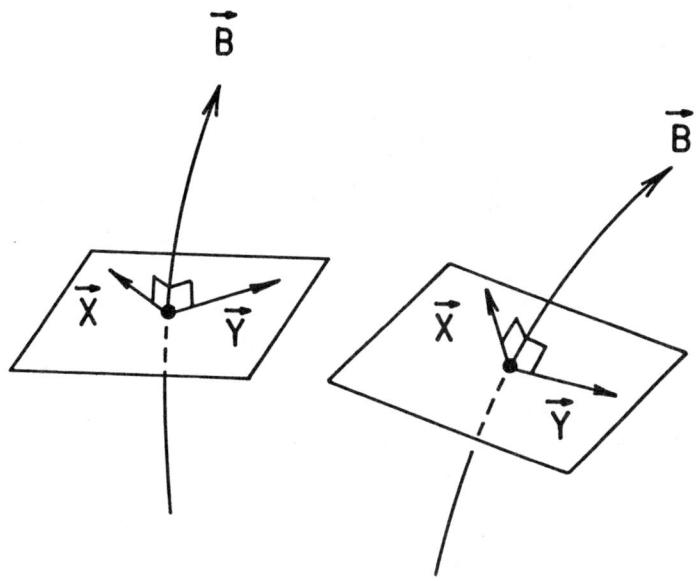

Fig. 1. The vector fields $\underset{\sim}{X}$ and $\underset{\sim}{Y}$ span the planes perpendicular to the magnetic field $\underset{\sim}{B}$.

independent at each point $\underset{\sim}{x}$ and perpendicular to $\underset{\sim}{B}$:

$$\underset{\sim}{X} \cdot \underset{\sim}{B} = \underset{\sim}{Y} \cdot \underset{\sim}{B} = 0. \tag{34}$$

Such vectors fields are easy to find and are far from unique, because any given $\underset{\sim}{X}$ and $\underset{\sim}{Y}$ can be replaced by any nonsingular pair of linear combinations of themselves. At each point of space, $\underset{\sim}{X}$ and $\underset{\sim}{Y}$ span a plane perpendicular to $\underset{\sim}{B}$, and if the scalar C of Eq. (33) exists, these planes will be tangent to the contour surfaces of C (see Fig. 1). Thus, the contour surfaces of C can be considered to be composed of a large number of small pieces, each one taken from one of the planes spanned by the local $\underset{\sim}{X}$ and $\underset{\sim}{Y}$; and the question of the existence of C is equivalent to seeing if the various planes spanned by $\underset{\sim}{X}$ and $\underset{\sim}{Y}$ can be pieced together into a smooth surface, with no mismatch.

A simple way to construct the contour surfaces of C (if they exist) is to follow along the integral curves generated by $\underset{\sim}{X}$ and $\underset{\sim}{Y}$, i.e. to solve the following differential equations:

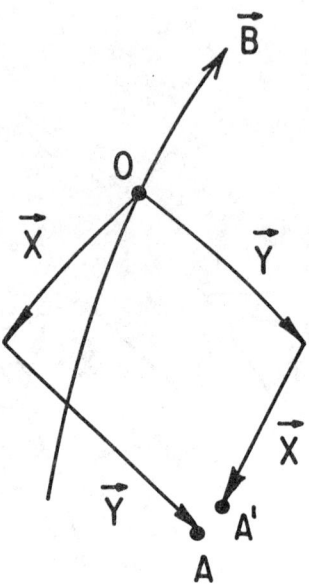

Fig. 2. The processes of following along the $\underset{\sim}{X}$ and $\underset{\sim}{Y}$ trajectories are not commutative, in general. But what is important for the Frobenius theorem is whether these processes stay in one surface. The condition for this can be expressed in terms of the commutator of the vector fields $\underset{\sim}{X}$ and $\underset{\sim}{Y}$.

$$\frac{d\underset{\sim}{x}}{ds} = \underset{\sim}{X}(\underset{\sim}{x}), \qquad \frac{d\underset{\sim}{x}}{dt} = \underset{\sim}{Y}(\underset{\sim}{x}). \qquad (35)$$

Then, starting from some point, such as O in Fig. 2, one can proceed a certain elapsed s parameter along the $\underset{\sim}{X}$ curves, and then a certain t parameter along the $\underset{\sim}{Y}$ curves, and arrive at a point A, which depends on s and t. Doing this for all values of s and t fills out a two-dimensional surface.

Of course, these operations are not commutative, so that if we reversed the order of the curves we moved along, we would arrive at a different point, A'. This is quite all right, as long as the reversed procedure fills out the same surface. Indeed, if the required contour surface exists, then we must always be confined to it, regardless of the order of the operations, because $\underset{\sim}{X}$ and $\underset{\sim}{Y}$ always lie in the surface. Conversely, if a unique surface is generated by following $\underset{\sim}{X}$ and $\underset{\sim}{Y}$ curves, regardless of order, then the surface is the one we want, because it is everywhere perpendicular to $\underset{\sim}{B}$.

So the existence of the function C of Eq. (33) reduces to the commutativity of the s and t advance operations. This can be expressed quantitatively in terms of the commutator of the vector fields $\underset{\sim}{X}$ and $\underset{\sim}{Y}$. The commutator is defined by

$$[\underset{\sim}{X},\underset{\sim}{Y}] = \underset{\sim}{X}\cdot\nabla\underset{\sim}{Y} - \underset{\sim}{Y}\cdot\nabla\underset{\sim}{X}. \qquad (36)$$

The bracketed quantity is a vector, and it is called a commutator because it is closely related to the commutator (in the usual sense of linear operators) of the operators $\underset{\sim}{X}\cdot\nabla$ and $\underset{\sim}{Y}\cdot\nabla$. Geometrically, it represents the displacement between the points A and A' of Fig. 2 when s and t are small. In order to generate a unique surface, regardless of the order of the operations, this commutator must lie in the surface itself (it need not vanish). That is, it must be a linear combination of $\underset{\sim}{X}$ and $\underset{\sim}{Y}$:

$$[\underset{\sim}{X},\underset{\sim}{Y}] = a\underset{\sim}{X} + b\underset{\sim}{Y}. \qquad (37)$$

This is the necessary and sufficient condition for the existence of the scalar C of Eq. (33). And it is easy to show the equivalence of this to the condition $\underset{\sim}{B}\cdot(\nabla\times\underset{\sim}{B})=0$, simply by noting that we must have $\underset{\sim}{B}=k\underset{\sim}{X}\times\underset{\sim}{Y}$ for some scalar $k\neq 0$, and using some vector algebra.

This has been a sketch of a simple version of Frobenius' theorem. This theorem can be stated in two complementary but equivalent ways, one using differential forms, and the other using vectors. The condition $\underset{\sim}{B}\cdot(\nabla\times\underset{\sim}{B})=0$ is essentially a statement about differential forms ($B \wedge dB=0$), whereas the commutator relation Eq. (37) takes the vector point of view. The latter point of view is somewhat more useful for what we shall do next.

CASIMIR FUNCTIONS EXIST

Let us return to Eq. (31). Each one of the K covariant vectors $\beta_{(k)i}$ in the D-dimensional phase space can be thought of as being orthogonal to some (D-1)-dimensional hyperplane in the space (really the tangent space). Thus, we are drawing an analogy between the $\beta_{(k)}$ and the magnetic field vector $\underset{\sim}{B}$ previously considered. Since the $\beta_{(k)}$ are linearly independent, the intersections of these K (D-1)-dimensional hyperplanes is a (D-K)-dimensional hyperplane, which is simultaneously orthogonal to all the $\beta_{(k)}$. The (D-K)-dimensional hyperplanes are analogous to the two-dimensional perpendicular planes of Fig. 1. Note that D-K=2N; we will show that these

2N-dimensional hyperplanes can be smoothly fitted together into 2N-dimensional surfaces. These are symplectic manifolds, to which all the $\beta_{(k)}$ are orthogonal.

Because the K covariant vectors $\beta_{(k)}$ are linearly independent, it is always possible to find at each point precisely D-K=2N linearly independent (contravariant) vectors $X^{(n)}$ which are orthogonal to them:

$$\sum_{i=1}^{D} X^{(n)i} \beta_{(k)i} = 0, \qquad (38)$$

for n=1,...,2N, k=1,...,K. These vectors $X^{(n)}$ span the 2N-dimensional hyperplanes orthogonal to the $\beta_{(k)}$, and are analogous to $\underset{\sim}{X}$ and $\underset{\sim}{Y}$ of Eq. (34).

By following along the integral curves of the vectors $X^{(n)}$, i.e. by integrating the 2N systems of differential equations

$$\frac{dz^i}{ds_n} = X^{(n)i}(\underset{\sim}{z}), \qquad (39)$$

for n=1,...,2N, we can form 2N-dimensional surfaces.

However, the 2N-dimensional surfaces we produce in this way will be everywhere orthogonal to the $\beta_{(k)}$ only if the surfaces themselves are invariant with respect to the order of following integral curves. And this will be the case if all of the commutators of the vector fields $X^{(n)}$ are linearly dependent on the $X^{(n)}$, i.e. if

$$[X^{(n)}, X^{(m)}] = \sum_{r=1}^{2N} \Gamma_r^{nm} X^{(r)}, \qquad (40)$$

for some quantities Γ_r^{nm}. Here we generalize Eq.(36), and define the commutator of two vectors X and Y, with components X^i and Y^i, by

$$[X,Y]^i = \sum_{j=1}^{D} \left(X^j Y^i_{,j} - Y^j X^i_{,j} \right). \qquad (41)$$

Now it is easy to find vectors orthogonal to all the $\beta_{(k)}$. Let us define D vectors $Y^{(i)}$ as the D rows of the matrix J^{ij}:

$$Y^{(i)j} = J^{ij}. \qquad (42)$$

Then from Eq. (19) we have

$$\sum_{j=1}^{D} Y^{(i)j} \beta_{(k)j} = 0. \tag{43}$$

The vectors $Y^{(i)}$ are not linearly independent, because rank(J^{ij}) =2N < D. But we can always form 2N linearly independent combinations of the $Y^{(i)}$, and these can serve as the $X^{(n)}$ in Eqs. (38)-(40). In any case, if we express the Jacobi identity, Eq. (16), in terms of the $Y^{(i)}$, then we see that the commutator of the $Y^{(i)}$ naturally appears:

$$[Y^{(i)}, Y^{(j)}] = \sum_{k=1}^{D} J^{ij}_{,k} Y^{(k)}. \tag{44}$$

Finally, it is easy to show that Eq. (40) follows from this, by using the fact that the $X^{(n)}$ and the $Y^{(i)}$ are linear combinations of one another.

The result of this application of the Frobenius theorem is that a 2N-dimensional surface passes through each point of phase space, and it is orthogonal to all the covariant vectors $\beta_{(k)}$. Since the whole space has dimensionality 2N+K, these 2N-dimensional subspaces can be parametrized by K quantities C_k, constant on each subspace. These are the Casimir functions, satisfying Eq. (31).

CANONICAL COORDINATES

Let us introduce a set of 2N coordinates, $x^1, \ldots, x^{2N}$, on each of the 2N-dimensional surfaces we have just constructed. These coordinates could be the quantities s_n of Eq. (39), if we are committed to a particular order of following vector fields when reaching a point of a given surface. Since the surfaces themselves are parametrized by the C_k, we can regard the entire collection of 2N+K=D quantities, $(x^1, \ldots, x^{2N}, C_1, \ldots, C_K)$ as a coordinate system on the original D-dimensional phase space. Let us consider the form which the Poisson tensor takes on when expressed in terms of these coordinates. Calling on Eqs. (12) and (23), we can partition J^{ij} into submatrices by D=2N+K:

$$J^{ij} = \left(\begin{array}{c|c} \tilde{J}^{nm} & 0 \\ \hline 0 & 0 \end{array} \right), \qquad (45)$$

where

$$\tilde{J}^{nm} = \{x^n, x^m\}. \qquad (46)$$

And since rank(J^{ij})=2N, we must have det($\tilde{J}^{nm}$)≠0.

Thus we conclude that each of the 2N-dimensional surfaces constructed by the Frobenius theorem is a symplectic manifold, with Poisson bracket specified by $\tilde{J}^{nm}$. By Darboux's theorem, we can always find q's and p's on each one of these manifolds, so it is finally possible to construct a coordinate transformation of the form

$$(z^1, \ldots, z^D) \to (q_1, \ldots, q_N, p_1, \ldots, p_N, C_1, \ldots, C_K). \qquad (47)$$

In this coordinate system, the Poisson bracket takes on its usual form. This is the principal conclusion of this talk.

It is instructive to see how this works out for the spin system. We already have identified the Casimir function S^2; and a possible choice for p and q is easy to guess. These are just S_z and the azimuthal angle ϕ, the latter being given implicitly by

$$S_x = (S^2 - S_z^2)^{1/2} \cos \phi,$$
$$S_y = (S^2 - S_z^2)^{1/2} \sin \phi. \qquad (48)$$

So the transformation (47) appears as

$$(S_x, S_y, S_z) \to (\phi, S_z, S^2), \qquad (49)$$

with $\{\phi, S_z\}=1$. In the case of the spin system, the Poisson bracket of Eq. (9) causes the original spin space to foliate into a one-parameter family of two-dimensional symplectic manifolds, these being the spheres S^2=const.

CONCLUSIONS

We conclude that the abstract properties of the Poisson bracket assumed by Morrison and Greene are sufficient to guarantee the existence of q's and p's. However, in general there will also be a certain number of Casimir functions, the C's, which have vanishing Poisson brackets with all other functions. Geometrically, the original phase space foliates into a family of symplectic manifolds, to which the physical motion is confined.

POSTSCRIPT

After the completion of this work it was brought to my attention that the central theorem on the existence of Casimir functions was previously demonstrated by Bayen et al,[9] and is further discussed by Sudarshan and Mukunda.[10] I would also like to call the reader's attention to the work of Bialynicki-Birula and Hubbard,[11] who have derived Poisson brackets and Hamiltonian structures for relativistic plasmas.

ACKNOWLEDGMENTS

It is a pleasure to acknowledge several fruitful discussions with T. Frankel, who suggested the strategy of Eqs. (42)-(44) for fulfilling the requirements of the Frobenius theorem, and also with A. Weinstein, who suggested the Casimir functions of Eq. (27). I am also grateful to N. Pomphrey for calling my attention to classical spin systems.

I would also like to thank the Aspen Center for Physics for their hospitality during the time in which some of this material was worked out.

This work was supported by the Office of Naval Research under contract No. N00014-79-C-0537 and the Office of Fusion Energy of the U.S. Department of Energy under contract No. DOE-DE-AM03-76500010 PA 26, Task I.

REFERENCES

1. P. J. Morrison and J. M. Greene, Phys. Rev. Lett. __45__, 790 (1980).
2. P. J. Morrison, Phys. Lett. __80A__, 383 (1980).
3. P. J. Morrison, Princeton Plasma Physics Laboratory Report No. PPPL-1783, 1981.
4. J. E. Marsden and A. Weinstein, Physica D, to appear, 1982.
5. Francis Bretherton, J. Fluid Mech. __44__, 19 (1970).
6. V. I. Arnold, Mathematical Methods of Classical Mechanics (Springer-Verlag, New York, 1978).
7. R. G. Littlejohn, J. Math. Phys. __20__, 2445 (1979).

8. M. Spivak, <u>Differential Geometry</u> (Publish of Perish, Berkeley, Cal., 1979).
9. F. Bayen, M. Falto, C. Fronsdal, A. Lichnerowicz, and D. Sternheimer, Ann. Phys. <u>111</u>, 61 (1978).
10. E. C. G. Sudarshan and N. Mukunda, <u>Classical Dynamics: A Modern Perspective</u> (Interscience, New York, 1974), p. 117.
11. I. Bialynicki-Birula and J. C. Hubbard, University of California at Irvine, Department of Physics Report No. 81-55, submitted to Phys. Rev. A.

DYNAMICAL SYSTEMS DEFINED ON INFINITE DIMENSIONAL LIE ALGEBRAS OF THE "CURRENT ALGEBRA" OR "KAC-MOODY" TYPE

Robert Hermann*
Association for Physical and Systems Mathematics
53 Jordan Road, Brookline, Massachusetts 02146

ABSTRACT

Recent work by Morrison, Marsden, and Weinstein has drawn attention to the possibility of utilizing the cosymplectic structure of the dual of the Lie algebra of certain infinite dimensional Lie gruops to study hydrodynamical and plasma systems. This paper treats certain models arising in elementary particle physics, considered by Lee, Weinberg, and Zumino; Sugawara; Bardacki, Halpern, and Frishman; Hermann; and Dolan. The Lie algebras involved are associated with the "current algebras" of Gell-Mann. This class of Lie algebras contains certain of the algebras that are called "Kac-Moody algebras" in the recent mathematics and mathematical physics literature.

1. INTRODUCTION

Let M be a manifold (assumed, for the moment, to be paracompact, finite dimensional, and C^∞) and let V be a C^∞ vector field on M. The study of many mechanical systems with a finite number of degrees of freedom amounts to studying the orbit curves of V of the one-parameter pseudogroup $t \to \exp(tV)$ of diffeomorphisms of M generated by V.

Often, the trajectories of the pseudogroup (hence, the "dynamics" of the physical system) can be related to a Lie group G, assumed for the purposes of this introductory discussion to be finite dimensional. (The point of this work will be to extend the ideas to infinite dimensional Lie groups.)

Let $T^d(G)$ be the cotangent bundle[1] of the Lie group G. There seem to be two ways of relating the dynamical system (M,V) to G:

a) Embedd M as a submanifold of $T^d(G)$ such that the vector field V is the restriction of a vector field V' on $T^d(G)$ which (as I shall explain in Section 2) arises from the universal enveloping algebra $U(\mathcal{G})$ of the Lie algebra $\mathcal{G}$ of G.

b) Construct[2,3] a submersion mapping $\alpha: T^d(G) \to M$ such that V is the projection under α of a vector field on $T^d(G)$ constructed from $U(\mathcal{G})$.

These geometric approaches to the structure of mechanical systems with a finite number of degrees of freedom suggest the possibility of generalization to infinite dimensional G's and M's. One such

---
*Supported by Ames Research Center (NASA), Grant NSG-2402, U.S. Army Research Office, Contract #ILI61102RH57-05MATH, NSF MCS8003227.

possibility was investigated in the 1960's under the name of the "σ of Sugawara model".[4] The situation here was that the Lie algebra $\mathcal{G}$ was an infinite dimensional Lie algebra of the type Gell-Mann called a "current algebra",[5] and suggested as an algebraic building block for elementary particle-quantum mechanical systems. The "Sugawara model" was especially pretty as a realization of Gell-Mann's ideas. It is also of great mathematical and physical interest: On the one hand, it is a field-theoretic generalization of the rotating rigid body model, and on the other hand, its classical equations of motion are those which differential geometers have studied under the name "harmonic maps". (If the underlying Lie algebra is abelian, the solutions of the equations of motion are just harmonic functions in the usual sense.) However, I believe that the relation between this model and the symplectic structure mentioned above was not recognized in the 1960's. In this note I want to sketch how this might go. Recently (e.g., in the work of Morrison[6] and Marsden-Weinstein[7]), it has been recognized that physical field theories might carry interesting structures of this type, following a pattern initiated by Arnold.[8]

2. THE SYMPLECTIC STRUCTURE ON THE COTANGENT BUNDLE OF A LIE GROUP

Continue, for the moment, with G as a finite dimensional Lie group, with $T^d(G)$ as its cotangent bundle. Let $\mathcal{G}$ be the Lie algebra of G, $\mathcal{G}^d$ its dual space. $T^d(G)$ may then be identified (say, by left-translation) with $G \times \mathcal{G}^d$. Let

$$\alpha: T^d(G) \to \mathcal{G}^d$$

be the projection map. Let $\mathcal{F}(\mathcal{G}^d)$ be the C^∞ real valued functions on $\mathcal{G}^d$, and let:

$$\alpha^*: \mathcal{F}(\mathcal{G}^d) \to \mathcal{F}(T^d(G))$$

be the pull-back map. One may then prove[9,10] that:

$$\{\alpha^*(\mathcal{F}(\mathcal{G}^d), \alpha^*(\mathcal{F}(\mathcal{G}^d))\} \subset \alpha^*(\mathcal{F}(\mathcal{G}^d)) \quad , \tag{2.1}$$

where $\{,\}$ denotes the Poisson bracket defined by the standard symplectic structure on the cotangent bundle. Equation (2.1) says that $\alpha^*(\mathcal{F}(\mathcal{G}^d))$ is a subalgebra of $\mathcal{F}(T^d(G))$ in both the associative and Lie algebra sense. ($\alpha^*(\mathcal{F}(\mathcal{G}^d))$ is one example of what Sophus Lie called[10,11] a "function group".)

Each $A \in \mathcal{G}$ determines an element f_A of $\mathcal{F}(\mathcal{G}^d)$:

$$f_A(\theta) = \theta(A) \tag{2.2}$$

for $\theta \in \mathcal{G}^d$ (the dual space to $\mathcal{G}$).

The mapping $A \to f_A$ is a Lie algebra isomorphism $A \to f_A$ of $\mathcal{G}$ into $\alpha^*(\mathcal{F}(\mathcal{G}^d))$. It extends to a map of $U(\mathcal{G})$ into $\mathcal{F}(\mathcal{G}^d)$. $U(\mathcal{G})$ is a filtered associative algebra: The corresponding graded algebra is commutative, and is sent isomorphically into the $\mathscr{F}(\mathscr{G}^d)$ (as an associative algebra) under this map.

Thus, an element Δ of $U(\mathscr{G})$ determines an element $f_\Delta \in \alpha^*(\mathscr{F}(\mathscr{G}^d))$, and a symplectic vector field V_Δ on $T^d(G)$. The orbit curves of V_Δ then project under α to curves in $\mathscr{G}^d$; these are the dynamical systems we want to study. Notice that we get "free" a source of constants of motion for this dynamical system: Elements $\Delta' \in U(\mathscr{G})$ such that

$$[\Delta, \Delta'] = 0 \qquad (2.3)$$

give rise to functions $f_{\Delta'}$ on $T^d(G)$ such that

$$V_\Delta(f_{\Delta'}) = 0 . \qquad (2.4)$$

In favorable cases[12] one can prove that there are "enough" such Δ' to prove "complete integrability".

3. THE SUGAWARA MODEL

Notice that the formalism of Section 2 is algebraic and makes sense for infinite dimensional G's. Since the dynamics projects down to $\mathscr{G}^d$, we can even work completely in $\mathscr{G}^d$, which is an advantage in infinite dimensional situations because $\mathscr{G}^d$ is a vector space. (Of course, the equations of motion are nonlinear, so that there are still enormous analytical problems to making everything rigorous.) I will now use as a starting point a paper by Sugawara.[4] He wrote down the equations of a quantum field theory described by a set of "currents" which form an infinite dimensional Lie algebra. A related version was given earlier by Lee, Weinberg, and Zumino,[4] which was called the *algebra of fields*. I will abstract from this and related work by physicists the following quasi-algebraic structure.[13-16]

Let V be a real, finite dimensional vector space, and let $C_0^\infty(R^n)$ be the C^∞, real-valued functions on R^n with compact support. Let $\underline{D}(R^n)$ denote the space of linear, C^∞ differential operators

$$D: C_0^\infty(R^n) \to C_0^\infty(R^n) .$$

Form the tensor product (over the reals as scalar field):

$$V \otimes C_0^\infty(R^n) .$$

Let us suppose the following data is given: A skew-symmetric bilinear map

$$\omega: V \times V \to V ,$$

a bilinear map

$$\beta: V \times V \to \underline{D}(R^n) .$$

Let $\mathscr{L}$ be given as a vector space by the following formula:

$$\mathscr{L} = V \otimes C_0^\infty(R^n) \oplus R \quad . \tag{3.1}$$

Define a bilinear bracket operation

$$[\ ,\ ]: \mathscr{L} \times \mathscr{L} \to \mathscr{L}$$

as follows:

$$[v_1 \otimes f_2,\ v_2 \otimes f_2] = \omega(v_1, v_2) \otimes f_1 f_2 + \int \beta(v_1, v_2)(f_1) f_2 \tag{3.2}$$

$$[V \otimes C_0^\infty(R^n),\ R] = 0 \tag{3.3}$$

Physically, each element $v \otimes f$ of V is to be thought of as a skew Hermitian operator in a Hilbert space, which can be written as:

$$v \otimes f = i \int A(v)(x) f(x)\ dx \tag{3.4}$$

$x \to A(v)(x)$ is to be thought of as a Hermitial operator valued "function" of $x \in R^n$, parameterized by the $v \in V$. Thus, the relations (3.2)-(3.3) translate into certain "commutation relations" between "operator-valued functions of x". The second term on the right hand side of (3.2) is the "Schwinger term",[17] and depends on the derivatives of these fields. The Hamiltonian is then a formal expression of the type

$$H = \int \sum_{a=1}^n A(v^a)(x) A(v^a)(x)\ dx \quad . \tag{3.5}$$

The equation of motion of the fields are those defined as:

$$\frac{\partial}{\partial t} A(v)(x,t) = [H,\ A(v)(x,t)] \tag{3.6}$$

for all $v \in V$.

Rather than attempting to make this quantum field theory lore precise, I will proceed as follows. Let $\mathscr{L}^d$ be the dual space of $\mathscr{L}$; for the moment, without any topological conditions. Let $\mathscr{F}(\mathscr{L}^d)$ be the real-valued functions on $\mathscr{L}^d$, which define a "Poisson bracket" as follows:

$$\{f_1, f_2\}(\theta) = <\theta,\ [df_1(\theta),\ df_2(\theta)]> \tag{3.7}$$

(In the language of Reference (10), this is the Lie algebra structure on functions on the dual of the Lie algebra.) The convention implied in (3.7) is as follows:

For $\theta \in \mathscr{L}^d$, $df_1(\theta)$, the value of the differential of f_1 at θ, is an element of the dual space of $\mathscr{L}^d$. Suppose this is in $\mathscr{L}$. (The dual of the dual naturally contains the original vector space.) Then, $[df_1(\theta), df_2(\theta)]$ is an element of $\mathcal{L}$, the value of θ on it is a real number, which is the meaning of the expression on the right hand side of (3.7).

Now, let $\mathcal{L}$ be the Lie algebra determined by Equations (3.1)-(3.3). $\mathcal{L}$ is a vector space. Give $C_0^\infty(R^n)$ the usual locally convex topological vector space structure, so that its dual space is the space of distributions in the sense of Schwartz. $\mathcal{L}$ inherits a topology. Its dual space is then

$$V^d \otimes \mathcal{D}(R^n) \oplus R \ , \qquad (3.8)$$

where $\mathcal{D}(R^n)$ are the distributions (defined on the compactly supported test functions). Denote an element of $\mathcal{D}(R^n)$ as f^d. Pretend that f^d is a function of x, hence has a "value" of $f^d(x)$ as $x \in R^n$. For $v \in V$, $x \in R^n$, let $A(v,x)$ be "symbolically" the following element of

$$A(v,x)(v^d \otimes f^d) = <v,v^d> f^d(x) \qquad (3.9)$$
$$\text{for } v^d \in V^d, \quad f^d \in \mathcal{D}(R^n) \ .$$

Consider $A(v,x)$ as an element of $\mathcal{F}(\mathcal{L}^d)$. One easily sees that the formal Poisson bracket structure (3.7) agrees with the Lie algebra structure defined precisely by formulas (3.2)-(3.3). We now have the possibility of defining objects like (3.5), and using them to calculate (within the Poisson bracket Lie algebra) equations of motion like (3.6) without mentioning quantum fields. One can construct such models by following the ideas of Reference (15). This work is in progress, and will be published elsewhere.

REFERENCES

1. R. Abraham and J. Marsden, *Foundations of Mechanics*, 2nd edition, Addison-Wesley, Reading, MA, 1978.

2. J. Marsden and A. Weinstein, "Reduction of Symplectic Manifolds with Symmetry", *Rep. Math. Phys.* 5, 121-130 (1974).

3. V. Guillemin and S. Sternberg, "On the Equations of Motion of a Classical Particle in a Yang-Mills Field and the Principle of General Covariance", *Hadronic J.* 1, 1-32 (1978).

4. M. Gell-Mann and M. Levy, *Nuovo Cimento* 16, 705 (1960); T.D. Lee, S. Weinberg, and B. Zumino, *Phys. Rev. Lett.* 18, 1029 (1967); K. Bardacki, Y. Frishman, and M.B. Halpern, *Phys. Rev.* 170, 1353 (1968); H. Sugawara, *Phys. Rev.* 170, 1659 (1968); K. Bardacki and M.B. Halpern, *Phys. Rev.* 172, 1542 (1968); H. Sugawara and M. Yoshimura, 173, 1419 (1968); C. Sommerfield, *Phys. Rev.* 176, 2019 (1968); L Dolan, *Phys. Rev. Lett.* 47, 1371 (1981).

5. M. Gell-Mann, *Phys. Rev.* 125, 1067 (1962).

6. J.E. Marsden and A. Weinstein, "The Hamiltonian Structure of the Maxwell-Vlasov Equations", Preprint, 1981; P.J. Morrison, "The Maxwell-Vlasov Equations as a Continuous Hamiltonian System, *Phys. Lett. A* (to appear); P.J. Morrison and J.M. Greene, "Noncanonical Hamiltonian Density Formulation of Hydrodynamics and Ideal Magnetohydrodynamics", *Phys. Rev. Lett.* 45, 790-794 (1980).

7. J.E. Marsden and A. Weinstein, "The Hamiltonian Structure of the Maxwell-Vlasov Equations" (to appear).

8. V. Arnold, "Sur la Geometrié Differentielle des Groupes de Lie de Dimension Infinite et ses Applications a l'Hydrodynamique des Fluids Parfaits", *Ann. Inst. Grenoble* 16, 319-361 (1966).

9. R. Hermann, "Left Invariant Geodesics and Classical Mechanics on Manifolds", *J. Math. Phys.* 13, 460 (1972).

10. R. Hermann, *Toda Lattices, Cosymplectic Manifolds, Backländ Transformations and Kinks, Part A*, Math Sci Press, Brookline, MA 1977.

11. S. Lie, *Behrührungstransformationen*, Chelsea, New York, 1965.

12. M.A. Olshanetski and A.M. Perelmonov, "Completely Integrable Hamiltonian Systems Connected with Semi-Simple Lie Algebras, *Inv. Math.* 37, 93-108 (1976).

13. R. Hermann, "Infinite Dimensional Lie Algebras and Current Algebra", *Proceedings of the 1969 Battelle Rencontre*, Springer-Verlag, Berlin, 1970.

14. R. Hermann, *Lie Algebras and Quantum Mechanics*, W.A. Benjamin, Reading, MA, 1970.

15. R. Hermann, "Current Algebras, the Sugawara Model, and Differential Geometry", *J. Math. Phys.* 11, 1825-1829 (1970); "Geometric Formula for Current-Algebra Commutation Relations", *Phys. Rev.* 177, 2449 (1969); "Quantum Field Theories with Degenerate Lagrangians", *Phys. Rev.* 177, 2453 (1969).

16. R. Hermann, *Tensor Products and Current Algebra*, Interdisciplinary Mathematics, Volume 6, Math Sci Press, Brookline, MA, 1973.

17. J. Schwinger, *Phys. Rev. Lett.* 3, 296 (1959).

GYROSCOPIC ANALOG FOR MAGNETOHYDRODYNAMICS

Darryl D. Holm
Los Alamos National Laboratory, Los Alamos, New Mexico

ABSTRACT

The gross features of plasma equilibrium and dynamics in the ideal magnetohydrodynamics (MHD) model can be understood in terms of a dynamical system which closely resembles the equations for a deformable gyroscope.

1. INTRODUCTION

In the ideal magnetohydrodynamics (MHD) model, electrically neutral plasma convects like an adiabatic fluid that carries an embedded magnetic field. During convection, induced electrical currents flow instantaneously to oppose change of magnetic flux through each comoving surface. The resultant magnetic stresses alter the convective motion of the plasma by opposing bending of magnetic field lines.

We shall seek motions in three-dimensional MHD for which the velocity varies linearly in space. For such flows, time dependence factorizes out from all of the fluid variables in the Lagrange representation. The dynamical system which then results from Hamilton's principle closely resembles the equations for a deformable gyroscope.

Reduction to gyroscopic motion of fluid flow with linear velocity profiles was noted already in 1879 by Greenhill for circulation of a fluid of constant density within an ellipsoidal cavity. Before Greenhill, fluid flows with linear profiles had also been studied by Dirichlet, Dedekind, and Riemann, in connection with ellipsoidal figures of fluid equilibrium. The history and development of the latter topic is given with complete references by Chandrasekhar (1969). Rotating ellipsoidal fluid solutions are also treated in the classical texts on fluid mechanics by Basset (1888) and Lamb (1932).

More recently Parker (1957) has studied the expansion of a magnetic gas cloud which undergoes homogeneous dilation with linear velocity profiles, but which does not rotate or circulate. Likewise, Dyson (1968) has studied isothermal expansion and circulation of an ideal fluid whose velocity profile is linear and whose density profile is of Gaussian shape. Before Dyson, compressible fluid flows with linear profiles had also been noted by Ovsjannikov (1956). Subsequently Anisimov and Lysikov (1970) have found special solutions to Dyson's equations, that involve elliptic integrals for $\lambda = 5/3$ ideal gas.

In the next section we explain how time dependence factorizes out for MHD fluid flows with linear profiles, in the Lagrange representation. We then derive the equations of motion from Hamilton's principle, and analyze the resultant dynamical system for the time dependence of the flow.

The results provide an analogy between circulation of a magnetic fluid and angular momentum of a gyroscope. In this analogy, magnetic stresses produce elastic-like forces within the fluid which tend to restore both the circulatory motion and expansion of the fluid. In fact the equations for MHD with linear velocity profiles separate into two gyroscope equations which in general are coupled to each other by both magnetic stresses and deformations of shape. In the planar case with fixed elliptical boundary the equations reduce to the equation for a simple pendulum.

2. THE MHD EQUATIONS

In the Lagrange representation the particle paths are fundamental objects, and partial derivatives of the particle paths are basic dependent variables. The paths of fluid particles through fixed Eulerian space are given by vector functions $\underline{x}(t,\underline{x}^0)$ with initial conditions $\underline{x}(0,\underline{x}^0) = \underline{x}^0$, the Lagrange coordinate. The partial derivatives of the particle paths $\underline{x}(t,\underline{x}^0)$ produce the kinematical variables, velocity $\underline{v}$ and displacement gradients $\underline{F}_j$ with components

$$v_i(t,\underline{x}^0) = \left.\frac{\partial x_i}{\partial t}\right|_{\underline{x}^0} = \dot{x}_i \qquad F_{ij}(t,\underline{x}^0) = \left.\frac{\partial x_i}{\partial x^0_j}\right|_t \qquad (1)$$

where subscripts t, $\underline{x}^0$ label the variables held constant in the partial derivatives.

In the Lagrange representation with Cartesian coordinates the equation of motion for ideal MHD is

$$\rho \frac{\partial^2 x_i}{\partial t^2} = -\frac{d}{\partial x^0_j}\left(P + \frac{B^2}{8\pi}\right) F^{-1}_{ji} + \frac{1}{4\pi} B^0_k \frac{\partial}{\partial x^0_k} B_i \qquad (2)$$

where ρ, p are fluid density and pressure; $B_i(t,\underline{x}^0)$ without a superscript is magnetic field along a particle trajectory, and $B^0_i = B_i(0,\underline{x}^0)$ is its initial distribution. The equation of motion (2) follows from Hamilton's principle

$$0 = \delta \int dt\, d^3x^0\, \rho^0 \left[\frac{1}{2}\dot{x}^2 - e(\rho,s) - \frac{B^2}{8\pi\rho}\right] \dot{x}^2 \qquad (3)$$

for variations of the particle paths δx_k that vanish on the boundaries of the Lagrange domain of integration. The added notation in Hamilton's principle (3) defines $\rho^0 = \rho(0,\underline{x}^0)$ as the initial density distribution, and $e(\rho,s)$ as the specific internal energy of the fluid which is a function of density, ρ, and specific entropy, s.

The variations of particle paths must be performed subject to the constraints of the following subsidiary conditions for MHD

$$\rho \det F = \rho^0 = \rho(0, \underline{X}^0) \quad (4)$$

$$B_i = F_{ij} B_j^0 / \det F \quad (5)$$

$$S(\rho, p) = S^0 = S(\rho^0, p^0) \quad (6)$$

$$e(\rho, S) = e(\rho, S^0) \quad (7)$$

These subsidiary conditions impose respectively conservation of mass, Faraday's Law of magnetic induction, and the equations of state for adiabatic convection with the prescribed specific internal energy. In Faraday's Law and in the motion equation one uses Ampere's Law, curl $\underline{B} = 4\pi \underline{J}/c$, and Ohm's Law for the case of infinite electrical conductivity, $\underline{E} + \underline{v} \times \underline{B}/c = 0$, in order to eliminate current density, $\underline{J}$, and electric field, $\underline{E}$, in favor of magnetic field, $\underline{B}$, and particle velocity, $\underline{v}$.

Faraday's Law implies preservation of the divergence equation div $\underline{B} = 0$, which thus may be regarded as an initial condition.

3. FACTORIZATION ANSATZ

By inspection of the subsidiary conditions for MHD one notes that time dependence factorizes in all of the variables, provided the displacement gradient is a function of time only,

$$F_{ij} = F_{ij}(t) \quad (8)$$

Once factorized, ideal MHD motion reduces to a dynamical system for the nine components of $F_{ij}(t)$. Hamilton's principle then acquires the matrix form

$$0 = \delta \int dt \left[\frac{1}{2} \text{Tr}\,(\dot{F} I^0 \dot{F}^T) - \pi^0 E(\det F) - \frac{\text{Tr}(F S^0 F^T)}{\det F} \right] \quad (9)$$

with variations $\delta F_{ij}(t)$ and constants I^0, Π^0, S^0 defined by integrals over the initial distributions of matter and magnetic fields,

$$I^0{}_{kl} = \int d^3x^0 \, \rho^0 \, x^0_k x^0_l \qquad \begin{pmatrix}\text{initial moment}\\ \text{of inertia}\end{pmatrix}$$

$$\pi^0 = \int d^3x^0 \, p^0 \qquad \begin{pmatrix}\text{initial, integrated}\\ \text{pressure}\end{pmatrix} \qquad (10)$$

$$S^0{}_{kl} = \int d^3x^0 \, \frac{B^0_k B^0_l}{8\pi} \qquad \begin{pmatrix}\text{initial, integrated}\\ \text{magnetic stress}\end{pmatrix}$$

Variation of the action with respect to generalized coordinates $F_{ij}(t)$ produces the following motion equation when one chooses the polytropic adiabat $p/\rho^\gamma = p_0/\rho_0^\gamma$,

$$\ddot{F} \, I^0 F^T = \mathbb{1}\left[\frac{\pi^0}{(\det F)^{\gamma-1}} + \frac{\text{Tr}(FS^0 F^T)}{\det F}\right] - 2\frac{FS^0 F^T}{\det F} \qquad (11)$$

In the case that magnetic stress tensor S^0_{kl} is absent and the initial moment of inertia is unity, $I^0{}_{kl} = \delta_{kl}$, one recovers Dyson's equations for the spinning gas cloud.

4. MATHEMATICAL REMARKS

Before discussion of the motion of the fluid in detail, let us remark briefly on a mathematical aspect of the factorization Ansatz. The particle trajectories $\underline{x}(t,\underline{x}^0)$ arise from a smooth one-to-one mapping g_t that depends on time.

$$g_t : R \times R^3 \to R^3 \qquad \underline{x}(t,\underline{x}^0) = g_t(\underline{x}^0) \qquad (12)$$

A smooth, one-to-one mapping whose inverse is also smooth is called a diffeomorphism, and such diffeomorphisms form a Lie group under functional composition.

The fluid velocity along each particle path is related to the curve of mappings g_t in the diffeomorphism group by

$$\dot{g}_t(\underline{x}^0) = \frac{d}{dt} g_y(\underline{x}^0) = \underline{v}\left(t, g_t(\underline{x}^0)\right) \qquad (13)$$

or equivalently

$$\underline{v}(t,\underline{x}) = \dot{g}_t \bullet g_t^{-1}(\underline{x}) \qquad (14)$$

Thus, the fluid velocity field along the particle trajectories is determined from the Lie algebra of vector fields associated with the diffeomorphism group.

Likewise, the displacement gradient is identified with $dg_t \cdot g_t^{-1}(\underline{x})$, the Jacobian of the map g_t. When this Jacobian is a function of only time, the particle paths become linear functions of the initial coordinates

$$x_i(t,\underline{x}^0) = F_{ij}(t) x^0_j \qquad (15)$$

where the time dependent matrix $F_{ij}(t)$ represents a general linear transformation of Cartesian coordinates, i.e., $F(t) \in GL(3,R)$. Thus, the diffeomorphism group for fluids specializes to the Lie Group GL(3,R) when the Jacobian of the evolution map depends only on time.

5. DYNAMICAL DESCRIPTION IN THREE DIMENSIONS

After the factorization Ansatz, the velocity along particle paths is given by

$$v_i(t,\underline{x}) = [\dot{F} F^{-1}(t)]_{ij} x_j \qquad (16)$$

or, in terms of Lagrange coordinates,

$$v_i(t,\underline{x}^0) = \dot{F}_{ij}(t) x^0_j \quad . \qquad (17)$$

The magnetic field also evolves by a linear transformation

$$B_i(t,\underline{x}^0) = F_{ij}(t) B^0 \qquad (18)$$

and the density undergoes a time-dependent scaling

$$\rho(t,\underline{x}^0) = \rho^0 / \det F \qquad (19)$$

while the specific entropy is constant along each particle trajectory.

When this time dependence for $\rho, s, \underline{v}, \underline{B}$, is substituted into Hamilton's principle for MHD, there results the dynamical system (11) for the linear transformations $F_{ij}(t)$. Such transformations stretch the initial configuration of particles, and rotate the particle configuration relative to both Eulerian and Lagrange coordinate frames. Accordingly the displacement gradient $F_{ij}(t)$ may be decomposed into a matrix product

$$F = R_1 D R_2 \tag{20}$$

where R_1 and R_2 are orthogonal and D is diagonal. Each matrix depends upon time, and the decomposition $F = R_1 D R_2$ turns out to separate the motion into Eulerian rotations (R_1), dilations (D), and Lagrange rotations (R_2), the last of which represent circulatory motions of the fluid.

Upon substitution of the triple product $F(t) = R_1 D R_2$ into the dynamical system (11) one obtains the following separated equations

$$\dot{J} = 0$$

$$\dot{K} = \frac{2}{\det F} [S^o, F^T F] \tag{21}$$

$$\ddot{D} = -\frac{\partial}{\partial D} U(D)$$

where the skew-symmetric matrices J, K represent fluid angular momentum and circulation respectively

$$J = \int d^3x \, \rho \left[x_i v_j - x_j v_i \right] = F \dot{F}^T - \dot{F} F^T \quad (I^o = \mathbf{1})$$
$$K = F^T (\text{curl } v) F = F^T \dot{F} - \dot{F}^T F \tag{22}$$

The bracket in the $\dot{K}$ equation is the matrix commutator, and the potential function $U(D)$ in the equation for the dilation matrix D is given by

$$U(D) = \frac{1}{4} \text{Tr} (\omega_1 L + \omega_2 N) + E(\det D) + \frac{\text{Tr} S * D^2}{\det D} \tag{23}$$

with dynamical quantities ω_1, ω_2, Ln, N, S* defined by

$$\omega_1 = -R_1^{-1} \dot{R}_1$$

$$\omega_2 = \dot{R}_2 R_2^{-1}$$

$$L = R_1^{-1} J R_1 = D^2\omega_1 + \omega_1 D^2 - 2 D\omega_2 D \qquad (24)$$

$$N = R_2^{-1} K R_2 = D^2\omega_2 + \omega_2 D^2 - 2 D\omega_1 D$$

$$S^* = R_2 S^0 R_2^{-1}$$

The quantities ω_1, ω_2 are angular velocities of rotation and circulation respectively. The quantities L,N represent the angular momentum and circulation expressed in fixed, Eulerian coordinates. Finally $S^* = R_2 S^0 R_2^{-1}$ is the magnetic stress tensor referred to the fixed Eulerian frame.

The equations of motion (21) for J, K, and D first of all express conservation of fluid angular momentum, J. The circulation K is also conserved provided the magnetic stress tensor S^0_{ij} can be simultaneously diagonalized with the initial mass distribution I^0. However when the commutator $[S^0, F^T F]$ does not vanish, the circulation experiences a restoring torque due to magnetic stresses which are developed as the lines of magnetic field wind around themselves during fluid circulation. Finally the last equation for the dilation matrix D expresses the coupling between expansion of the fluid and its circulation and rotation. With the expansion potential U(D) the centrifugal, thermodynamical, and magnetic forces are each conservative, so energetic trade offs among them are clear.

6. COMPARISON WITH THE GYROSCOPE EQUATIONS

The reduced equations for J, K, and D express the fluid motion in the co-moving Lagrange frame. When transferred to the fixed Eulerian frame the resultant equations for $L = R_1^{-1} JR_1$ and $N = R_2^{-1} KR_2$ closely resemble the gyroscope equations expressed in body coordinates.

In body coordinates within a gyroscope that spins with angular velocity ω in a uniform gravitational field (g = (0,0,-g) in fixed coordinates) the equations of motion for angular momentum $M = I\omega$ take the matrix form

$$\dot{M} + [\omega, M] = [g, C]$$

$$M = D^2\omega + \omega D^2 \qquad (25)$$

$$\dot{g} + [\omega, g] = 0$$

where C corresponds to the center-of-mass vector in the body, and the well-known correspondence, e.g., $C_{ij} = \epsilon_{ijk} C_k$ between $O(3)$ and R^3 has been used.

In order to compare with the gyroscope equations (25) one expresses the factorized equations for J,K in terms of their fixed-frame representatives L,N as

$$\dot{L} + [L, \omega_1] = 0$$

$$\dot{N} + [N, \omega_2] = \frac{2}{\det D}[S^*, D^2] \qquad (26)$$

$$\dot{S}^* + [S^*, \omega_2] = 0$$

Thus when $\omega_1 = 0 = \dot{D}$, for MHD fluid circulations with fixed shape and Eulerian orientation, the equation for fluid circulation in the fixed Eulerian frame, N, has an analog with the gyroscope equation for angular momentum in the moving frame, M, under the following identifications

$$M \leftrightarrow N$$

$$\omega \leftrightarrow -\omega_2 \qquad (27)$$

$$g \leftrightarrow S^*$$

$$C \leftrightarrow 2D^2/\det D$$

The analog is not exact though, because S^*, D^2 are symmetric matrices while g, C are skew-symmetric.

Thus, the equations for MHD motion with linear velocity profiles separate into two gyroscope equations which are coupled to each other by magnetic stresses and by deformations of shape. When the magnetic and material distributions can be simultaneously diagonalized, the angular motion becomes torque-free motion on $O(3) \times O(3)$, which can be further combined into geodesic motion on $O(4)$ by standard methods, see Holm (1982). In that case for motions with fixed shape the equations are completely integrable.

7. PLANAR FLOW: PENDULUM EXAMPLE

When constrained to rotate in a single plane the gyroscope reduces to a simple pendulum. Likewise the MHD circulation equation for N in fixed coordinates reduces to the equation for a simple pendulum in the case of planar MHD flow with a fixed elliptical boundary.

Consider planar circulation of a MHD fluid within a fixed ellipse whose principal axes (d_2, d_3) are aligned with the coordinate axes of the vertical (x_2, x_3) plane. Because of the problem statement $\omega_1 = 0 = \dot{D}$, and the dynamical equation that remains is

$$\dot{N} + [N, \omega_2] = \frac{2}{\det D} [S^*, D^2] \tag{28}$$

where the quantities N, ω_2, D, S^* are given by

$$N = D^2 \omega_2 + \omega_2 D^2 = \dot{\phi}(t) \; (d_2^2 + d_3^2) \begin{pmatrix} 0 & 1 \\ -1 & 0 \end{pmatrix}$$

$$\omega_2 = \dot{R}_2 R_2^{-1}(\phi) + N / (d_2^2 + d_3^2) \tag{29}$$

$$D = \begin{pmatrix} d2 & 0 \\ 0 & d3 \end{pmatrix}$$

$$S^* = R_2(\phi+\alpha) \begin{pmatrix} S_2 & 0 \\ 0 & S_3 \end{pmatrix} R_2^{-1}(\phi+\alpha)$$

with α the angle of rotation whereby S^o is diagonalized,

$$R_2(\alpha) = \begin{pmatrix} \cos\alpha & \sin\alpha \\ -\sin\alpha & \cos\alpha \end{pmatrix}$$

$$\tan\alpha = 2 B_2^o B_3^o / \left[\left(B_2^o\right)^2 - \left(B_3^o\right)^2 \right] \tag{30}$$

Upon substitution of these definitions into the circulation equation, the pendulum equation emerges

$$\ddot{\phi} = -A \sin(\phi+\alpha) \qquad (31)$$

with natural frequency-squared A given by

$$A = \frac{d_2^2 - d_3^2}{d_2^2 + d_3^2} \frac{S_2-S_3}{d_2 d_3} \qquad (32)$$

Thus, the gyroscopic analog for MHD provides an interpretation of planar circulation in terms of pendulum motion. The particle trajectories for fluid circulation are defined by

$$x_i = \left[D\, R_2\left(\phi(t)\right)\right]_{ij} x_j^0 \qquad (33)$$

from which it follows that div $\underline{v}$ - div $\underline{\dot{x}}$ = 0. For this flow the density of the fluid is constant, and the magnetic field varies according to

$$\begin{pmatrix} B_1 \\ B_2 \end{pmatrix} = \frac{1}{d_2 d_3} \begin{pmatrix} d_2 & 0 \\ 0 & d_3 \end{pmatrix} \begin{pmatrix} \cos\phi & \sin\phi \\ -\sin\phi & \cos\phi \end{pmatrix} \begin{pmatrix} B_1^0 \\ B_2^0 \end{pmatrix} \qquad (34)$$

Thus the co-ordinates of the fluid particles and the magnetic field coordinates undergo pendulum motion within the ellipse.

At the boundary of the ellipse

$$\text{Tr}\,(X^T D^{-2} X) - 1 = A(\underline{x}) = 0 \qquad (35)$$

the normal components of both velocity and magnetic field vanish, provided the initial magnetic field B_i^0 is linearly related to Lagrange coordinates by

$$\begin{pmatrix} B_1^0 \\ B_2^0 \end{pmatrix} = b \begin{pmatrix} 0 & 1 \\ -1 & 0 \end{pmatrix} \begin{pmatrix} x_1^0 \\ x_2^0 \end{pmatrix} \qquad (36)$$

where b is a constant of proportionality. Thus the condition of impermeability at the wall is satisfied for this solution, and both the velocity and magnetic field are divergenceless.

$$\underline{v} \cdot \nabla A = 0 = \underline{B} \cdot \nabla A \qquad (37)$$

$$\text{div } \underline{v} = 0 = \text{div } \underline{B}$$

The ensuing motion within the ellipse for a uniformly dense, magnetic field is a type of "angular sloshing", for which the embedded magnetic field provides a pendular restoring force.

REFERENCES

1. S.I. Anisimov and Yu.I. Lysikov (1970) Expansion of a Gas Cloud in a Vacuum. PMM $\underline{34}$, 926.
2. A.B. Basset (1888) A Treatise on Hydrodynamics, Volume II. Reprinted by Dover Publications, Inc., New York, 1961.
3. S. Chandrasekhar (1969) Ellipsoidal Figures of Equilibrium, Yale University Press, New Haven.
4. F.J. Dyson (1968) Dynamics of a Spinning Gas Cloud. J. Math. Mech. $\underline{18}$, 91.
5. D.D. Holm (1981) Magnetic Tornadoes: Three-dimension Affine Motions in Ideal Magnetohydrodynamics. Physica D, to appear.
6. H. Lamb (1932) Hydrodynamics, Sixth edition. Reprinted by Dover Publications, Inc., New York, 1945.
7. L.V. Ovsjannikov (1956) New Solution of Hydrodynamic Equations. Dokl. Acad. Nauk SSSR $\underline{111}$, 47.
8. E.N. Parker (1956) The Gross Dynamics of a Hydromagnetic Gas Cloud. Astrophys. J. Suppl $\underline{3}$, 51.
9. E.N. Parker (1979) Cosmical Magnetic Fields. Clarendon Press, Oxford.

GAUGE GROUPS AND NOETHER'S THEOREM FOR CONTINUUM MECHANICS

Frank S. Henyey
CSND, La Jolla Institute, P. O. Box 1434, La Jolla, CA 92038

ABSTRACT

The group of "gauge" transformations of variables which arise in the Hamiltonian formulation of dissipation-free continuum systems is discussed. These transformations leave all physical quantities invariant. The quantities which are thereby conserved as a result of Noether's theorem are discussed in the cases that the group is Abelian or non-Abelian. Magnetohydrodynamics and stratified fluids are examples of these two cases.

Variational formulations, especially in Hamiltonian form, of dissipation-free continuum systems, are useful for a number of reasons. Perturbation theory can be systematically developed in a Hamiltonian formulation, variational techniques can be applied, other approximations which preserve important features of the system can be made, the eikonal expansion can easily be carried out with automatic action conservation, etc.

There is a motivated recipe for constructing a variational principle[1,2,3]. A Lagrangian is written down as kinetic energy minus potential energy. Variables are then constrained to, for example, conserve mass. Additional constraints ensure the part of Hamilton's principle that says that the initial and final configurations are not free to vary.

In the construction of the variational principle for a dissipation-free continuum system, three types of terms may occur in the Lagrangian density. 1) Energy terms, such as $\rho v^2/2$; 2) Physical constraints with Lagrange multipliers, such as $\lambda(\partial_t \rho + \nabla \cdot (\rho \vec{v}))$; and 3) Lin constraints[4] with their multipliers such as $\alpha(\partial_t \beta + \vec{v} \cdot \nabla \beta)$.

It automatically turns out that λ is canonically conjugate to ρ, and α to β, due to the presence of $\lambda \partial_t \rho$ and $\alpha \partial_t \beta$ in the Lagrangian density. The remainder is the negative of the Hamiltonian density:

$$L + \Sigma p_i \partial_t q_i - H$$

The variables λ, α, β, and other similar variables are unphysical; i.e., their value cannot be determined in terms of $\vec{v}$, ρ, etc. The velocity $\vec{v}$ does not appear as a dynamical variable. Its time derivative nowhere appears, and it has no conjugate variable whose time derivative it multiplies. Therefore, it can be eliminated algebraically and can be written in terms of other variables, including the unphysical variables. There must be at least three unphysical variables in order to allow arbitrary initial conditions. If there are fewer without a Lin constraint, one must be imposed. Assuming multivalued unphysical variables are allowed, only enough Lin constraints to allow arbitrary initial conditions are needed. The expression for the velocity often resembles the Clebsch[5] representation, $\vec{v} = \nabla a + b\nabla c$.

Depending on the physics under consideration, there may be enough physical variables with unphysical conjugate variables so that Lin constraints are superfluous, or there may be too few. A case where there are enough is magnetohydrodynamics, where the physical variables are the density, the entropy density, and the various components of the magnetic field. In contrast, a case in which Lin constraints are required is a stratified fluid in which the density and the entropy density or their equivalents are the only physical variables with conjugate variables. (For an incompressible fluid, $\nabla \cdot \vec{v} = 0$ takes the place of the entropy transport equation.)

If one follows the recipe that has been briefly sketched, for any dissipation-free continuum mechanical system, one obtains:
1) Pairs of canonical variables, p_j, q_j. The p_j's are all unphysical and q_j's are physical except when Lin constraints are needed, where both p_j and q_j are unphysical.
2) An expression for the velocity $\vec{v} = \vec{v}(p_j, q_j)$, involving at least three unphysical variables.
3) A Hamiltonian density, which is just the energy density with the velocity expressed in terms of the p_j's and q_j's.

The dynamical equations are Hamilton's canonical equations. There is a close analogy with the Hamiltonian formulation of Maxwell's equations. The canonical pairs are the components of the electric field, E_j, and the vector potential, A_j. The physical magnetic field $\vec{B}$ is given by $\nabla \times \vec{A}$. The Hamiltonian is just the energy, $\frac{1}{2}\epsilon_0 E^2 + \frac{1}{2}\mu_0 B^2$, with $\vec{B}$ expressed in terms of $\vec{A}$. Only two components of $\vec{B}$ are independent, since $\nabla \cdot \vec{B} = 0$. Yet $\vec{A}$ has three components. As a consequence, the theory has (local) gauge invariance $\vec{A} \rightarrow \vec{A} + \nabla \Lambda$. Noether's theorem on Hamiltonian systems asserts that the generator of any invariance is a conserved quantity. In the case of electromagnetism, the generator is $\nabla \cdot \vec{E}$, the electric charge. Thus, a very physical conservation law is associated with an invariance under transformations of unphysical variables.

Within the last ten years, other gauge theories have become accepted as a description of nuclear particles. These Yang-Mills theories[6] also have unphysical potentials. The nature of the gauge invariance is quite different, however. The transformations are described by a non-Abelian group rather than an Abelian group, as is the case with electromagnetism. It has been found that the phenomena arising in the theory have very important differences due to the non-Abelian nature of the group.

The charges conserved by virtue of Noether's theorem are also different in Yang-Mills theories. The charge densities are unphysical and change under a gauge transformation. This property is a direct consequence of the non-Abelian character of these theories.

We may also expect a gauge group for continuum systems if the number of unphysical quantities exceeds three, the number required to describe the velocity. The word "gauge" is being used here only to describe the transformations of unphysical variables that leave all physical variables invariant. Electromagnetism and Yang-Mills theories share other features which are not shared by the continuum theories. These additional features are considered by many to be an essential part of the definition of the word "gauge."

We may further expect that there are important consequences of the nature of the gauge group. In particular, the Abelian or non-Abelian property and the Noether relationship between gauge transformations and conserved quantities are significant. It turns out that these differences hinge on the existence of Lin constraints.

We first examine the case without Lin constraints. MHD is an example of this case. The generator must be independent of the p_j's, since all q_j's are physical, so $0 = \delta q_j = -\frac{\delta\phi}{\delta p_j}$. The dependence of ϕ on q_j is only constrained by $0 = \delta \vec{v}(p_j, q_j)$. The dependence of $\vec{v}$ on p_j comes from terms in the Lagrangian like $\lambda \vec{v} \cdot \nabla\rho$, which leads to a term in $\vec{v}$ like $-\frac{\lambda}{\rho}\nabla\rho$. Therefore, the velocity depends linearly on the p_j's. The condition on ϕ thus is $0 = \Sigma \frac{\partial \vec{v}}{\partial p_j} \frac{\delta\phi}{\delta q_j}$. This is a linear system of PDE's for ϕ, solutions superpose, and the generating function for a composition of two gauge transformations is the sum of the individual generators. Since addition is commutative, so is the composition of gauge transformations. This is precisely the definition of an Abelian group. The generators $\phi(q_j)$ are the Noetherian conserved quantities. Since they are independent of p_j, they are all physical quantities.

For MHD, the general generator is

$$\phi = \int d^3\vec{r} \left[F(\nabla \cdot \vec{B}, \vec{r}) + \rho G(s, \frac{\vec{B}}{\rho} \cdot \nabla s, \frac{\vec{B}}{\rho} \cdot \nabla(\frac{\vec{B}}{\rho} \cdot \nabla s), \ldots) \right],$$

where ρ is the density, s is the entropy per unit mass, and $\vec{B}$ is the magnetic field. F and G are arbitrary functions.

The F term originates because $\nabla \cdot \vec{B} = 0$ is imposed as an initial condition in the formulation of MHD I use. The first term implies $\partial_t(\nabla \cdot \vec{B}) = 0$.

The conservation of the G term is equivalent to:

1. $d_t s \equiv \partial_t s + \vec{v} \cdot \nabla s = 0$
2. If $d_t a = 0$ then $d_t(\vec{B}/\rho \cdot \nabla a) = 0$; starting with a = s and iterating, an infinite set of such quantities can be generated.

3. ρ times any function G of such variables obeys the equation $d_t(\rho G) + G\nabla \cdot \vec{v} = 0$ and is therefore a conserved density.

The first of these is conservation of entropy and the second expresses the "freezing in" of magnetic field lines to the plasma.

In the case that Lin constraints are required, the situation is quite different. The velocity now has terms which are quadratic in the unphysical Lin variables, such as

$$-\alpha \nabla \beta / \rho.$$

The PDE's are now nonlinear, their solutions do not superpose, and the gauge group is non-Abelian.

An example of this case, of interest in oceanography, is an incompressible stratified fluid in a rotating reference frame. In this problem, as I formulate it, the variables are the vertical Lagrangian coordinate C, defined in reference to the stable equilibrium solution of unmoving horizontal stratified layers, its conjugate variable which is unphysical, and variables A and B, which are essentially the conjugate pair of unphysical Lin variables. Given the quantities of fluid of each density (or equivalently, the Brunt-Väisälä frequency $N = \sqrt{(g/\rho)\partial\rho/\partial C}$), one can uniquely relate the density and C.

A, B, and C are "Pseudo-Lagrangian" coordinates, i.e.,

$$(\partial_t + \vec{v} \cdot \nabla)(A, B, C) = 0.$$

They are not true Lagrangian coordinates since their Jacobian with respect to Eulerian coordinates

$$J = \nabla C \cdot (\nabla A \times \nabla B) \neq 1$$

in general.

The general (infinitesimal) generator of the gauge transformation is an arbitrary function of A, B, and C but is independent of the variable conjugate to C. The "conservation" law is

$$(\partial_t + \vec{v} \cdot \nabla)F(A, B, C) = 0.$$

Of these, only those F's which are independent of A and B are physical. Most conservation laws are unphysical just as with the Yang-Mills non-Abelian theories. The group generated by F is an infinite direct product of identical groups, one for each value of C. For fixed C, the group is the canonical group in one pair of variables A and B. The physical interpretation is that this is the group of area-preserving deformations of layers of the static stable stratified fluid.

As a consequence of $(\partial_t + \vec{v} \cdot \nabla)(A, B, C) = 0$, the Jacobian J also obeys $(\partial_t + \vec{v} \cdot \nabla)J = 0$. It turns out that J is a physical variable, $fJ = \nabla C \cdot (\nabla \times \vec{v} + \vec{f})$, where $\vec{f}$ is the vorticity due to the coordinate system rotation, which is called the inertial frequency. fJ is known as the potential vorticity, and its integral is the vertical angular momentum. J is, however, not a gauge generator, but generates an invariance which changes physical quantities.

It is a pleasure to thank K. Watson for his interest in and comments on the work of which this is a part. Discussions with J. Greene and P. Morrison are appreciated. Research support was provided by La Jolla Institute.

References

1. R. L. Seliger and G. R. Whitham
   Proc. Roy. Soc. $\underline{A305}$, 1 (1968)

2. F. Henyey, "Canonical Construction of a Hamiltonian for Dissipation-Free MHD" La Jolla Inst. Preprint. LJI-R-81-150

3. F. Henyey, "Hamiltonian Description of Stratified Fluid Dynamics" La Jolla Inst. preprint, LJI-R-81-162.

4. C. C. Lin "Hydrodynamics of Helium II" in Proc. Int. School Phys. "Enrico Fermi," course XXI (1963)

5. H. Lamb, Hydrodynamics, (Cambridge, 1932) Sec. 167

6. C. N. Yang and R. Mills Phys. Rev. $\underline{96}$, 191 (1954)

7. P. J. Morrison and J. M. Greene, Phys. Rev. Lett $\underline{45}$, 790 (1980)

NONCANONICAL HAMILTONIAN MECHANICS

John M. Greene
Plasma Physics Laboratory, Princeton University
Princeton, New Jersey 08544

A number of books and papers contain the best formulations of noncanonical Hamiltonian mechanics [1-3]. Here, instead, the object is to present the essential ideas in their most simple-minded form. The root idea is that quantities are "real" if they are independent of the coordinate system used to describe them. In orthodox Hamiltonian mechanics a restricted set of coordinate systems is used and interesting quantities are invariant under the canonical transformations that relate these coordinates. Here more general coordinate systems will be used, with a reality condition taken from tensor calculus.

To start at the beginning, the Hamiltonian equations are customarily written

$$\frac{dq_i}{dt} = \frac{\partial H(\vec{q},\vec{p})}{\partial p_i} \quad , \quad \frac{dp_i}{dt} = -\frac{\partial H(\vec{q},\vec{p})}{\partial q_i} \tag{1}$$

where $\vec{q} = (q_1, \ldots, q_n)$ and $\vec{p} = (p_1, \ldots, p_n)$ represent respectively n position coordinates and n momenta. In a notation that does not distinguish between position and momentum coordinates, this can be rewritten

$$\frac{dz^i}{dt} = J^{ij} \frac{\partial H(\vec{z})}{\partial z^j} \tag{2}$$

where the 2n phase space coordinates z^i are $\vec{z} = (\vec{q},\vec{p})$, the cosymplectic form J^{ij} is given by

$$J^{ij} \equiv \begin{pmatrix} 0 & I \\ -I & 0 \end{pmatrix} \quad , \tag{3}$$

I is the unit n × n matrix, and the Einstein summation convention is used.

Consider a time independent transformation to new coordinates $\hat{z}^i$, and its inverse,

$$\hat{z}^i = \hat{f}^i(\vec{z}) \quad , \quad z^i = f^i(\vec{\hat{z}}) \quad . \tag{4}$$

Then, by the chain rule,

$$\frac{d\hat{z}^i}{dt} = M_{ij} \frac{dz^j}{dt} \tag{5}$$

where

$$M_{ij} \equiv \frac{\partial \hat{f}^i}{\partial z_j} \tag{6}$$

Quantities that transform under change of coordinates according to Eq. (5) are called contravariant vectors, or contravariant rank one tensors, and will be characterized by a superscript. Similarily,

$$\frac{\partial H}{\partial \hat{z}^i} = M^*_{ij} \frac{\partial H}{\partial z^j} \tag{7}$$

where differentiating Eq. (4) shows that M^* is the inverse of the transpose of M. Quantities that transform this way are called covariant vectors, and will be denoted by a subscript.

If Eq. (2) is to be independent of coordinates, the matrix J must transform according to

$$\hat{J}^{ij} = M_{ik} M_{jl} J^{kl} = MJM^\dagger \; . \tag{8}$$

in matrix notation, where $M^\dagger$ is the transpose of M. That is, it must transform as a contravariant second rank tensor.

In short, since $M^\dagger M^*$ is the unit matrix, multiplication of a contravariant component by a covariant component yields a coordinate independent result, which is the fundamental idea of this note.

It is natural to define a scalar inner product based on J,

$$[F,G] \equiv \left(\frac{\partial F}{\partial z^i}\right) J^{ij} \left(\frac{\partial G}{\partial z^j}\right) \; . \tag{9}$$

This is known as the Poisson bracket. It is particularly useful since any function $F(\vec{z})$ satisfies, from Eq. (2),

$$\frac{dF(\vec{z})}{dt} = [F,H] \; . \tag{10}$$

The inverse of J is called the symplectic form, and is denoted by ω.

It is interesting to draw an analogy between metric geometry and this Hamiltonian structure, which is called symplectic geometry. Here the matrix J plays the role of the metric tensor. The analogy is of no use in providing results, but it does give a structure for thinking about symplectic geometry. One thing to note is that no distinction is made between the metric tensor and its inverse, but J and its inverse are given different names and symbols. Also, in metric geometry $\partial H/\partial z^i$ and $J\partial H/\partial z^i$ would not be considered to be physically different quantities, but would be called the covariant and contravariant components of a more fundamental entity. Canonical coordinates are those for which J retains the form of Eq. (3). This is roughly analogous to orthogonal coordinates for which the metric tensor is diagonal. In either case it is not necessary to carefully distinguish covariant and contravariant components, but in canonical symplectic geometry it is necessary to mind the p's and q's since the canonical J mixes them.

The next step is to identify those characteristics of J that are invariant under transformation. The canonical J of Eq. (3) is antisymmetric and nonsingular, Det $J \neq 0$, and so it will have these properties in all coordinate systems. The story of singular J's is given by Littlejohn in another paper in these proceedings [4].

In crude analogy to the curvature tensor of metric geometry, a third rank contravariant tensor can be formed from J and its derivatives,

$$S^{ijk} \equiv J^{i\ell}\frac{\partial J^{jk}}{\partial z^\ell} + J^{j\ell}\frac{\partial J^{ki}}{\partial z^\ell} + J^{k\ell}\frac{\partial J^{ij}}{\partial z^\ell} \ . \tag{11}$$

This is shown to be a tensor by subsituting Eq. (8) and using the antisymmetry of J to cancel all terms containing derivatives of M. The tensor S is antisymmetric under odd permutations of its indices. The scalar formed from gradient vectors and S is

$$\frac{\partial F}{\partial z^i}\frac{\partial G}{\partial z^j}\frac{\partial H}{\partial z^k} S^{ijk} = [F,[G,H]] + [G,[H,F]] + [H,[F,G]] \tag{12}$$

whose vanishing is the Jacobi identity. Thus S will be called the Jacobi tensor.

Note that the Jacobi tensor vanishes in canonical coordinates, and thus must vanish in any system obtained by coordinate transformation. On the other hand, it is quite possible to find cosymplectic tensors J whose Jacobi tensor does not vanish. They lead to a non-Hamiltonian dynamics.

Fortunately, we have Darboux's theorem that any antisymmetric, nonsingular J with vanishing Jacobi tensor can be brought into canonical form by a coordinate transformation. Thus any set of equations of the form of Eq. (2) with such a J is Hamiltonian.

Just as a curved space can be embedded in a flat space of more dimensions, it is easy to make the Jacobi tensor vanish by adding extraneous independent variables to Eq. (2). It is not clear that this really simplifies any problem.

The covariant components of the Jacobi tensor have a particularly simple form,

$$S_{ijk} = \omega_{i\ell} \omega_{jm} \omega_{kn} S^{\ell mn} = \frac{\partial \omega^{ij}}{\partial z^k} + \frac{\partial \omega^{jk}}{\partial z^i} + \frac{\partial \omega^{ki}}{\partial z^j} \qquad (13)$$

which is linear in ω. This equation is the mathematical pole of the theory, as Eq. (2) is the physical pole. This accounts for some of the notation.

For a number of reasons one is generally interested in the behavior of a bundle of orbits in the neighborhood of a given orbit. For one thing, it is often instructive to know the stability properties of a chosen orbit. More fundamentally, the essential character of Hamiltonian systems is the absence of attracting orbits. In fact, in a 2n dimensional phase space, there is at most an n dimensional manifold of orbits that asymptotically approach a given orbit at infinite positive or negative time. This is a strong restriction. Some aspects of this will be discussed in the remainder of this note.

The linearized equation for the displacement between two nearby orbits is

$$\frac{d}{dt} \delta z^i = \frac{\partial (J^{ij} \partial H / \partial z^j)}{\partial z^k} \delta z^k \qquad (14)$$

where the quantity in square brackets is a function of time obtained from a solution of Eq. (2), and $\vec{\delta z}$ is the contravariant vector difference between two orbits. It is convenient to define a linear Hamiltonian matrix by

$$H^{"}_{m,n} \equiv \left(\frac{\partial^2 H}{\partial z^m \partial z^n}\right) + \omega_{mk}\left(\frac{\partial J^{k\ell}}{\partial z^n}\right) \frac{\partial H}{\partial z^\ell} \qquad . \qquad (15)$$

Then Eq. (14) becomes

$$\frac{d}{dt} \vec{\delta z} = JH'' \vec{\delta z} \quad . \tag{16}$$

There is a formal problem here that $d\,\vec{\delta z}/dt$ and H'' are not tensors. In fact, for example,

$$\frac{d\hat{\delta z}^i}{dt} = M_{ij} \frac{d\delta z^j}{dt} + \delta z^j \frac{\partial M_{ij}}{\partial z^k} \frac{dz^k}{dt} \tag{17}$$

and the second term on the right is not consistent with the transformation law of Eq. (5). However, H'' is also not a tensor in just such a way that $\vec{\delta z}$ calculated from Eq. (16) is a tensor. For reference, note that the coodinates z^i of Eq. (2) do not form a vector, since the matrix M of Eq. (6) is not relevant to their transformation, but dz^i/dt is a vector. That is, the time derivative operation changes the tensor character.

It is useful to introduce a time advancement tensor U that contains information about the whole bundle of orbits in the neighborhood of a given orbit,

$$\vec{\delta z}(t) = U(t)\,\delta z(0) \quad . \tag{18}$$

Then U is contravariant with respect to coordinate changes around $\vec{z}(t)$, and covariant with respect to coordinate changes around $\vec{z}(0)$, where these values of $\vec{z}$ are points on an orbit that satisfies Eq. (2). It evolves in time according to

$$\frac{dU}{dt} = JH''U \tag{19}$$

with the initial condition that U(0) is the identity.

To obtain information about the eigenvalues of U we first establish the relation

$$\omega[\vec{z}(t)]\,U(t)\,J[\vec{z}(0)] = U^*(t) \tag{20}$$

where U^* is the inverse of the adjoint of U. This is a generalization of the definition of a symplectic matrix given by Arnold and Avez [5]. To show this, note that the linear

Hamiltonian matrix can be used to derive an equation for the evolution of J,

$$\frac{dJ^{ij}}{dt} = \left(\frac{\partial J^{ij}}{\partial z^k}\right)\frac{dz^k}{dt} = \left(\frac{\partial J^{ij}}{\partial z^k}\right) J^{k\ell} \frac{\partial H}{\partial z^\ell}$$

$$= J^{im}(H''_{mn} - H''_{nm})J^{nj} - S^{ij\ell}\frac{\partial H}{\partial z^\ell} \qquad (21)$$

and

$$\frac{d\omega_{ij}}{dt} = H''_{ji} - H''_{ij} + S_{ijk} J^{k\ell}\frac{\partial H}{\partial z^\ell} \qquad . \qquad (22)$$

Then, when $S = 0$,

$$\frac{d[\omega(t)U(t)J(0)]}{dt} = H''^\dagger J(t) [\omega(t)U(t)J(0)] \qquad . \qquad (23)$$

From the definition of the star operation as the inverse of the transpose we find

$$\frac{dU^*}{dt} = -U^*\left(\frac{dU^\dagger}{dt}\right)U^*$$

$$= H''^\dagger JU^* \qquad (24)$$

with the aid of the transpose of Eq. (19). Thus $\omega(t)U(t)J(0)$ and $U^*(t)$ satisfy the same equation with the same initial condition, and are equal.

Now consider the eigenvalue problem

$$U - \lambda K = 0 \qquad (25)$$

where K is some matrix satisfying $\omega(t)KJ(0) = K^*$. This vague statement is useful either for coordinates for which J is independent of $\vec{z}$, or for closed periodic orbits so that $\vec{z}(t) = \vec{z}(0)$. In either case we take K to be the identity. Then

$$\text{Det}(U - \lambda K) = \frac{\text{Det } \omega(t)(U - \lambda K) J(0)}{\text{Det } \omega(t) J(0)}$$

$$= \frac{\text{Det}(U^* - \lambda K^*)}{\text{Det } \omega(t) \ J(0)}$$

$$= \frac{\text{Det}(U^{-1} - \lambda K^{-1})}{\text{Det } \omega(t) \ J(0)}$$

$$= \frac{\text{Det}(K - \lambda U)}{\text{Det } U \ \omega(t) \ J(0) \ K} \tag{26}$$

Thus if λ is an eigenvalue of Eq. (25) then λ^{-1} is also. No more than half the eigenvalues can be larger than unity, or smaller than unity, in absolute value. This is a beginning toward the assertion above that the orbits of Hamiltonian systems are not attractors. It has some of the information contained in the Poincare invariants of the canonical formulation.

In conclusion, if the solution of some practical problem is most simply expressed in a noncanonical coordinate system, there is no cause for regret. An example is given in Ref. [6].

ACKNOWLEDGMENTS

I am grateful to a number of people for teaching me the ideas contained in this note, of whom I should mention Drs. R. C. Churchill, A. N. Kaufman, F. Estabrook, and especially R. G. Littlejohn. The hospitality of Dr. Ken Watson during my year at the LaJolla Institute was greatly appreciated.

This work was supported by the U.S. Department of Energy contract No. DE-AC02-76-CHO3073.

REFERENCES

1. R. Abraham and J. E. Marsden, Foundations of Mechanics (Benjamin, Reading, Mass., 1978).
2. V. I. Arnold, Mathematical Methods of Classical Mechanics (Springer, New York, 1978).
3. R. G. Littlejohn, to be published in Physics Reports.
4. R. G. Littlejohn, these proceedings.
5. V. I. Arnol'd and A. Avez, Ergodic Problems of Classical Mechanics (Benjamin, New York, 1968) p. 221.
6. R. G. Littlejohn, Phys. Fluids 24, 1730 (1981).

PERFECT INVISCID FLUIDS AND GAUGE THEORY

Edward Detyna*
Institute for Plasma Research
Stanford University, Stanford, CA 94305, U.S.A.

ABSTRACT

The classical description of matter is based on two very different pictures: it is viewed as a collection of interacting particles (point-like objects in empty space) or as a continuous medium. The author believes that only the latter picture is appropriate in classical physics--particles should be obtained by quantizing classical continuous matter. For that reason it is important to study Euler's equations--the simplest form of equations of motion of a continuous medium.

In the first part of this paper the variational principle for these equations is developed. It is shown that the ideal fluid posseses spin and if spin zero solutions are sought, the standard Clebsch representation of velocity is obtained.

The Lagrangian depends on several gauge potentials and it is indicated in the last part of this paper that the induced gauge fields can be identified as electromagnetic and gravitational fields.

1. VARIATIONAL PRINCIPLE

The general method of constructing a variational principle when equations of motion are known was proposed in the earlier paper[1]. It states that equations of motion can be found by seeking a stationary value of the difference of kinetic and pottential energies subject to all gauge conservation laws. In other words, the Lagrangian L is

$$L = L_H + \sum_{\alpha = 1}^{N} \lambda_\alpha (\partial_t Q_\alpha + \underline{\nabla} \cdot \underline{F}_\alpha) , \qquad (1.1)$$

where L_H is the Lagrangian, used in Hamilton's principle λ_α are N Lagrange multipliers and the expressions in the bracket are gauge conservation laws: Q_α and $\underline{F}_\alpha$ are the N gauge-conserved quantities and their fluxes, respectively.

---

* On leave of absence from: Mathematics Department
Reading University
Whiteknights
Reading RG6 ZAX
U.K.

The Lagrangian (1.1) should be invariant under a group of infinitesimal transformations of space and time: translations of space and time and rigid rotations of space. By Noether's theorem[2] we can deduce that the Lagrangian (1.1) then induces kinematic conservation laws, namely conservation of momentum, energy and angular momentum.

This method can be viewed as follows: let a physical system be described by a set of equations of motion. From these equations of motion we can construct all the conservation laws admitted by the system. Now we separate them into two families--kinematic and gauge conservations laws. A properly constructed Lagrangian will, by Noether's theorem, induce both families of conservation laws but for different reasons: the kinematic laws are due to invariance of L under space and time transformations while gauge laws are due to gauge invariance of field functions. Therefore, if the system admits gauge conservation laws, the Lagrangian cannot be constructed from physical observables alone and must involve "gauge potentials". The role of these potentials is played by functions λ_α in (1.1).

The ideal inviscid fluid admits all the kinematic conservation laws, namely conservation of energy, momentum (Euler's equation) and angular momentum. The last one is in degenerate form stating that the stress tensor is symmetric.

But we also have conservation of mass

$$\partial_t \rho + \underline{\nabla} \cdot (\rho \underline{v}) = 0 \qquad (1.2)$$

conservation of entropy

$$\partial_t (\rho S) + \underline{\nabla} \cdot (\rho S \underline{v}) = 0 \qquad (1.3)$$

and conservation of "particles" (fluid lines), i.e., we can construct a vector $\underline{a}(\underline{x},t)$ which is constant along fluid lines[3]

$$\partial_t (\rho \underline{a}) + \underline{\nabla} (\cdot \underline{v} \, \rho \underline{a}) = 0 \quad . \qquad (1.4)$$

It is worth noting that the vector $\underline{a}$ may be, but need not be, identified with Lagrangian coordinates of particles.

Now we can construct Lagrangian according to formula (1.1)

$$L = \frac{1}{2} \rho \underline{v}^2 - \rho E(\rho,S) - \phi[\partial_t \rho + \underline{\nabla} \cdot (\rho \underline{v})] \qquad (1.5)$$

$$- \lambda[\partial_t (\rho S) + \underline{\nabla} \cdot (\rho S \underline{v})] - b_j [\partial_t (\rho a_j) + \nabla_i (\rho a_j v_i)]$$

where $E(\rho,S)$ is the internal energy and ϕ, λ and $\underline{b}$ are Lagrange multipliers; $\frac{1}{2}\rho v^2 - \rho E$ is the Lagrangian for Hamilton's principle.

Taking variations with respect to $\underline{v}$, ρ, S and $\underline{a}$ in turn gives equations of motion which are algebraic with respect to these variables. Therefore one could solve them and simplify the Lagrangian by substituting the solutions into (1.5). This would result in the Lagrangian dependent only on the potentials ϕ, λ and $\underline{b}$. This is similar to a method of classical mechanics where it is possible to simplify the description by introducing new variables, if the motion is subject to constraints.

Unfortunately, although elimination of all physical variables leads to a Lagrangian depending on very few variables, it is rather cumbersome. But the elimination of just velocity instead simplifies the description. Taking the variation with respect to $\underline{v}$ gives

$$v_i = -(\nabla_i \phi + S\nabla_i \lambda + a_j \nabla_i b_j) \qquad (1.6)$$

By substituting (1.6) into (1.5) and integrating by parts gives a simplified Lagrangian

$$L = -\rho[\frac{1}{2}(\underline{\nabla}\phi + S\underline{\nabla}\lambda + a_j\underline{\nabla}b_j)^2 + E(\rho,S) - \partial_t\phi - S\partial_t\lambda - \underline{a}\cdot\partial_t\underline{b}] \qquad (1.7)$$

It is worth noting here that the definition of velocity, equation (1.6), is not really needed, since the Lagrangian (1.7) induces conservation of momentum, which by Noether's theorem is

$$\underline{m} = -\sum_{A}^{10} \frac{\partial L}{\partial \partial_t y_A} \underline{\nabla} y_A = -\rho(\underline{\nabla}\phi + S\underline{\nabla}\lambda + a_j\underline{\nabla}b_j) \qquad (1.8)$$

where y_A stands for ten field variables: ρ, ϕ, S, λ, $\underline{a}$, $\underline{b}$. Since $\underline{m}$ is the momentum density it should be written as

$$\underline{m} = \rho\underline{v} \qquad (1.8a)$$

hence we have the definition of velocity identical to (1.6) directly from the Lagrangian.

The equations of motion can now be derived from the variational principle using Lagrangian (1.7) and taking variations with respect to each variable in turn. They are not given here since it is an elementary procedure. But it is worth noting that only time derivatives of potentials are present in the Lagrangian (1.7). That means the field variables can be split into two groups[6]

a) potentials
$$q_\nu \equiv (\phi, \lambda, \underline{b}) \qquad \nu = 1, 2, \ldots, 5 \qquad (1.9a)$$

b) conjugate momenta
$$\pi_\nu \equiv \frac{\partial L}{\partial \partial_t q_\nu} = (\rho, \rho S, \rho \underline{a}) \qquad (1.9b)$$

Thus we can formulate a Hamiltonian H
$$H = \sum_\nu \pi_\nu \partial_t q_\nu - L \qquad (1.10)$$

which with the aid of equations (1.7) and (1.9) turns out to be a standard expression of total energy as expected.

The equations of motion can now be written in Hamiltonian form
$$\partial_t q_\nu = \frac{\partial H}{\partial \pi_\nu} ,$$
$$\partial_t \pi_\nu = -\frac{\partial H}{\partial q_\nu} . \qquad (1.11)$$

2. CONSERVATION LAWS

The Lagrangian (1.7) is invariant under a group of transformations of time and space as well as under the group of gauge transformations:
$$\phi \rightarrow \phi + \delta\phi$$
$$\lambda \rightarrow \lambda + \delta\lambda$$
$$\underline{b} \rightarrow \underline{b} + \delta\underline{b} \qquad (2.1)$$

where $\delta\phi$, $\delta\lambda$ and $\delta\underline{b}$ are constants. This invariance induces three conservation laws of mass (1.2) entropy (1.3) and vector $\underline{a}$ (1.4). This is not surprising since we put these equations into the Lagrangian (1.5) as constraints.

The invariance of L under a group of time and space translations also induces the expected conservation laws of energy and momentum in their standard forms (Euler's equation for the momentum).

But the invariance of L under a group of rigid rotations induces the conservation law of angular momentum in an interesting form. The conserved angular momentum $\underline{J}$ and its flux $\underline{\underline{M}}$ are respectively.

$$\underline{J} = \rho \underline{v} \times \underline{x} + \rho \underline{a} \times \underline{b} \qquad (2.2)$$

and

$$\underline{\underline{M}} = \underline{\underline{T}} \times \underline{x} + \rho (\underline{a} \times \underline{b}) \underline{v}$$

where $\underline{x}$ is the position vector and $\underline{\underline{T}} = \rho \underline{v}\, \underline{v} + \underline{\underline{I}}\, p$ the stress tensor. The two parts of the angular momentum, the kinematic $\rho \underline{v} \times \underline{x}$ part and the intrinsic part $\rho\, \underline{a} \times \underline{b}$ are conserved independently since momentum is conserved and the stress $\underline{\underline{T}}$ is symmetric.

The conserved intrinsic angular momentum or spin

$$\underline{s} = \rho\, \underline{a} \times \underline{b} \qquad (2.3)$$

did not appear in the original formulation of ideal inviscid fluids and has only emerged when additional potentials were introduced into the variational principle. This situation is identical to that of Maxwell's electrodynamics: the equations for the electric and magnetic fields $\underline{E}, \underline{B}$ allow conservation of electromagnetic energy and momentum. Also, since the stress tensor is symmetric, the kinematic angular momentum is trivially conserved. But Maxwell's equations provide no information about the intrinsic angular momentum, which can only be deduced from variational principles <u>after</u> introduction of a suitable electromagnetic potential. We note in passing that the intrinsic angular momentum vanishes identically if the Lagrangian depends on scalar functions only and its value depends on the type of non-scalar functions (vectors, spinors, etc.) Maxwell's equations can be derived from a variety of variational principles involving scalar potentials only or vector potential only. In the former case, the intrinsic angular momentum would be identically zero. This is in direct contradiction with experiments which can measure the angular momentum of, say, circularly polarized waves.

Therefore, a conclusion has to be drawn that the gauge potentials have a deeper physical meaning.

The ideal inviscid fluid conserves spin $\underline{s}$ hence if initially we set $\underline{s} = 0$ the intrinsic angular momentum will be zero at all future times. It is seen that the most general solution in this case is

$$\underline{a} = \alpha \underline{b} \qquad (2.4)$$

where α is an arbitrary scalar function. Noting that the Lagrangian (1.7) depends only on terms of the type

$$\underline{a} \cdot \partial_t \underline{b} = \alpha \partial_t (\tfrac{1}{2} \underline{b}^2)$$

$$\underline{a} \cdot \nabla_i \underline{b} = \alpha \nabla_i (\tfrac{1}{2} \underline{b}^2)$$

We can set $\beta = \tfrac{1}{2} \underline{b}^2$ which is another scalar and then the Lagrangian depends only on these two scalars α and β in the same manner it depended on vectors $\underline{a}$ and $\underline{b}$. Now the velocity representation (1.6) becomes

$$\underline{v} = -(\nabla \phi + S \nabla \lambda + \alpha \nabla \beta) \qquad (2.5)$$

which is the Clebsch representation[4,5].

It is clear that the scalars ϕ, α and β are sufficient to represent any velocity field $\underline{v}$ and the choice of this representation or the more general one (1.6) can only be made on the grounds that the fluid does not or does possess spin. This question cannot be resolved by studying Euler's equations alone, since we already mentioned these equations do not deal with the problems of spin at all.

3. GAUGE THEORY

We have noted earlier that the Lagrangian (1.7) is invariant under the gauge transformation

$$\phi \rightarrow \phi + \delta\phi \qquad (3.1)$$

where $\delta\phi$ is a constant throughout space and time. This induces continuity equation (1.2) by Noether's theorem. In order to make the Lagrangian (1.7) invariant under a local gauge transformation[7]

$$\delta\phi = \mu(\underline{x},t) \qquad (3.2)$$

we have to add to it a vector $\underline{A}$ and scalar ψ

$$L = -\rho[\frac{1}{2}(\underline{\nabla}\phi - k\underline{A} + S\underline{\nabla}\lambda + a_j\underline{\nabla}b_j)^2 + E(\rho,S)$$

$$- \partial_t\phi + k\psi - S\partial_t\lambda - \underline{a}\cdot\partial_t\underline{b}] \qquad (3.3)$$

where k is a constant. Now if the new functions $\underline{A}$ and ψ transform as

$$\underline{A} \rightarrow \underline{A} + k^{-1}\underline{\nabla}\mu$$

$$\psi \rightarrow \psi + k^{-1}\partial_t\mu \qquad (3.4)$$

the new Lagrangian (3.3) is invariant under local gauge transformation (3.2). We can construct fields from functions $\underline{A}$ and ψ

$$\underline{F} = \nabla \times \underline{A}$$

$$\underline{G} = \underline{\nabla}\psi - \partial_t\underline{A} \qquad (3.5)$$

where vectors $\underline{F}$ and $\underline{G}$ are invariant under the local gauge transformation (3.4).
Now there is a choice: either we set these vectors identically to zero or we can assume that there is a new gauge field coupled to the matter and its energy has to be added to the Lagrangian (3.3). Although it is not obvious from the non-relativistic theory, the only scalar invariant made from $\underline{G}$ and $\underline{F}$ is $\underline{G}^2 - \underline{F}^2$.

We also note that if we choose the gauge $\mu = -\phi$, all the terms involving ϕ in the Lagrangian (3.3) vanish. The Lagrangian with the gauge field is now

$$L = -\rho[\frac{1}{2}(-k\underline{A} + S\underline{\nabla}\lambda + a_j\underline{\nabla}b_j)^2 + E(\rho,S)$$

$$+ k\psi - S\partial_t\lambda - \underline{a}\cdot\partial_t\underline{b}] + c(\underline{F}^2 - \underline{G}^2) \qquad (3.6)$$

where c is a constant.

Taking variations with respect to ψ and $\underline{A}$ in turn gives

$$\nabla\cdot\underline{G} = \frac{k}{c}\rho$$

$$-\partial_t\underline{G} + \nabla\times\underline{F} = \frac{k}{c}\rho\underline{v} \qquad (3.7)$$

Therefore, the continuity equation (1.2) can be obtained from (3.7) by adding the time derivative of the first equation to the divergence of the second. In other words by requiring local gauge invariance (3.2) we had to introduce a new gauge field and the gauge conservation law (1.2) is then replaced by a dynamical equation imposed by the equations of motion of the new field. It will be obvious to the reader that the new field can be identified with the electromagnetic field provided that appropriate constants k and c are chosen. The scalar kρ/c and vector k $\rho\underline{v}$/c are the gauge charge and gauge current respectively and act as the sources of the gauge field. That would indicate that the gauge field can only exist if the sources are present and, consequently, the so-called free field solutions have no meaning.

The vector $\underline{b}$ is also invariant under the gauge transformation

$$\underline{b} \to \underline{b} + \delta\underline{b} \qquad (3.8)$$

where $\delta\underline{b}$ is a constant. In order to make the Lagrangian (3.6) invariant under a local transformation

$$\delta\underline{b} = \underline{R}(\underline{x},t) \qquad (3.9)$$

we have to introduce a new tensor field in a manner similar to that for introducing the vector field (3.4). Although it is possible to do so in the present non-relativistic treatment the arguments are cumbersome. The introduction of the gauge field induced by the local gauge transformation (3.9) will be given in a future paper dealing with fully relativistic perfect fluid. It is only worth noting here that such a gauge field can be identified with gravity.

The invariance of the Lagrangian (3.6) under the gauge

transformation $\lambda \to \lambda\ \delta\lambda$ can also induce a vector gauge field with entropy as its source. But its physical meaning remains unclear since there may be surfaces in space where entropy is not conserved (shocks). This poses the question whether entropy is an elementary physical quantity or a convenient approximation to more elementary quantities.

ACKNOWLEDGEMENTS

The author wishes to acknowledge the warm hospitality of Professor Oscar Buneman and his colleagues at Stanford University who clarified the author's ideas in many valuable discussions.

REFERENCES

1. E. Detyna, Fluid Dynamics Report 2/81, Mathematics Dept., Reading University (1981). Submitted to Proc. Roy. Soc.
2. E. Noether, Goettingen Nachriehten, 235 (1918)
3. C. C. Lin, Proc. Int. School of Physics, Course XXI, New York, Academic Press (1963).
4. A. Clebsch, J. Reine Angew. Math. 56, 1 (1859).
5. R. L. Seliger and G. B. Whitham, Proc. Roy. Soc. A305, 1 (1968).
6. O. Buneman, Phys. Fluids, 23, 1716 (1980).
7. D. Sciama, in "Recent Developments in General Relativity", Pergamon Press (1963).

THE LIE ALGEBRAIC INTERPRETATION OF THE COMPLETE INTEGRABILITY OF THE ROSOCHATIUS SYSTEM

Tudor Ratiu*)
University of Michigan, Ann Arbor, MI 48109

ABSTRACT

The present note answers a question posed by A.G. Reyman [5] as to the Lie algebraic reasons of the complete integrability of a system studied by E. Rosochatius [6].

1. The Rosochatius System

Consider the motion of a particle on the sphere $S^{n-1} \subset \mathbb{R}^n$ under the influence of the potential

$$U(\underline{x}) = \frac{1}{2}(A\underline{x} \cdot \underline{x} - \|C/\underline{x}\|^2)$$

where $A = \text{diag}(a_1,\ldots,a_n)$, $C = \text{diag}(c_1,\ldots,c_n)$ and $C/\underline{x}$ denotes the vector $(c_1/x_1,\ldots,c_n/x_n)$. The equations of motion are

$$\ddot{x}_i = -\partial U/\partial x_i + \lambda x_i = -(a_i x_i + \frac{c_i^2}{x_i^3}) + \lambda x_i \qquad (1.1)$$

where λ is the Lagrange multiplier defined by the condition that $\underline{x} \in S^{n-1}$ during the motion. To find λ, multiply (1.1) by x_i and sum over i, taking into account that $\|\underline{x}\|^2 = 1$, and therefore $\underline{x} \cdot \underline{\dot{x}} = 0$, $\|\underline{\dot{x}}\|^2 + \underline{x} \cdot \underline{\ddot{x}} = 0$. Thus $\lambda = A\underline{x} \cdot \underline{x} + \|C/\underline{x}\|^2 - \|\underline{\dot{x}}\|^2$. Set $\underline{y} = \underline{\dot{x}}$; with λ just found (1.1) becomes

$$\begin{cases} \dot{x}_i = y_i, \quad \|\underline{x}\|^2 = 1, \quad \underline{x} \cdot \underline{y} = 0 \\ \dot{y}_i = -a_i x_i - c_i/x_i^3 + (A\underline{x} \cdot \underline{x} + \|C/\underline{x}\|^2 - \|y\|^2)x_i \end{cases} \qquad (1.2)$$

This Hamiltonian system has been shown to be completely integrable by E. Rosochatius [6]. Moser [2] finds a Lax pair for these equations, which however is not equivalent to (1.2) but only implied by it. Moreover, Moser proves that the integrals in involution of (1.2) are eigenvalues of a matrix obtained by a rank two perturbation. Reyman asks what is the Lie algebraic interpretation of Moser's Lax pair, since it does not fit into a general framework developed by him [5].

---
*) Partially supported by NSF Grant MCS 81-01642

We shall prove the following: (1.2) is equivalent to a degenerate Euler-Poisson equation [3], [4] on a minimal dimensional orbit of the semidirect product of so(n) with the space of symmetric traceless matrices. Introducing a parameter in Moser's Lax pair, it is shown that this Euler-Poisson equation is equivalent to a Hamiltonian system on an invariant submanifold of a subalgebra in the Kac-Moody extension of $s\ell(n)$. This bihamiltonian formulation of the same problem, yields as usual, the complete integrability of the problem; see also Adler-van Moerbeke [1].

This system is very similar to the C. Neumann system [1], [2], [4], [5], and many results here are implied by facts already proved for the Neumann system. What is new, however, is the fact that it seems to be the first completely integrable Euler-Poisson system with non-linear potential - at least to the knowledge of the author.

I would like to thank H. Fleschka for drawing my attention to this problem. The exposition that follows is due to length considerations quite dense and hereby based on [4].

2. The Euler-Poisson Equations

We start by reviewing a few known facts about orbits in semidirect products.

2.1 Let $\mathcal{G}$ be a semisimple Lie algebra with κ some constant multiple of the Killing form. The semidirect product of $\mathcal{G}$ with itself by the adjoint representation is a Lie algebra with underlying vector space $\mathcal{G} \times \mathcal{G}$ and bracket

$$[(\xi_1,\eta_1),(\xi_2,\eta_2)] = ([\xi_1,\xi_2], [\xi_1,\eta_2]-[\xi_2,\eta_1]) \quad (2.1)$$

There is a bilinear, symmetric, bi-variant, non-generate two form on this semidirect product induced by κ and it is given by

$$\kappa_s((\xi_1,\eta_1),(\xi_2,\eta_2)) = \kappa(\xi_1,\eta_2) + \kappa(\xi_2,\eta_1) . \quad (2.2)$$

2.2 In the following considerations, the coadjoint orbit theory plays a central role. If $\mathcal{G}$ is any Lie algebra, the Kirillov-Kostant-Souriau theorem states that its coadjoint orbits are symplectic manifolds. If $\mathcal{G}$ is semisimple, the equivariant diffeomorphism induced by κ makes the adjoint orbits symplectic manifolds. Suppose now that $\mathcal{G} = \mathcal{h} \oplus \mathcal{n}$ with $\mathcal{n}$ a Lie subalgebra. Then $\mathcal{n}^* \cong \mathcal{h}^{\perp}$ ($\perp$ with respect to κ) so that the coadjoint orbits of $\mathcal{n}^*$ can be identified with orbits in $\mathcal{h}^{\perp}$ which by the above considerations are symplectic manifolds. Tracing through all the above diffeomorphisms, it can be shown [4] that the Poisson bracket in $\mathcal{h}^{\perp}$ is given by

$$\{f,g\}(\xi) = -\kappa([\Pi_{\mathcal{n}}(\text{grad } f)(\xi), \Pi_{\mathcal{n}}(\text{grad } g)(\xi)], \xi) \quad (2.3)$$

for $\xi \in \mathcal{h}^\perp$, $\Pi_{\mathcal{n}}: \mathcal{g} \to \mathcal{n}$ the comonical projection along $\mathcal{h}$ and "grad" the gradient with respect to κ.

2.3 Euler-Poisson equations are Hamiltonian systems on adjoint orbits of the semidirect product $\mathcal{g} \times \mathcal{g}$ with Hamiltonian of the form $H(\xi,\eta) = \frac{1}{2}\kappa(L\eta,\eta) + V(\xi)$ for $V: \mathcal{g} \to \mathbb{R}$ a smooth function, and $L: \mathcal{g} \to \mathcal{g}$ a linear map. They are of the form

$$\dot{\xi} = [\xi, L\eta], \quad \dot{\eta} = [\eta, L\eta] + [\xi, (\text{grad } V)(\xi)] ; \quad (2.4)$$

see [3] for more information and the connection with reduction. The most famous example of such equations are the equations of the heavy rigid body with a fixed point; in this case $\mathcal{g} = so(3)$ which is identified as Lie algebra with $\mathbb{R}^3$ endowed with the cross-product. In this case V is a linear function of ξ. One sees thus, that in general, such Hamiltonian systems are not completely integrable.

2.4 For $x, y \in \mathbb{R}^n$, $\|x\| = 1$, $x \cdot y = 0$, set $X = x \otimes x - \frac{1}{n}\text{Id}$, $P = x \wedge y$, where $(x \otimes x)_{ij} = x_i x_j$, $(x \wedge y)_{ij} = x_i y_j - y_i x_j$. Let sym denote the vector space of symmetric traceless matrices; then $X \in$ sym, $P \in so(n)$. Split the semidirect product $\mathcal{g} = s\ell(n) \times s\ell(n) = \mathcal{h} \oplus \mathcal{n}$, $\mathcal{h} =$ sym $\times so(n)$, $\mathcal{n} = so(n) \times$ sym. The underlying Lie groups of $\mathcal{g}$ and $\mathcal{n}$ respectively are $G = SL(n) \times s\ell(n)$, $N = SO(n) \times$ sym. If $\kappa(A,B) = -\frac{1}{2}\text{Tr}(AB)$, remark that $\mathcal{h}^\perp = \mathcal{h}$ where $\perp$ is taken with respect to κ_s. The following has been proved in [4].

2.1 Theorem. The N-orbit through $(z \otimes z - \frac{1}{n}\text{Id}, 0) \in \mathcal{h}$, $z = (1,\ldots,1)/\sqrt{n}$, consists of pairs $(X,P) \in \mathcal{h}$. With the Kirillov-Kostant-Souriau symplectic form, this $(2n-2)$-dimensional orbit is symplectically diffeomorphic via $(X,P) \mapsto (x,y)$ to the cotangent bundle of S^{n-1} with the symplectic structure induced from $\mathbb{R}^{2n}$ by $\sum_{i=1}^{n} dx_i \wedge dy_i$.

2.5 Let us return to the system (1.2). Put $N_{ij} = \dfrac{c_i^2 x_j}{x_i^3} + \dfrac{c_j^2 x_i}{x_j^3}$,
$N_{ii} = \dfrac{2c_i^2}{x_i^2} - \dfrac{2}{n}\sum_{i=1}^{n}\dfrac{c_i^2}{x_i^2}$ and form $N \in$ sym with entries N_{ij}. From now on denote by $A = \text{diag}(a_1,\ldots,a_n) - \dfrac{1}{n}(\text{Tr}(A))\text{Id} \in$ sym.

2.2 Proposition. The system (1.2) is equivalent to

$$\dot{X} = [X,P], \quad \dot{P} = [A + N, X], \quad \|\underset{\sim}{x}\| = 1, \quad \underset{\sim}{x}\cdot\underset{\sim}{y} = 0. \quad (2.5)$$

The proof is a direct, somewhat lengthy, verification.

A comparison between the equations (2.4) and (2.5) shows that (2.5) are degenerate Euler-Poisson equations. Indeed, (2.5) if one chooses L = identity and $V(X) = -\kappa(A,X) + \dfrac{1}{2}\text{Tr}\phi(X)$, where $\phi(X)$ is chosen such that $(\text{grad } V)(X) = -A-N$, (2.5) becomes a special case of (2.4). The form of ϕ is irrelevant; only its existence matters. Specifically, $V(X)$ can be obtained by tracing through the diffeomorphism given by Theorem 2.1 and computing the expression of the push-forward of Rosochatius' potential. We proved the following.

2.3 Theorem The Rosochatius system (1.2) is a Hamiltonian system on the N-orbit in $so(n) \times$ sym given by Theorem 2.1. Hamilton's equations coincide with the degenerate Euler-Poisson equations (2.5).

3. Hamilton's equations in Kac-Moody Lie algebra setting and complete integrability.

3.1. Introduce the matrices $M \in$ sym, $M_{ij} = (c_i x_j/x_i) + (c_j x_i/x_j)$,

$M_{ii} = 2c_i - 2(\sum_{i=1}^{n} c_i)/n$, $D = \text{diag}(c_1/x_1^2,\ldots,c_n/x_n^2) - \dfrac{1}{n}(\sum_{i=1}^{n} c_i/x_i^2)\text{Id} \in$ sym.

3.1 Lemma. If $\|\underset{\sim}{x}\| = 1$, $\underset{\sim}{x}\cdot\underset{\sim}{y} = 0$, then

$$[M,X] = [D,X], \quad [D,M] = [N,X]$$

If $\dot{\underset{\sim}{x}} = \underset{\sim}{y}$, then $\dot{M} = [D,P]$.

The proof is a straight forward verification. Using these relations it is easy to show the following.

3.2 Proposition. If $\|\underset{\sim}{x}\| = 1$, $\underset{\sim}{x} \cdot \underset{\sim}{y} = 0$, the degenerate Euler-Poisson equations (2.5) are equivalent to

$$(-X + (P+M)\lambda + A\lambda^2)^{\bullet} = [-X + (P+M)\lambda + A\lambda^2, P - D + M + A\lambda] \quad (3.1)$$

If $\lambda = 1$, the above Lax equation coincides with the restriction of Moser's to the cotangent bundle of S^{n-1}. Equation (3.1) has also a Hamiltonian interpretation, but in the Kirillov-Kostant-Souriau structure of an infinite dimensional Lie algebra.

3.2 Here is a quick review of how (3.1) is Hamiltonian in the Kac-Moody extension of $s\ell(n)$. Let

$\widetilde{s\ell(n)} = \{\sum_{n \in \mathbb{Z}} \xi_n \lambda^n \mid \text{finite sum}\}$ be the Kac-Moody extension of $s\ell(n)$; the bracket is defined by

$$[\sum_{k \in \mathbb{Z}} \xi_k \lambda^k, \sum_{n \in \mathbb{Z}} \eta_n \lambda^n] = \sum_{n \in \mathbb{Z}} (\sum_{i+j=n} [\xi_i, \eta_j]) \lambda^n.$$

The form $\widetilde{\kappa}$ defined by $\widetilde{\kappa}(\sum_{k \in \mathbb{Z}} \xi_k \lambda^k, \sum_{n \in \mathbb{Z}} \eta_n \lambda^n) = \sum_{n+k=-1} \kappa(\xi_k, \eta_n)$, for $\kappa(A,B) = -\frac{1}{2} \text{Tr}(AB)$, $A, B \in s\ell(n)$ is bilinear, symmetric, bi-invariant and weakly non-degenerate. Let $K = \{\sum_{n=0}^{\infty} \xi_n \lambda^n\}$, $N = \{\sum_{n=-\infty}^{-1} \eta_n \lambda^n\}$; then $\widetilde{s\ell(n)} = K \oplus N$, K, N are Lie subalgebras, $K^{\perp} = K$, $N^{\perp} = N$. Denote by $\mathcal{G}(\widetilde{s\ell(n)})$ those real values functions on $\widetilde{s\ell(n)}$ which have gradient with respect to $\widetilde{\kappa}$. The submanifold $Q_A = \{\xi + \eta\lambda + A\lambda^2 \mid \xi, \eta \in s\ell(n)\}$ is an invariant submanifold of $K^{\perp} = K$ with respect to the Kirillov-Kostant-Souriau structure given by $\mathcal{G}(s\ell(n))$. The submanifold $\{-X + (P+M)\lambda + A\lambda^2\}$ is not invariant but there exists an ad-invariant Hamiltonian whose $\widetilde{\kappa}$-gradient projected onto K equals $P + M - D + A\lambda$. Thus (3.1) is of the form $\dot{\widetilde{\xi}} = [\pi_K(\text{grad } H)\widetilde{\xi}, \widetilde{\xi}]$, for $\widetilde{\xi} = -X + (P+M)\lambda + A\lambda^2$ which is the expression of a Hamiltonian vector field on $K^{\perp}$; see [1], [3], [4], [5] for more details and related problems.

3.3. From (3.1) it follows that the functions $\phi_K(X,P) = \frac{1}{2(k+1)} \text{Tr}(-X + (P+M)\lambda + A\lambda^2)^{k+1}$ are conserved along the flow of (3.1) and thus the coefficients $f_K(X,P)$ of λ^{2k+1} in this expansion are conserved along the flow of (2.5).

These functions are in involution. Instead of carrying out here the long computations, we simply remark that the general outline in [4] will work. First, one shows that $\{\phi_K, \phi_\ell\} = 0$ for every λ in the semidirect product $s\ell(n) \times s\ell(n)$ by using the proof of Theorem 3.2 in [4] and a direct computation in which one expresses M as a function of X. Specifically, using the expression of the Poisson bracket in the semidirect product for ϕ_K, ϕ_ℓ, one gets the terms appearing in the proof of Theorem 3.2 of [4] plus three other terms. The first group is handled as in the above mentioned theorem and the last three terms are easily seen to vanish by using the explicit expression of M as a function of X. Unfortunately, there is no simple involution theorem at hand which could simplify this step. Thus $\{\phi_K, \phi_\ell\} = 0$ in the Poisson bracket of $s\ell(n) \times s\ell(n)$ for every λ. In particular, the highest non-zero λ-coefficients in this expansion must also vanish, i.e. $\{f_K, f_\ell\} = 0$ in $s\ell(n) \times s\ell(n)$. Next, one uses Theorem 3.3 of [4] to conclude $\{f_K, f_\ell\} = 0$ on the N-orbit. This somewhat laborious - but direct - proof shows again how the bihamiltonian formulation of the same problem in Lie algebras is ultimately responsible for involution.

The next step should be the proof of the generic independence of these integrals. We simply refer the reader to [4] for a proof that works - with minor technical modifications - also in this case. Finally, the Kac-Moody Lie algebraic formulations shows that the hyperelliptic curve $\det(-X + (P+M)\lambda + A\lambda^2 - z\text{Id}) = 0$ of genus $n-1$ is isospectral. Applying here the standard methods of Adler-van Moerbeke [1], or Moser [3], the flow linearizes on the Jacobian of this curve.

We proved hence purely Lie algebraically the following.

3.3 Theorem. The Rosochatius system (2.5) is a completely integrable Hamiltonian system with non-linear potential of degenerate Euler-Poisson equations on a minimal dimensional orbit of the semidirect product $so(n) \times sym$. Its flow linearizes in the Jacobian of a hyperelliptic curve given by its interpretation as a Hamiltonian vector field on a submanifold in a subalgebra of the Kac-Moody extension of $s\ell(n)$.

REFERENCES

1. M. Adler, P. van Moerbeke, Linearization of Hamiltonian Systems, Jacobi varieties, and Representation Theory, Advances in Math. $\underline{38}$, (1980), 318 - 379.
2. J. Moser, Geometry of Quadrics and Spectral Theory, in "The Chern Symposium, Berkeley, June 1979", Springer Verlag (1980), 147 - 188.
3. T. Ratiu, Euler-Poisson Equations on Lie Algebras and the N-Dimensional Heavy Rigid Body, Proc. Nat. Acad. Sci. USA, $\underline{78}$, No. 3 (1981), 1327 - 1328.
4. T. Ratiu, The C. Neumann Problem as a Completely Integrable System on an Adjoint Orbit, Trans. Amer. Math. Soc. $\underline{264}$ (1981), 321 - 329.
5. A. G. Reyman, Integrable Hamiltonian Systems Connected with Graded Lie Algebras, Zap. Nauk Sem. Leningrad O.D.T.L., Math. Inst. Steklov (L.O.M.I.), Vol. $\underline{95}$, (1980) 3-54.
6. E. Rosochatius, Über die Bewegung eines Punktes, Inaugural Dissertation, Universität Göttingen, Gebr. Unger, Berlin (1877).

HAMILTONIAN KINETIC THEORY OF PLASMA PONDEROMOTIVE PROCESSES

Steven W. McDonald and Allan N. Kaufman
Physics Department and Lawrence Berkeley Laboratory
University of California, Berkeley 94720

ABSTRACT

The nonlinear nonresonant interaction of plasma waves and particles is formulated in a Hamiltonian kinetic theory which treats the wave-action and particle distributions on an equal footing, thereby displaying reciprocity relations. In the quasistatic limit, a nonlinear wave-kinetic equation is obtained. The generality of the formalism allows for applications to arbitrary geometry, with the nonlinear effects expressed in terms of the linear susceptibility.

Plasma dynamics utilizes a number of models which are self-consistent and obey conservation laws. Heretofore these properties have been deduced from the evolution equations of each model, rather than from the underlying structure of the model. The recent realization that a Hamiltonian field theory could be based on a Poisson structure, without the need for conjugate pairs of fields, has led to the Hamiltonian formulation of several standard models, namely MHD[1], Vlasov-Coulomb[2], Vlasov-Maxwell[3], and two-fluid[4].

One great advantage of a Hamiltonian formulation is that it points the way to new self-consistent models. In the present work, we extend the Poisson structure of the Vlasov-Coulomb model to the kinetic ponderomotive model[5], i.e., the nonresonant nonlinear interaction of high-frequency plasma waves with the low-frequency dynamics of a Vlasov plasma. The resulting evolution equations extend previous studies to arbitrary geometry and waves, and are inherently self-consistent.

We recall[2] that a Vlasov distribution $f(z)$, i.e., a six-dimensional phase-space density of particles, allows a Poisson structure [,] for functionals $A(f)$:

$$[A_1(f), A_2(f)] = \int d^6z \, f(z) \, \{\delta A_1/\delta f(z), \, \delta A_2/\delta f(z)\}, \quad (1)$$

where $\{ , \}$ is the Poisson bracket for functions on z-space. With the evolution equation for any functional:

$$\dot{A}(f) = [A(f), H(f)], \quad (2)$$

for a given Hamiltonian functional $H(f)$, we find that f itself evolves as

$$\dot{f}(z) = -\{f(z), h(z;f)\}, \quad (3)$$

where

$$h(z;f) = \delta H(f)/\delta f(z) \quad (4)$$

is the self-consistent particle Hamiltonian. We recognize (3) as the standard Vlasov equation; note that the over-dot means evolution at fixed z. (We suppress species labels for simplicity.)

It is well-known that waves evolve as a Hamiltonian system, in the eikonal (or ray) approximation. Here the six-dimensional phase-space is $y = (\underline{x},\underline{k})$, and the Hamiltonian is the dispersion relation for frequency $\omega(\underline{x},\underline{k})$. By analogy to the formulas above, the standard wave-kinetic equation

$$\dot{J}(y) = - \{J(y),\omega(y)\}, \qquad (5)$$

for the action distribution $J(\underline{x},\underline{k})$, can be obtained from a Hamiltonian functional $H(J)$, with $\omega(y) = \delta H(J)/\delta J(y)$. In (5) $\{\,,\,\}$ is the standard Poisson bracket for functions on y-space; for simplicity, wave-branch labels are suppressed.

To couple particles and waves, we consider functionals $A(f,J)$ of distributions $f(z)$ and $J(y)$. Their Poisson bracket is taken as

$$[A_1(f,J), A_2(f,J)] = \int d^6z\, f(z)\, \{\delta A_1/\delta f(z),\, \delta A_2/\delta f(z)\}_z \\ + \int d^6y\, J(y)\, \{\delta A_1/\delta J(y),\, \delta A_2/\delta J(y)\}_y\,; \qquad (6)$$

note that there are now three kinds of Poisson bracket: $[\,,\,]$ $\{\,,\,\}_z$ $\{\,,\,\}_y$. With the evolution equation

$$\dot{A}(f,J) = [A(f,J), H(f,J)], \qquad (7)$$

we obtain the coupled equations:

$$\dot{f}(z) = - \{f(z), h(z;f,J)\}_z, \qquad (8a)$$
$$\dot{J}(y) = - \{J(y), \omega(y;f,J)\}_y, \qquad (8b)$$

with

$$h(z;f,J) = \delta H(f,J)/\delta f(z), \qquad (9a)$$
$$\omega(y;f,J) = \delta H(f,J)/\delta J(y). \qquad (9b)$$

The choice of Hamiltonian $H(f,J)$ (and of Poisson structure) determines the physical processes included in the model. We wish to study the dynamics of particles acted on by the ponderomotive forces of the waves, and the dynamics of the waves resulting from changes in the particle distribution. From previous work in this area[6], we identify z as oscillation-center variables, and f as the distribution of oscillation centers.

Thus, in (9a), h represents the oscillation-center Hamiltonian in the presence of the wave field J, while, in (9b), ω is the local wave dispersion relation for oscillation-center distribution f. A remarkable reciprocity relation follows on cross-differentiation of (9):

$$\delta h(z)/\delta J(y) = \delta\omega(y)/\delta f(z). \qquad (10)$$

The left side is the ponderomotive contribution to the Hamiltonian, while the right side is essentially the linear susceptibility, as we show below. This relation was discovered several years ago by one of us and Cary[7], by comparing explicit expressions for the ponderomotive Hamiltonian and the susceptibility. The present derivation uncovers its significance: each expression is the same functional second derivative of $H(f,J)$.

To be more explicit, we choose

$$H(f,J) = \int d^6z\, h_0(z) f(z) + \int d^6y\, \omega(y;f) J(y), \tag{11}$$

where $h_0(z)$ is the unperturbed particle Hamiltonian, and $\omega(y;f)$ is the dispersion relation, determined as a root of the dispersion function:

$$D(y,\omega;f) = 0. \tag{12}$$

From (12), we have

$$\delta\omega(y)/\delta f(z) = -[\delta D/\delta f(z)][\partial D/\partial \omega]^{-1}, \tag{13}$$

with ω replaced by $\omega(y)$ on the right after differentiation. The dispersion function D of (12) is obtained from the Hermitian part of the dispersion tensor $\underline{D}$ and the local polarization $\hat{e}$:

$$D = \hat{e}^* \cdot \underline{D} \cdot \hat{e}, \tag{14}$$

while

$$\underline{D} = \underline{I} - c^2(k^2 \underline{I} - \underline{kk})/\omega^{-2} + \underline{\chi}(y;f), \tag{15}$$

where $\underline{\chi}$ is the linear susceptibility. Thus (13) becomes

$$\delta\omega(y)/\delta f(z) = -[\partial D/\partial \omega]^{-1}\, \hat{e}^* \cdot \delta\underline{\chi}(y)/\delta f(z) \cdot \hat{e}. \tag{16}$$

Returning to (9a) and (10), and using (16), we have

$$h = h_0 - \int d^6y\, J(y) [\partial D/\partial \omega]^{-1}\, \hat{e}^* \cdot \delta\underline{\chi}/\delta f(z) \cdot \hat{e}. \tag{17}$$

Finally we express J in terms of the electric field[8]:

$$J(y) = (\partial D/\partial \omega)|\hat{e}^* \cdot \widetilde{\underline{E}}(y)|^2/4\pi, \tag{18}$$

to obtain the previously known relation:

$$h(z) = h_0(z) - \int d^6y\, \widetilde{\underline{E}}^*(y) \cdot \delta\underline{\chi}(y)/\delta f(z) \cdot \widetilde{\underline{E}}(y)/4\pi, \tag{19}$$

somewhat generalized.

Before proceeding, we should stress that the evolution equations obtained here are not yet complete, as they ignore the effects of the low-frequency fields. A fuller treatment, incorporating them, is in preparation, and will be published elsewhere.

Except for this omission, the present equations may be considered as the generalization of ponderomotive kinetic theory[5] to arbitrary geometry, e.g., to nonuniform plasma and magnetic field. As our next step, we consider the quasistatic limit for Eq. (8a), in order to eliminate f and to obtain a nonlinear equation for J, thereby generalizing the "nonlinear Schrödinger equation."

In (8a), we set $\dot{f} = 0$, and satisfy $\{f,h\} = 0$ by choosing

$$f(z) = a\, \exp(-\beta) h(z;f,J). \tag{20}$$

Thus f is a functional of J, and we have (ignoring the weak dependence of h on f):

$$\delta f(z)/\delta J(y) = -\beta f(z)\, \delta h(z)/\delta J(y) \qquad (21)$$
$$= -\beta f(z)\, \delta\omega(y)/\delta f(z),$$

by Eq. (10). Thus $\omega(y;f)$ has an implicit dependence on J:

$$\delta\omega(y)/\delta J(y') = \int d^6 z\, [\delta\omega(y)/\delta f(z)]\, [\delta f(z)/\delta J(y')]$$
$$= -\beta \int d^6 z\, f(z)\, [\delta\omega(y)/\delta f(z)]\, [\delta\omega(y')/\delta f(z)]. \qquad (22)$$

Noting the symmetry in (22), we obtain another reciprocity relation:

$$\delta\omega(y)/\delta J(y') = \delta\omega(y')/\delta J(y), \qquad (23)$$

governing nonlinear frequency shifts. This relation enables us to formulate the Hamiltonian functional for wave-action:

$$H(J) = \int d^6 y\, \omega_0(y) J(y) + \tfrac{1}{2} \iint d^6 y\, d^6 y'\, J(y)J(y')\omega_2(y,y'), \qquad (24)$$

with the coupling coefficient $\omega_2(y,y')$ given by the right side of (22). This Hamiltonian yields the desired nonlinear frequency, by (9b):

$$\omega(y;J) = \omega_0(y) + \int d^6 y'\, J(y')\omega_2(y,y'). \qquad (25)$$

To illustrate these formulas, we may consider the simplest case, Langmuir waves in an unmagnetized plasma. Here (16) reads

$$\delta\omega(\underline{k},\underline{x})/\delta f(\underline{r},\underline{p}) = -(\omega/2)(4\pi e^2/m)\delta(\underline{x}-\underline{r})(\omega - \underline{k}\cdot\underline{v})^{-2}, \qquad (26)$$

and the wave-wave coupling is

$$\omega_2(\underline{k},\underline{x};\underline{k}',\underline{x}') = -\delta(\underline{x}-\underline{x}')\beta(\omega\omega'/4)(4\pi e^2/m)^2 \cdot$$
$$\cdot \int d^3 p\, f(\underline{x},\underline{p})[(\omega - \underline{k}\cdot\underline{v})(\omega' - \underline{k}'\cdot\underline{v})]^{-2}, \qquad (27)$$

with $\omega = \omega_0(\underline{k},\underline{x})$, $\omega' = \omega_0(\underline{k}',\underline{x}')$.

Applications of this formulation to more interesting situations will be the subject of future publications.

This research was supported in part by the Office of Fusion Energy, U.S. Department of Energy.

REFERENCES

1. P. Morrison & J. Greene, Phys. Rev. Lett. 45, 790 (1980).
2. P. Morrison, Phys. Lett. 80A, 383 (1980); J. Gibbons, Physica 3D, 503 (1981).
3. J. Marsden & A. Weinstein, Physica 4D (in press).
4. R. Spencer & A. Kaufman, submitted to Phys. Rev. A.
5. J. Cary & A. Kaufman, Phys. Fluids 24, 1238 (1981).
6. R. Dewar, Phys. Fluids 16, 1102 (1973); S. Johnston, Phys. Fluids 19, 93 (1976).
7. J. Cary & A. Kaufman, Phys. Rev. Lett. 39, 402 (1977).
8. S. McDonald, to be published.

The Hamiltonian Structure of Multi-Species Fluid Electrodynamics

Richard G. Spencer

Physics Department and Lawrence Berkeley Laboratory
University of California, Berkeley, CA 94720

ABSTRACT

The phase space for multi-species fluid electrodynamics is the function space of fluid variables and Maxwell field variables. The Poisson bracket on phase functionals is constructed as a Lie algebra product following general methods of infinite dimensional symplectic geometry.

Recently, there has been a considerable amount of interest in the underlying Hamiltonian structures of the three standard models of non-dissipative plasma physics. Morrison and Greene [1] and Holm and Kupershmidt [2] have considered ideal magnetohydrodynamics, while Morrison [3], Weinstein and Morrison [4], and Marsden and Weinstein [5] have treated the Maxwell-Vlasov system, the special relativistic generalization of which has been studied by Bialynicki-Birula and Hubbard [6]. We present here the Hamiltonian structure for multi-species fluid electrodynamics [7].

One conclusion that has emerged from this activity is that it is advantageous to derive the appropriate Poisson bracket from geometric considerations which will insure that it is bilinear, skew-symmetric, and satisfies the Jacobi identity. Our work follows the approach of Marsden and Weinstein in that the main ingredients in the construction of the Poisson structure for functionals of the fluid variables and the Maxwell field variables are the symplectic manifold structure of the co-adjoint orbits of a Lie group [8], and the reduction of phase spaces with symmetry [9], [10].

We define the physical system of charged fluids under consideration. Label fluid species with subscript s; each is composed of particles of mass m_s and charge q_s. Let $a_s = q_s/m_s$; $q_s = 0$ is allowed. Our treatment holds for an arbitrary number of species, but two (oppositely charged) species is the situation most commonly discussed. Then, in terms of the fluid velocities $\underline{u}_s$, mass densities ρ_s, specific entropies σ_s, electric field $\underline{E}$, and magnetic field $\underline{B}$, the equations of ideal multi-fluid dynamics, in rationalized units, are

$$\dot{\underline{E}} = \nabla \times \underline{B} - \sum_s a_s \rho_s \underline{u}_s \qquad \dot{\underline{B}} = -\nabla \times \underline{E} \qquad (1a)$$

$$\nabla \cdot \underline{E} = \sum_s a_s \rho_s + \rho_{ext} \qquad \nabla \cdot \underline{B} = 0 \qquad (1b)$$

$$\dot{\rho}_s = -\nabla \cdot (\rho_s \underline{u}_s) \tag{2a}$$

$$\dot{\sigma}_s = -\underline{u}_s \cdot \nabla \sigma_s \tag{2b}$$

$$\dot{\underline{u}}_s = -(\underline{u}_s \cdot \nabla)\underline{u}_s + a_s(\underline{E} + \underline{u}_s \times \underline{B}) - \rho_s^{-1} \nabla P_s \tag{2c}$$

where the specific internal energy $U_s(\rho_s, \sigma_s)$, expressed as an equation of state, yields the (partial) pressure P_s according to

$$P_s = \rho_s^2 \, \partial U_s / \partial \rho_s \; . \tag{3}$$

Eqs. (1) are the Maxwell equations, including an external time-independent charge density ρ_{ext}, and eqs. (2) and (3) are the laws of compressible ideal fluid dynamics. We neglect heat flow, and therefore express entropy convection by the adiabatic equation (2b).

It is natural in our construction to replace the velocity field variable $\underline{u}_s$ with the momentum density $\underline{M}_s \equiv \rho_s \underline{u}_s$. Then phase space consists of the set of dynamical variables $(\underline{M}_s, \rho_s, \sigma_s, \underline{E}, \underline{B})$, while the energy of the system is

$$H(\underline{M}_s, \rho_s, \sigma_s, \underline{E}, \underline{B}) = \sum_s \int (\tfrac{1}{2} \rho_s^{-1} |\underline{M}_s|^2 + \rho_s U_s) d^3x \\ + \int (\tfrac{1}{2} |\underline{E}|^2 + \tfrac{1}{2} |\underline{B}|^2) d^3x. \tag{4}$$

(We shall sometimes use the notation of writing, e.g., $\underline{M}_s \equiv (\underline{M}_1, \ldots, \underline{M}_k)$ for k species. Whether this is the case or whether $\underline{M}_s$ refers to the single species s will always be clear from the context.)

We wish to express eqs. (1a) and (2) as Hamiltonian evolution equations

$$\dot{Z} = \{Z, H\} \; , \tag{5}$$

where Z represents one of the dynamical variables, and the Hamiltonian H is given by the energy, eq. (4). Because we are abandoning canonical coordinates in favor of these physical variables, the bracket in eq. (5) is not expected to have the form of a standard Poisson bracket. Here, we present only a sketch of the construction of the Poisson bracket; a detailed exposition will be presented in a later publication.

We regard the system defined by eqs. (1), (2), (3), and (4) as the coupling of the vacuum Maxwell equations to ordinary ideal fluid dynamics. Therefore, we shall briefly review the Hamiltonian structures of these theories.

The equations of motion for a single fluid species composed of uncharged particles are eqs. (2) and (3), with s=1 and $a_s=0$. The Hamiltonian is the first integral in eq. (4). The Poisson bracket for this has been given by Morrison and Greene, and rederived by Marsden and Weinstein (private communication) from the symplectic structure on co-adjoint orbits of a certain Lie group G. In order to motivate a guess of what G might be, we reason as follows. The dual to the Lie algebra g of G, denoted g^*, must contain the phase space variables $(\underline{M}, \rho, \sigma)$. This is achieved here by noting that momentum densities are just vector field densities on $\mathbb{R}^3$, the set of which we write as X^*, and that mass and entropy densities are scalar field densities on $\mathbb{R}^3$, written F^*. Thus the fluid phase space is, with x denoting the direct product,

$$\{(\underline{M}, \rho, \sigma)\} = g^* = X^* \times F^* \times F^* \quad .$$

The duality between g^* and g is via the L^2 pairing, "multiply and integrate". It is therefore immediate that

$$g = X \times F \times F$$

where X denotes vector fields and F denotes scalar fields on $\mathbb{R}^3$. Now X, which may be thought of as containing the velocity fields on $\mathbb{R}^3$, is the set of generators of displacements of the fluid. Hence the part of G which corresponds to the factor X in g is the group of diffeomorphisms on $\mathbb{R}^3$, denoted $\mathcal{D}$ (see (11) for a discussion of $\mathcal{D}$ as a Lie group). The group product is composition, since that is how two successive displacements of a fluid element combine to yield a single displacement. On the other hand, F is a vector space, so that the parts of G corresponding to these factors in g are F itself.

Hence G may be regarded as sets of triples of parameters labeling positions, densities, and entropies of fluid elements. These are physically related in the following way. Let η denote a diffeomorphism representing the change in position of fluid elements from the set of their initial positions $\{x_0\}$ to their positions $\{x\}$ at some later time. Then for the values of the density and entropy of the fluid element at x, we have, respectively, $\rho(x) = \rho_0 \circ \eta(x_0)$ and $\sigma(x) = \sigma_0 \circ \eta(x_0)$. In other words, we keep track of these quantities by composition with η. Hence the group $\mathcal{D}$ acts on $F \times F$ according to

$$\eta \bullet (f_1, f_2) = (f_1 \circ \eta, f_2 \circ \eta), \text{ where } (f_1, f_2) \in F \times F.$$

Finally, then, we conjecture that

$$G = \mathcal{D} \odot (F \times F) \quad .$$

The semi-direct product structure, indicated by $\odot$, is specified by the action above.

Then, by constructing explicitly the Kostant-Kirillov-Souriau symplectic structure on co-adjoint orbits of G, one obtains Morrison and Greene's Poisson structure on all of g^*. Suppressing the species label and using a dynamical variable as a subscript to denote the functional derivative with respect to that variable, it is

$$\{F,G\} (\underline{M}, \rho, \sigma) = -\int \left[F_{\underline{M}} \cdot (\nabla G_{\underline{M}}) - G_{\underline{M}} \cdot (\nabla F_{\underline{M}}) \right] \cdot \underline{M} \, d^3x \qquad (6)$$

$$- \int \left[F_{\underline{M}} \cdot (\nabla G_\rho) - G_{\underline{M}} \cdot (\nabla F_\rho) \right] \rho d^3x \; - \int \left[F_{\underline{M}} \cdot (\nabla G_\sigma) - G_{\underline{M}} \cdot (\nabla F_\sigma) \right] \sigma d^3x \quad .$$

The equations of motion for an ideal fluid now follow from eq. (5).

The vacuum Maxwell equations have been treated as a Hamiltonian system by Pauli (12) and Born and Infeld (13), but the discussion of Marsden and Weinstein (5) from the viewpoint of the reduction procedure has additional noteworthy aspects. The phase space is the cotangent bundle T^*A to the group A of vector potential fields $\underline{A}$ on $\mathbb{R}^3$. For $(\underline{A},\underline{Y}) \in T^*A$, we define the electric and magnetic fields $\underline{E} = -\underline{Y}$, $\underline{B} = \nabla \times \underline{A}$. Then the canonical symplectic structure on T^*A yields the following bracket of two functionals F, G of $\underline{E}$ and $\underline{B}$:

$$\{F, G\} (\underline{E}, \underline{B}) = \int \left[F_{\underline{E}} \cdot (\nabla \times G_{\underline{B}}) - G_{\underline{E}} \cdot (\nabla \times F_{\underline{B}}) \right] d^3x \quad . \tag{7}$$

This, together with the Maxwell Hamiltonian (the last term of eq. (4)), yields the evolution equations (1a) in the vacuum case. To obtain the vacuum versions of eqs. (1b) one uses the symmetry of the theory under the gauge transformation

$$\underline{A} \longmapsto \underline{A} + \nabla \Lambda \quad , \tag{8}$$

$\Lambda \in F$. We denote the gauge group $\{\Lambda\}$ by G. The Maxwell momentum map found in the reduction is $J_M (\underline{A}, \underline{Y}) = - \nabla \cdot \underline{Y}$.

We now consider the combined system of charged species plus the Maxwell equations. The Hamiltonian, written in terms of the canonical momenta $\underline{N}_s = \underline{M}_s + a_s \rho_s \underline{A}$ and the other pre-reduction variables, is

$$H(\underline{N}_s, \rho_s, \sigma_s, \underline{A}, \underline{Y}) = \sum_s \int \left[\tfrac{1}{2} \rho_s^{-1} |\underline{N}_s - a_s \rho_s \underline{A}|^2 + \rho_s U_s \right] d^3x \tag{9}$$
$$+ \tfrac{1}{2} \int \left[|\nabla \times \underline{A}|^2 + |\underline{Y}|^2 \right] d^3x \quad .$$

To apply reduction to the full phase space $P = \{(\underline{N}_s, \rho_s, \sigma_s, \underline{A}, \underline{Y})\} \equiv g_s^* \times T^*A$, the action of G on T^*A, eq. (8), must be extended to an action Φ of G on all of P. We require that $H \circ \Phi = H$, and that Φ preserves Poisson brackets of functionals F, G on P, i.e., $\{F \circ \Phi, G \circ \Phi\} = \{F, G\} \circ \Phi$. It is obvious from eq. (9) that the action

$$\Phi_\Lambda (\underline{N}_s, \rho_s, \sigma_s, \underline{A}, \underline{Y}) = (\underline{N}_s + a_s \rho_s \nabla \Lambda, \rho_s, \sigma_s, \underline{A} + \nabla \Lambda, \underline{Y}) \tag{10}$$

satisfies the first requirement, and one can show that it also satisfies the second.

In order now to obtain the momentum map $J: P \to G^*$ for Φ, it suffices to calculate $J_s: g_s^* \to G^*$, the momentum map on $g_s^* \equiv \{(\underline{N}_s, \rho_s, \sigma_s)\}$. We first define a map $\alpha_s: G \to F(P_s)$, from the Lie algebra of G to functions on P_s. Letting $\Lambda[g_s^*]$ be the infinitesimal generator of the action on g_s^* corresponding to $\Lambda \in G$, and letting $X[H]$ denote the Hamiltonian vector field on co-adjoint orbits in g_s^* with Hamiltonian function $H: P_s \to \mathbb{R}$, α_s is given by the relation $X[\alpha_s (\Lambda)] \equiv \Lambda[g_s^*]$. J_s is then defined by $J_s(\underline{N}_s, \rho_s, \sigma_s)(\Lambda) = \alpha_s(\Lambda)(\underline{N}_s, \rho_s, \sigma_s)$; one finds $J_s(\underline{N}_s, \rho_s, \sigma_s) = -a_s \rho_s$.

Therefore, the momentum map on P, obtained by summing

$$J(\underline{N}_s, \rho_s, \sigma_s, \underline{A}, \underline{Y}) = J_M(\underline{A}, \underline{Y}) + \sum_s J_s(\underline{N}_s, \rho_s, \sigma_s), \tag{11}$$

is $\quad J(\underline{N}_s, \rho_s, \sigma_s, \underline{A}, \underline{Y}) = - \nabla \cdot \underline{Y} - \sum_s a_s \rho_s$.

With $\underline{E} = -\underline{Y}$, reduction at the external charge density ρ_{ext} then specifies that the dynamics takes place on the level set of constant external charge

$$J^{-1}(\rho_{ext}) = \{(\underline{N}_s, \rho_s, \sigma_s, \underline{A}, \underline{Y}) \in P | \nabla \cdot \underline{E} = \rho_{ext} + \sum_s a_s \rho_s, \underline{E} = -\underline{Y}\} \quad .$$

Coordinates on the reduced phase space $J^{-1}(\rho_{ext})/G$ are now given by:

Proposition 1:

$$J^{-1}(\rho_{ext})/G = \{(\underline{M}_s, \rho_s, \sigma_s, \underline{B}, \underline{E}) | \nabla \cdot \underline{E} = \rho_{ext} + \sum_s a_s \rho_s, \nabla \cdot \underline{B} = 0\} .$$

Proof: To elements $(\underline{N}_s, \rho_s, \sigma_s, \underline{A}, \underline{Y})$ of P, associate quintuples $(\underline{M}_s, \rho_s, \sigma_s, \underline{B}, \underline{E})$, where $\underline{M}_s = \underline{N}_s - a_s \rho_s \underline{A}$, $\underline{B} = \nabla \times \underline{A}$, and $\underline{E} = -\underline{Y}$. Then the proposition follows from the momentum map constructed above, and from a simple verification that two elements of $J^{-1}(\rho_{ext})$ are associated to the same quintuple if and only if they are related by the gauge transformation eq. (10).

It remains now to compute the Poisson structure on $J^{-1}(\rho_{ext})/G$.

Proposition 2: For two functionals F, G of the field variables $(\underline{M}_s, \rho_s, \sigma_s, \underline{E}, \underline{B})$, the Poisson bracket is given by

$$\{F,G\}(\underline{M}_s, \rho_s, \sigma_s, \underline{E}, \underline{B}) = \sum_s \{F,G\}(\underline{M}_s, \rho_s, \sigma_s) + \{F,G\}(\underline{E}, \underline{B}) \qquad (12)$$
$$+ \sum_s \int (F_{\underline{M}} \cdot G_{\underline{E}} - G_{\underline{M}} \cdot F_{\underline{E}} + \underline{B} \cdot F_{\underline{M}} \times G_{\underline{M}}) a_s \rho_s \, d^3x$$

where the first and second terms are defined in eqs. (6) and (7), and species subscripts have been suppressed in the functional derivatives.

Proof: Given F and G, define $\bar{F}$ on P according to $\bar{F}(\underline{N}_s, \rho_s, \sigma_s, \underline{A}, \underline{Y}) = F(\underline{M}_s, \rho_s, \sigma_s, \underline{B}, \underline{E})$. Define $\bar{G}$ similarly. Then $\{F,G\}$ is found by computing $\{\bar{F},\bar{G}\}$ as the sum of eq. (6), written for unreduced variables, and the canonical bracket on T^*A, and by expressing the result in terms of the variables on $J^{-1}(\rho_{ext})/G$.

We observe that the first term of eq. (12) involves only the fluid variables and that the second is purely electromagnetic, while the third provides the coupling of the fluids to the electric and magnetic fields. Bilinearly, skew symmetry, and the Jacobi identity all follow for eq. (12) by the methods used in its derivation. In addition it is readily verified that the correct evolution equations for the phase space variables, in the form of eq. (5), follow from eqs. (12) and (4). Additional body forces, such as gravity, can easily be incorporated into eq. (2c) by the inclusion of an appropriate term in the Hamiltonian. Finally, eqs. (1b), rather then being postulated separately as initial conditions, follow from the gauge invariance of electromagnetism.

ACKNOWLEDGEMENTS

It is a pleasure to acknowledge valuable discussions with Professors Jerrold Marsden and Alan Weinstein, and with Professor Allan Kaufman, who suggested this problem.

This work was supported by the U.S. Department of Energy, under contract No. W-7405-ENG-48.

REFERENCES

1. P. Morrison and J. Green, Phys. Rev. Lett. $\underline{45}$, 790 (1980).
2. D. Holm and B. Kupershmidt, preprint.
3. P. Morrison, Phys. Lett. $\underline{80A}$, 383 (1980).
4. A. Weinstein and P. Morrison, Phys. Lett. $\underline{86A}$, 235 (1981).
5. J. Marsden and A. Weinstein, Physica $\underline{4D}$ (in press).
6. I. Bialynicki-Birula and J. Hubbard, submitted to Phys. Rev. A.
7. R. Spencer and A. Kaufman, submitted to Phys. Rev. A (Rapid Comm.).
8. R. Abraham and J. Marsden, Foundations of Mechanics (W. A. Benjamin Co., Reading, Mass, 1978).
9. J. Marsden and A. Weinstein, Rep. Math. Phys $\underline{5}$, 121 (1974).
10. J. Marsden, A. Weinstein, R. Schmid, and R. Spencer, Hamiltonian Systems and Symmetry Groups with Applications to Plasma Physics (in preparation).
11. J. Marsden, D. Ebin, and A. Fischer, Proceedings of the thirteenth biennial seminar of the Canadian Mathematical Congress, ed. J. Vanstone, pg. 135 (1972).
12. W. Pauli, General Principles of Quantum Mechanics (1933), (reprinted in English translation by Springer-Verlag, New York, New York, 1981).
13. M. Born and L. Infeld, Proc. Roy. Soc. A $\underline{150}$, 141 (1935).

HAMILTON'S PRINCIPLE AND ERTEL'S THEOREM

Rick Salmon*
Scripps Institution of Oceanography, La Jolla, Ca. 92093

1. Introduction

Variation principles for the equations governing the motion of perfect fluids are of two types. In the first type, which corresponds to Hamilton's principle in particle mechanics, the positions of marked fluid particles are varied at fixed times.[1,2] In the second type of variation principle, appropriately chosen field variables are varied at fixed locations and times.[3,4] The field variables include a set of scalar potentials which represent the fluid velocity. It has recently been shown that these two types of variation principle are really the same: they are related by canonical transformations.[5,6]

This note has two objectives. The first is to demonstrate a simple and particularly illuminating connection between variation principles of the two types mentioned above. The second objective is to give a new and direct derivation of Ertel's theorem of hydrodynamics based upon a symmetry property of the fluid Lagrangian. The results reported here were noticed in the course of an application of Hamiltonian methods to a study of the ocean's main thermocline.

2. Hamilton's Principle

Consider first a classical system composed of N discrete particles. Let i be a subscript index which identifies the particle, and let m_i and $\vec{x}_i(\tau)$ be the mass and the Cartesian position of the i-th particle at time τ. Let $V(\vec{x}_1,\ldots,\vec{x}_N)$ be the potential energy of the system. Then the Lagrangian is

---

*Work supported by the National Science Foundation (contract # OCE78-25670) as a part of POLYMODE.

$$L = \left(\sum_{i=1}^{N} \tfrac{1}{2} m_i \vec{\dot{x}}_i \cdot \vec{\dot{x}}_i \right) - V(\vec{x}_1, \ldots, \vec{x}_N) \qquad (2.1)$$

and the dynamical equations result from Hamilton's principle in the form

$$\delta \int L \, d\tau = 0 \qquad (2.2)$$

where δ corresponds to arbitrary variations $\delta\vec{x}_i(\tau)$ in the particle trajectories, and $\delta\vec{x}_i(\pm\infty)=0$. Alternatively, one can define the conjugate momenta,

$$\vec{p}_i = \partial L / \partial \vec{\dot{x}}_i , \qquad (2.3)$$

and invoke Hamilton's principle in the "extended" form,

$$\delta \int d\tau \left\{ \sum_{i=1}^{N} \vec{p}_i \cdot \vec{\dot{x}}_i - H(\vec{p}_j, \vec{x}_j) \right\} = 0 , \qquad (2.4)$$

where

$$H = \sum_{i=1}^{N} \vec{p}_i \cdot \vec{\dot{x}}_i - L \qquad (2.5)$$

and δ now corresponds to arbitrary independent variations $\delta\vec{p}_i(\tau)$, $\delta\vec{x}_i(\tau)$ in the momenta and positions of the particles.

Consider next the fluid continuum. Let the positions $\vec{x}=\vec{x}(\vec{a},\tau)$ of marked fluid particles be considered as functions of curvilinear labeling coordinates $\vec{a}$, amd the time τ. The labeling coordinates remain constant following fluid particles, and they are analogous to the subscript i above. It is convenient to assign these labeling coordinates so that equal volumes in $\vec{a}$-space contain equal masses. Then

$$\rho = \frac{\partial(a_1, a_2, a_3)}{\partial(x_1, x_2, x_3)} . \qquad (2.6)$$

is the mass density of the fluid. It follows directly from (2.6), and the fact that $\vec{a}$ is conserved on particles, that

$$\partial \rho / \partial \tau + \rho \nabla_x \cdot \vec{u} = 0 \qquad (2.7)$$

where

$$\vec{u} = \partial \vec{x}/\partial \tau \qquad (2.8)$$

and

$$\nabla_x = (\partial/\partial x_1, \partial/\partial x_2, \partial/\partial x_3) \qquad (2.9)$$

is the gradient operator in $\vec{x}$-space. Equation (2.7) is the usual equation of mass conservation.

Let the fluid Lagrangian be

$$L = \iiint d\vec{a} \, \{ \tfrac{1}{2} \partial \vec{x}/\partial \tau \cdot \partial \vec{x}/\partial \tau - V(\rho) \} \qquad (2.10)$$

where the potential V is a specified function of the density ρ. In (2.10) and below, the symbol ρ is merely an abbreviation for the Jacobian in (2.6). Hamilton's principle now states that

$$\delta \int L \, d\tau = 0 \qquad (2.11)$$

where L is given by (2.10) and δ corresponds to arbitrary variations $\delta \vec{x}(a,\tau)$ in the position of particle $\vec{a}$ at time τ. Assume for convenience that the fluid is infinite and that $\delta \vec{x}$ vanishes at large $|\vec{a}|$ and $|\tau|$. Then by the ordinary rules of variational calculus,

$\delta x_1:$
$$0 = \delta \int L \, d\tau =$$

$$= \int d\tau \iiint d\vec{a} \, \{ \partial x_1/\partial \tau \, \partial \delta x_1/\partial \tau + V'(\rho)\rho^2 \frac{\partial(\delta x_1, x_2, x_3)}{\partial(a_1, a_2, a_3)} \}$$

$$= \int d\tau \iiint d\vec{a} \, \{ -\partial^2 x_1/\partial \tau^2 - \frac{\partial(\rho^2 V', x_2, x_3)}{\partial(a_1, a_2, a_3)} \} \delta x_1 \qquad (2.12)$$

implies that

$$\partial^2 x_1/\partial \tau^2 = - \frac{\partial(x_1, x_2, x_3)}{\partial(a_1, a_2, a_3)} \frac{\partial(\rho^2 V', x_2, x_3)}{\partial(x_1, x_2, x_3)} = -\frac{1}{\rho} \partial(\rho^2 V')/\partial x_1$$

$$(2.13)$$

plus similar equations for δx_2 and δx_3. The definition

$$p = \rho^2 \, dV/d\rho \qquad (2.14)$$

brings (2.13) and its counterparts into the familiar form

$$\partial \vec{u}/\partial \tau = -\frac{1}{\rho} \nabla_x p . \qquad (2.15)$$

If p is required to be the thermodynamic pressure, and the flow is isentropic, then by (2.14) $V(\rho)$ must be the internal energy per unit mass. However, no such interpretation is actually required, because the laws of particle mechanics do not depend on the axioms of thermodynamics. The extended principle analogous to (2.4) is

$$\delta \int d\tau \{ \int\int\int d\vec{a}\ \vec{u} \cdot \partial \vec{x}/\partial \tau - H \} = 0 \qquad (2.16)$$

where

$$H = \int\int\int d\vec{a} \{ \tfrac{1}{2} \vec{u} \cdot \vec{u} + V(\rho) \} . \qquad (2.17)$$

Independent variations $\delta \vec{u}(\vec{a},\tau)$, $\delta \vec{x}(\vec{a},\tau)$ yield (2.8) and (2.15).

From a slightly different point of view, the fluid motion is a time-dependent map,

$$\vec{x} = \vec{x}(\vec{a},\tau) , \qquad (2.18)$$

from $\vec{a}$-space into $\vec{x}$-space, and Hamilton's principle requires that $\int L\ d\tau$ be stationary for arbitrary variations in this map. Since each forward map (2.18) uniquely determines an inverse map,

$$\vec{a} = \vec{a}(\vec{x},t) , \qquad (2.19)$$

from $\vec{x}$-space into $\vec{a}$-space, it is obvious that variations in this inverse map would serve as well. Here $t=\tau$, but $\partial/\partial\tau$ implies that $\vec{a}$ is held constant, while $\partial/\partial t$ implies constant $\vec{x}$. Rewrite (2.10) as

$$L = \int\int\int d\vec{x} \rho \{ \tfrac{1}{2} \vec{u} \cdot \vec{u} - V(\rho) \} \qquad (2.20)$$

and substitute for $\vec{u}$ from the identities

$$\rho\ u_1 = - \frac{\partial(a_1, a_2, a_3)}{\partial(t, x_2, x_3)} \quad , \text{etc.} \qquad (2.21)$$

The result is

$$L = \iiint d\vec{x} \frac{1}{\rho} \left\{ \tfrac{1}{2} \{ \frac{\partial(a_1,a_2,a_3)}{\partial(t,x_2,x_3)} \}^2 + \tfrac{1}{2} \{ \frac{\partial(a_1,a_2,a_3)}{\partial(x_1,t,x_2)} \}^2 \right.$$

$$\left. + \tfrac{1}{2} \{ \frac{\partial(a_1,a_2,a_3)}{\partial(x_1,x_2,t)} \}^2 - \rho^2 V(\rho) \right\} . \quad (2.22)$$

Hamilton's principle now requires that $\int L \, d\tau$ be stationary with respect to variations $\delta \vec{a}(\vec{x},t)$ in the labeling coordinates. This variation principle is Eulerian in the sense that $\vec{x}$ and t are the independent variables. To obtain the extended form, define momenta conjugate to $\vec{a}$, viz.

$$\vec{\Pi} = \delta L / \delta(\partial \vec{a}/\partial t) , \quad (2.23)$$

and eliminate $\partial \vec{a}/\partial t$ from (2.21). The result is

$$\rho u_1 = - \Pi_1 \partial a_1/\partial x_1 - \Pi_2 \partial a_2/\partial x_1 - \Pi_3 \partial a_3/\partial x_1 , \text{ etc. } (2.24)$$

and the extended principle is therefore

$$\delta \int dt \{ \iiint d\vec{x} \; \vec{\Pi} \partial \vec{a}/\partial t - H \} = 0 \quad (2.25)$$

where now δ corresponds to independent variations $\delta \vec{\Pi}(\vec{x},t)$ and $\delta \vec{a}(\vec{x},t)$. It is convenient to define

$$\vec{A} = -\vec{\Pi}/\rho \quad (2.26)$$

which can be freely varied in place of $\vec{\Pi}$. Then (2.25) takes the form

$$\delta \int dt \{ \iiint d\vec{x} \; \rho \vec{A} \cdot \partial \vec{a}/\partial t + H \} = 0 \quad (2.27)$$

where

$$H = \iiint d\vec{x} \; \rho \{ \tfrac{1}{2} \vec{u} \cdot \vec{u} + V(\rho) \} , \quad (2.28)$$

$$\vec{u} = A_1 \nabla_x a_1 + A_2 \nabla_x a_2 + A_3 \nabla_x a_3 , \quad (2.29)$$

and δ corresponds to variations $\delta\vec{A}(\vec{x},t)$ and $\delta\vec{a}(\vec{x},t)$. The variation principle (2.27-2.29) was obtained by Seliger and Whitham[4] by a rather different approach. The present derivation emphasizes the close connection between (2.27) and the Hamilton's principle of particle mechanics, and it puts a clear interpretation on $\vec{a}$ and $\vec{A}$. The $\vec{a}$ are mass labeling coordinates which can be assigned in numerous ways to satisfy (2.6). But once the a_i have been chosen, the A_i are uniquely determined from (2.29) as the projections of $\vec{u}$ on the curvilinear basis vectors $\nabla_x a_i$. The $\nabla_x a_i$ form a basis provided only that ρ is nonzero. Thus $\vec{a}$ and $\vec{A}$ are always single-valued. I note for future use that the reciprocal of (2.29) is

$$\vec{A} = u_1 \nabla_a x_1 + u_2 \nabla_a x_2 + u_3 \nabla_a x_3 \qquad (2.30)$$

where

$$\nabla_a = (\partial/\partial a_1, \partial/\partial a_2, \partial/\partial a_3) \qquad (2.31)$$

is the gradient operator in $\vec{a}$-space.

3. Ertel's Theorem[10]

As remarked in section 2, the Lagrangian (2.10) is unaffected by any transformation of the labeling coordinates which leaves the Jacobian (2.6) unchanged. This symmetry property leads to a conservation law discovered by Ertel[7] using wholly different methods. The following derivation by way of Noether's theorem is considerably more direct. Suppose that $\delta\vec{a}(x,t)$ is indeed such that

$$\delta \frac{\partial(a_1,a_2,a_3)}{\partial(x_1,x_2,x_3)} = 0 . \qquad (3.1)$$

This implies that

$$\partial\delta a_1/\partial a_1 + \partial\delta a_2/\partial a_2 + \partial\delta a_3/\partial a_3 = 0 \qquad (3.2)$$

provided that ρ is nonzero. Thus

$$\delta\vec{a} = \nabla_a \times \vec{T} \qquad (3.3)$$

for some $\vec{T}=\vec{T}(\vec{a},\tau)$. For such a variation,

$$\delta \int L\, d\tau = \delta \int d\tau \iiint d\vec{a}\ \tfrac{1}{2}\, \partial\vec{x}/\partial\tau \cdot \partial\vec{x}/\partial\tau$$

$$= - \int d\tau \iiint d\vec{a}\ \vec{A} \cdot \partial\vec{a}/\partial\tau \qquad (3.4)$$

since it can easily be shown that

$$\delta(\partial\vec{x}/\partial\tau) = - \frac{\partial\vec{x}}{\partial a_1}\frac{\partial \delta a_1}{\partial \tau} - \frac{\partial\vec{x}}{\partial a_2}\frac{\partial \delta a_2}{\partial \tau} - \frac{\partial\vec{x}}{\partial a_3}\frac{\partial \delta a_3}{\partial \tau} \qquad (3.5)$$

The vector $\vec{A}$ is that given by (2.30). Substitution from (3.3) and integrations by parts bring (3.4) into the form

$$\delta \int L\, d\tau = \int d\tau \iiint d\vec{a}\ \vec{T} \cdot \partial/\partial\tau\ \{\nabla_a \times \vec{A}\} \ . \qquad (3.6)$$

But $\vec{T}$ is arbitrary and (3.6) must vanish by Hamilton's principle. It follows that

$$\partial \vec{Q}/\partial \tau = \vec{0} \qquad (3.7)$$

where

$$\vec{Q} = \nabla_a \times \vec{A} \ . \qquad (3.8)$$

The vector $\vec{Q}$ is conserved on particles. Now let $\Phi = \Phi(a_1, a_2, a_3)$ be any quantity which is also conserved on particles. Then

$$q = \vec{Q} \cdot \nabla_a \Phi \qquad (3.9)$$

is also conserved. With help from (2.30), q may also be written

$$q = \frac{1}{\rho}\, \nabla_x \Phi \cdot (\nabla_x \times \vec{u}) \ . \qquad (3.10)$$

The statement

$$\partial q/\partial \tau = 0 \qquad (3.11)$$

is Ertel's theorem. Since Φ is arbitrary (3.11) and (3.7) are equivalent. With $\Phi = \rho$ (and assuming $\partial \rho/\partial \tau = 0$) the quantity q is called the potential vorticity.

The foregoing procedure extends to include rotating coordinates and it provides an elegant unification for all forms of Ertel's theorem which is lacking in the conventional derivations: for any continuum system, Ertel's theorem is simply the conservation law which results from the most general transformation of labeling coordinates that leaves every term in the Lagrangian unchanged. This approach also provides a motivation for Ertel's theorem: the conservation law is known to exist as soon as inspection of the Lagrangian reveals a symmetry property. One need not depend on unguided manipulations.

If the potential in (2.10) is replaced by

$$V(\rho, S(a_1, a_2, a_3)) ,\qquad (3.12)$$

where S is a function of the labeling coordinates only (usually considered to be the specific entropy), then the results of section 2 generalize easily, but the general conservation law is destroyed. This is obvious because, if S is completely arbitrary, as it must be to accomodate arbitrary initial entropy distributions, then no general transformation of the labeling coordinates leaves V unchanged.

Eckart[2] derived the conservation law (3.7) using the energy-momentum tensor formalism, which is related to the procedure followed here, but he did not notice the connection with Ertel's theorem. See also Bretherton[8,9] for some closely related results.

References

1. J. W. Herivel, Proc. Camb. Phil. Soc., $\underline{51}$, 344 (1955).

2. C. Eckart, Phys. Fluids, $\underline{3}$, 421 (1960).

3. A. Clebsch, Crelle's Journal, $\underline{56}$, 1 (1859).

4. R. L. Seliger and G. B. Whitham, Proc. Roy. Soc., $\underline{A305}$, 1 (1968).

5. L. J. F. Broer and J. A. Kobussen, Appl. Sci. Res., $\underline{29}$, 419 (1974)

6. W. van Saarloos, submitted to Physica (1981).

7. H. Ertel, Meteor. Zeit., $\underline{59}$, 277 (1942).

8. F. P. Bretherton, J. Fluid Mech., $\underline{44}$, 19 (1970).

9. F. P. Bretherton (manuscript).

10. While this note was in press, its author became aware that the proof of section 3 has previously been given by P. Ripa, A.I.P. proceedings, $\underline{76}$ (1981).

ADVANTAGES OF HAMILTONIAN FORMULATIONS IN COMPUTER SIMULATIONS
O. Buneman
Institute for Plasma Research, Stanford University, Stanford, CA 94305

ABSTRACT

Hamilton's equations for ideal vortical compressible flow, based on Clebsch's velocity representation, are well suited to numerical integration. Checkerboard data arrangement and leap-frogging in time guarantee second-order accuracy and stability. Incipient turbulence developing from a shear layer can be simulated by such a scheme.

1. The "Hamiltonian" Computer

In the age of computers the merits of advanced and elegant mathematical methods are often being questioned: what good is a closed solution to a problem in terms of--say--Bessel functions when it takes longer to calculate the sequence of function values than to integrate numerically the underlying equation?

In answer to such scepticism we can quote the modern justification for the mathematical effort to cast one's physical problem into a variational principle, namely, the finite element method. That combination of finite element which makes the action integral an extremum will give the best approximation to reality achievable with the finite resources of the computer.

It turns out that casting one's problem into a set of Hamilton's equations brings similar benefits. This is, perhaps, because the computer is itself a "Hamiltonian device": It has a memory--denote its state by Ψ -- and a central processor, H, programmed to operate on that memory and to produce $H(\Psi)$, the update $\delta\Psi$ of the memory in each cycle δt:

$$\delta\Psi/\delta t = H(\Psi)$$

The simulation of the evolution of a system which follows Hamilton's equations is therefore an ideal task for a computer. Even a digital computer becomes an analog device in this manner. Note that when solving elliptic problems numerically we often use iteration. This amounts to introducing an evolutionary process to approach the desired equilibrium.

Variational and Hamiltonian methods were originally developed in application to celestial and rigid body mechanics, their merits being (1) ease of transformation to new co-ordinates, (2) simplicity of perturbation calculus.

The superficial analogy between a working computer and an evolving Hamiltonian system can be refined in many practical applications where one distinguishes between conjugate variables, p and q. These variables can be "leap-frogged" over each other: the update of p is given by the q-configuration, the update of q by the p-configuration:

$$\frac{dp}{dt} = -\frac{\partial H}{\partial q} \qquad \frac{dq}{dt} = \frac{\partial H}{\partial p}$$

Records of p and q are therefore conveniently kept at alternating half time-steps. There is a dual advantage: the accuracy is that of central differencing, i.e. of order δt^2, and reversibility is preserved.

2. Hamilton's Equations from the Clebsch Representation

The details of this philosophy become apparent in the application to gas dynamics. Variational principles to describe ideal gas dynamics have been known for some time[1,2]. They tend to rely on the Clebsch representation of the flow field[3,4].

$$\vec{v} = \lambda \nabla \mu - \nabla \phi$$

(with additional terms to describe entropy changes[2]). λ and μ are identifiable as vortex labels which are transported by the fluid unchanged.

Recently, this author[5] was able to derive a set of Hamilton's equations from such variational descriptions, using (ρ,ϕ), (σ,μ) as conjugate variables where $\sigma \equiv -\rho\lambda$. Davidov[1] had recognized that σ and μ are conjugate. From the energy density

$$H = C\rho^\gamma + (\sigma\nabla\mu + \rho\nabla\phi)^2/2\rho ,$$

as the Hamiltonian, one obtains the functional derivatives

$$\frac{\partial H}{\partial \phi} = \nabla \cdot \rho\vec{v}$$

$$= -\frac{\partial \rho}{\partial t} \text{ (conservation of mass),}$$

$$\frac{\partial H}{\partial \rho} = C\gamma\rho^{\gamma-1} - \frac{1}{2}\left(\frac{\sigma}{\rho}\right)^2 (\nabla\mu)^2 + \frac{1}{2}(\nabla\phi)^2$$

$$= \frac{\partial \phi}{\partial t} \text{ (Bernoulli, alternative form),}$$

$$\frac{\partial H}{\partial \sigma} = -\vec{v} \cdot \nabla\mu$$

$$= \frac{\partial \mu}{\partial t} \text{ (vortex label } \mu \text{ follows the flow),}$$

$$\frac{\partial H}{\partial \mu} = \nabla \cdot \sigma\vec{v}$$

$$= -\frac{\partial \sigma}{\partial t} \text{ (vortex label } \lambda \text{ follows the flow)}$$

and these equations constitute the laws of compressible flow dynamics.

3. Checkerboard Data Arrangement

It is remarkable that in the temporal advance of any variable one requires the local values of the conjugate variables and only spatial derivatives of the variables themselves. The leap-frog feature seems to apply in space as it does in time! This space-time symmetry is not surprising since the Clebsch scheme and variational method can be made relativistically covariant as was realized by Dirac[6].

As a result, one can, and should, interleave the spatial records of mutually conjugate variables. For two dimensional simulations, this means a checkerboard arrangement of (ρ,σ) information on the dark squares with (ϕ,μ) information on the light squares at one half time-step, with the reverse arrangement at the next half time-step:

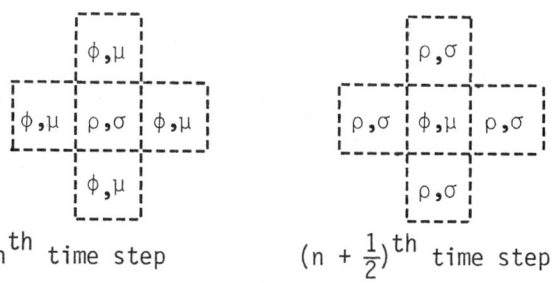

n^{th} time step $\qquad (n + \frac{1}{2})^{th}$ time step

Potential gradients and hence velocity components and fluxes are available, by spatial central differencing, at the density (ρ,σ) locations. That is where they are needed for updating potentials ϕ and μ in time. Divergences of fluxes are available at the potential (ϕ,μ) locations--where needed for updating the densities ρ and σ.

σ is a density of a conserved attribute (the vortex label λ moves with the flow unchanged). A more specific identification of this attribute is made difficult by the fact that the Clebsch potentials can be subjected to a large choice of gauge transformations: the initial labeling of vortices is highly arbitrary. In fact, as a simulation proceeds, a linear growth in time of potentials can occur. An occasional gauge transformation may be advisable. In three dimensions, one records density and potential information over complementary octahedral grids, data for each being arranged like the Na and Cl atoms in salt, respectively. But the two interchange roles every half time step. In a cubic mesh, four data are kept per grid point, two of the data referring to the present time step, the other two to the previous or next half time step.

The simulation is strictly causal: it can be made relativistic if necessary. The differential equations are hyperbolic, not elliptic. There is no need for predictor corrector methods or Crank-Nicolson algorithms. Numerical stability is subject to the Courant condition only, and there is no odd-even instability.

An important feature is that the procedure is local, not global. Updating can be pipelined through a small central processing unit, or in a parallel machine with communication only between close neighbors. The size of the random access memory of a computer presents no limitation on the volume of information to be handled, since not all the information needs to reside in core memory at once. For instance, a 256^3 simulation is being planned for a CRAY-1, using layer processing.

4. Preliminary Numerical Experiments

Some modest exploratory two-dimensional simulations have been carried out at Stanford by Y. Shiran. He initialized a continuous shear layer of finite width (Figure 1) by the appropriate choice of λ and μ (see contours in Figures 2 and 3).

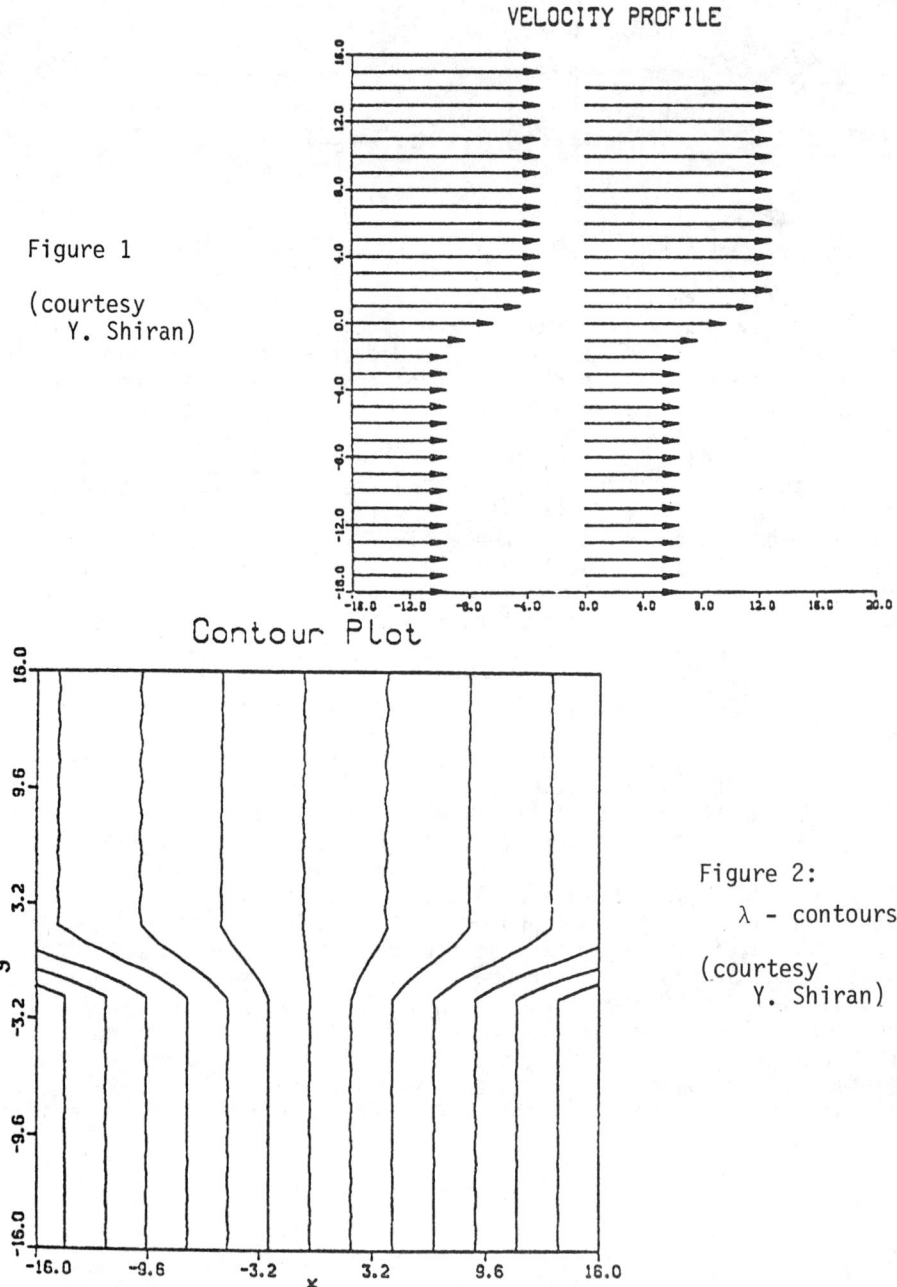

Figure 1

(courtesy Y. Shiran)

Figure 2:

λ - contours

(courtesy Y. Shiran)

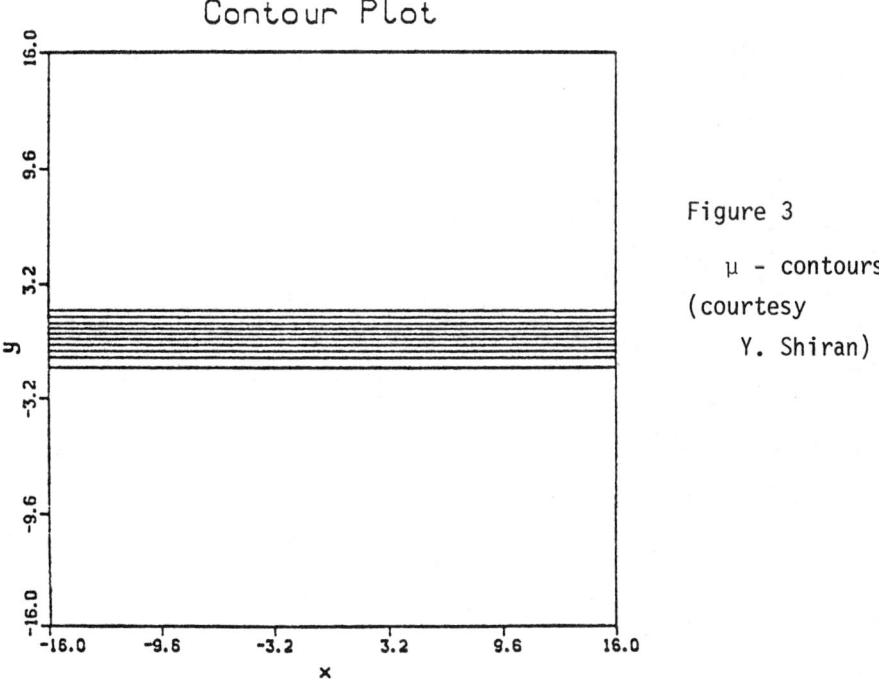

Figure 3

μ - contours

(courtesy Y. Shiran)

The vorticity, a scalar in two dimensions, is then initially as shown by its contour diagram, Figure 4.

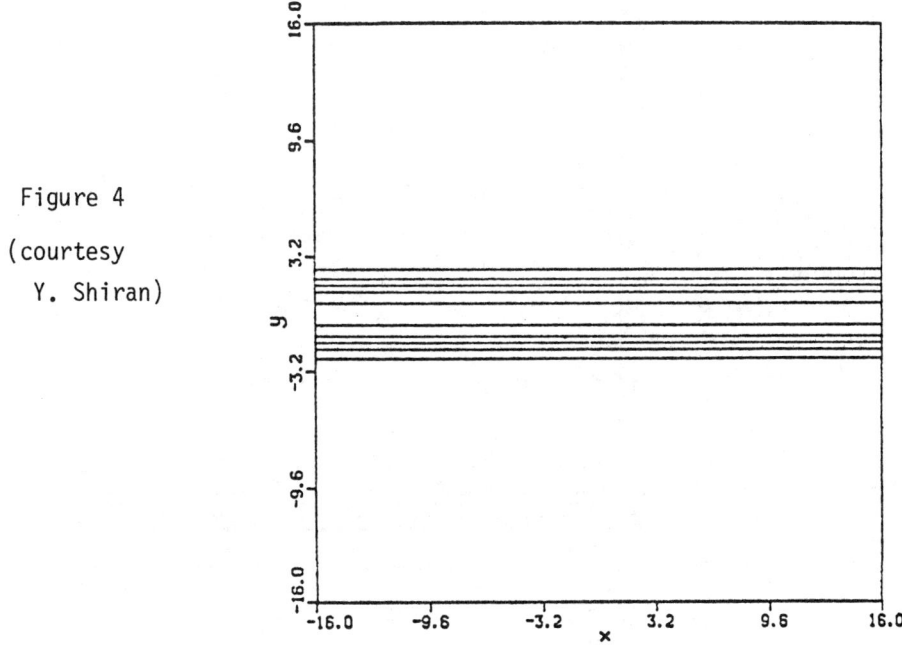

Figure 4

(courtesy Y. Shiran)

142

A 32 by 32 grid is used. After a number of time steps the vorticity contours appear as in Figure 5, with Figure 6 showing a vertical blow-up of the central part. The "eyes" are the result of the

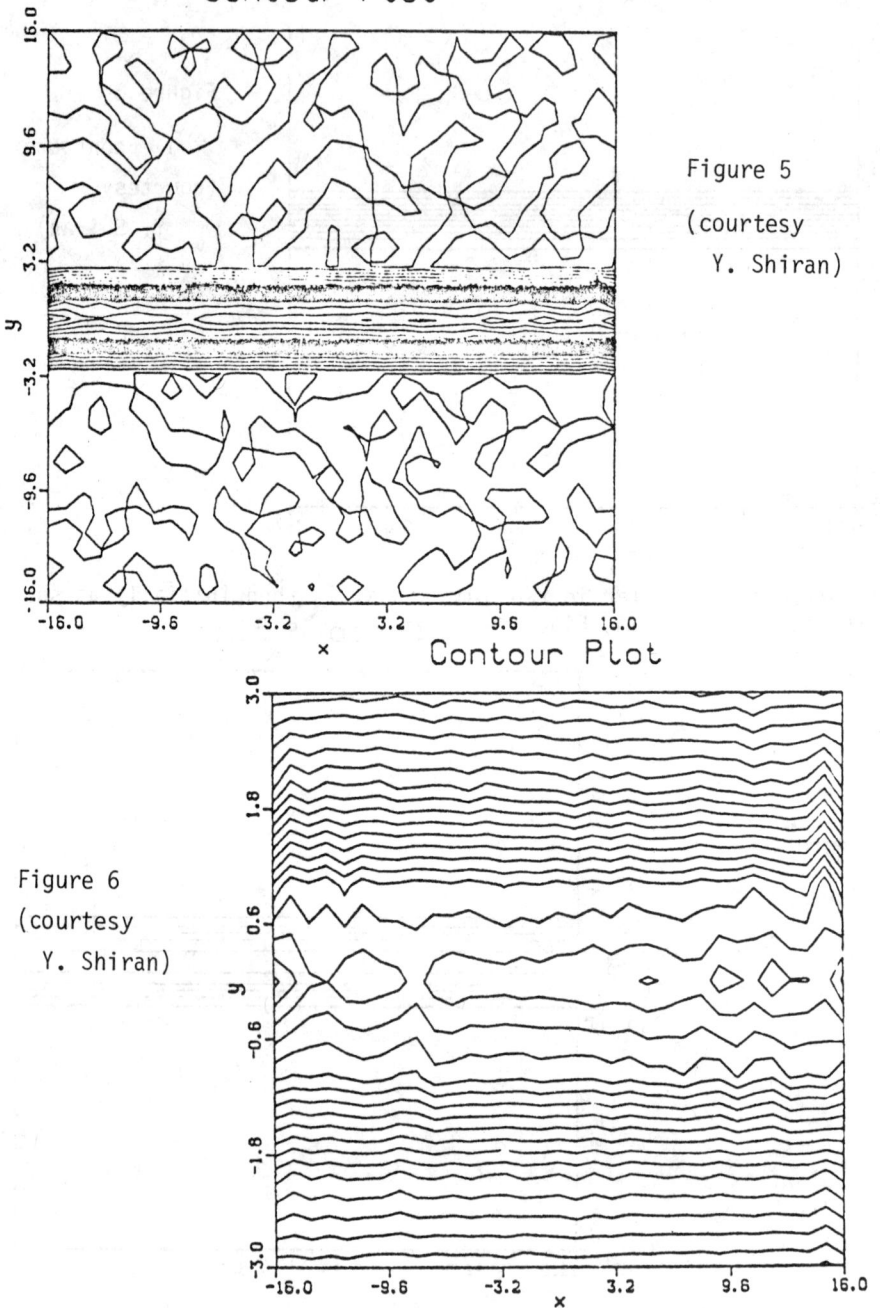

Figure 5
(courtesy
Y. Shiran)

Figure 6
(courtesy
Y. Shiran)

Kelvin-Helmholtz instability which has developed from noise level. They are the forerunners of the Karman street. (Y. Shiran had previously checked that the Clebsch representation does reproduce this genuine physical instability when subjected to a perturbation analysis.)

Conclusion

The formulation of the laws of compressible flow dynamics as a set of Hamilton's equations has more than aesthetic appeal. It leads to a numerical simulation scheme with many merits, and the first experiments with such a scheme have been successful.

References

1. B. I. Davidov, Dokl. Akad. Nauk SSSR, 69 No. 2, 165 (1949).
2. R. L. Seliger and G. B. Whitham, Proc. R. Soc. London Ser. A 305, 1 (1968).
3. H. Lamb, Hydrodynamics (Dover, New York, 1932), Article 167.
4. A. Clebsch, J. Reine Angew. Math. 56, 1 (1859).
5. O. Buneman, Phys. Fl. 23(8), 1716 (1980).
6. P. A. M. Dirac, Proc. R. Soc. London Ser. A 212, 330 (1952).

THE SIMPLE PARTICLE AND THE PERFECT FLUID

E. A. Spiegel
Columbia University, New York, New York 10027

ABSTRACT

A review of the analogy between the classical mechanics of a point particle and the dynamics of an ideal fluid is presented. The nonlocal forces in an irrotational fluid needed to convert its equation to a standard nonlinear Schrödinger equation are introduced. The changes needed to allow for vorticity are discussed.

1. INTRODUCTION

In the Hamilton-Jacobi description of a perfect fluid one considers a number of possible orbits at the same time. This picture of many representative points moving in the same physical space also applies to a fluid without internal stresses. The formal similarity was seen in the nineteenth century in the resemblance between the Hamilton - Jacobi equation and the Bernoulli integral [1]. The only difference, in the case of a barotropic fluid, is that the Bernoulli integral contains a specific enthalpy as well as a mechanical potential. The generalized equation [2,3], including the vorticities of the two flows, was already considered in that period.

The similarity between the descriptions of particle and fluid motions becomes more striking when the Schrödinger description of a simple particle is considered. This observation of Madelung [4] was however not particularly appreciated at first [5]. The linear Schrödinger equation was converted to nonlinear fluid equations and the analogue of the specific enthalpy term looked somewhat strange. Nonetheless, beginning in the early fifties, a steady trickle of papers on the fluid dynamical form of the Schrödinger equations began [6-12]. The earliest of these papers were motivated by an interest in nonstandard interpretations of quantum mechanics, which did nothing for the popularity of the fluid dynamical viewpoint among the majority of physicists. Later, the motivation came from superconductivity and the general interest in quantum fluids. Finally, once the nonlinear Schrödinger equation became interesting, there was no longer the fear of rendering the problem nonlinear.

In what follows, I give an elementary discussion of this analogy between the dynamics of a simple particle (a point endowed with only scalar properties) and of a perfect fluid (that conserves entropy). Since many contributions to this volume examine the Hamiltonian approach to fluid dynamics, I shall not follow that route, though I find it interesting [13]. Rather, I shall attempt to describe some fluid dynamical questions suggested by the analogy.

2. KINEMATICAL PRELIMINARIES

Orbits

The orbit of a particle is described by a set of three coordinates as a function of time. In Hamilton - Jacobi theory, one considers a set of possible realizations of the particle, each with a different orbit. To make things simple, suppose that at time $t = 0$ there is one representative point of the particle at each point in space. The orbits are given by:

$$x_i = X_i(t;\alpha_j) \qquad (2.1)$$

where the x_i are Cartesian coordinates of the particle at t, and

$$\alpha_j = X_i(0;\alpha_j) \qquad (2.2)$$

are parameters used to label the orbits.

Suppose that there is a constant T such that, for $0 \leq t < T$, no orbits have intersected, at least in some finite region of space. Then, in that region and for $0 \leq t < T$,

$$D \equiv \det \frac{\partial X_i}{\partial \alpha_j} \neq 0, \qquad (2.3)$$

and we can solve for the α_i:

$$\alpha_i = A_i(t; x_j). \qquad (2.4)$$

Velocity

The velocity of a representative particle labeled by α_i, or $\boldsymbol{\alpha}$, is

$$v_i \equiv \dot{x}_i \equiv \left.\frac{\partial x_i}{\partial t}\right|_{\boldsymbol{\alpha} \text{ fixed}} . \qquad (2.5)$$

Fluid dynamicists use the notation

$$\frac{D}{Dt}(\ ) \equiv (\ )^{\cdot} . \qquad (2.6)$$

Another fluid dynamical usage (exressed somewhat baldly) is obtained on noting that

$$v_i(t;\alpha_j) = v_i(t; A_j(t;x_j)), \qquad (2.7)$$

whence it is natural to define

$$u_i(x_j, t) \equiv v_i(t; A_j(t; x_j)). \quad (2.8)$$

If we know **v** we know the velocity of any particle **α** at t; **u** tells us the velocity at t of whatever particle happens to be at **x** just then. Here, I write **u** for u_i, and so on.

Acceleration

The acceleration of a particle is

$$\ddot{x}_i = \frac{D}{Dt} u_i = \frac{Dt}{Dt} \frac{\partial u_i}{\partial t}\bigg|_{\mathbf{x} \text{ fixed}} + \frac{Dx_j}{Dt} \frac{\partial u_i}{\partial x_j}\bigg|_{t \text{ fixed}}. \quad (2.9)$$

This is written conventionally as

$$\frac{Du_i}{Dt} = \frac{\partial u_i}{\partial t} + u_j \frac{\partial u_i}{\partial x_j} = \frac{\partial u_i}{\partial t} + (\mathbf{u} \cdot \nabla) u_i, \quad (2.10)$$

where the variable that is held fixed is to be deduced from the context.

Particle conservation

For $0 \leq t < T$ let

$$\rho(x_i, t) = m/D, \quad (2.11)$$

where m is the particle's mass. A straightforward calculation [14] gives

$$\partial_t \rho + \nabla \cdot (\rho \mathbf{u}) = 0. \quad (2.12)$$

where ∂_t is written for $\partial/\partial t$. The Jacobian D measures the volume of a swarm of representative points, hence ρ is the density of such a swarm. This conservation law holds so long as the orbits do not cross. When orbits do cross, D can be zero, and ρ diverges.

3. SIMPLE DYNAMICS

The material of the previous section could have been used in either an elementary fluids or a dynamics course. The distinction usually comes at the dynamical level when we specify whether the particles **α** are in interaction or not. If we speak of the elementary dynamics of a simple particle, we omit interactions between the particles since each is a representation of the one particle we have in mind. Motion of such a particle under the influence of a prescribed potential per unit mass $V(x_i, t)$, is governed by

the equation
$$\frac{D\mathbf{u}}{Dt} = -\nabla V. \tag{3.1}$$

Let
$$\boldsymbol{\omega} = \nabla \times \mathbf{u}, \tag{3.2}$$

and recall the relation
$$(\mathbf{u}\cdot\nabla)\mathbf{u} = \nabla(\tfrac{1}{2}\mathbf{u}^2) - \mathbf{u}\times\boldsymbol{\omega}.$$

Then (3.1) is
$$\partial_t \mathbf{u} = \mathbf{u}\times\boldsymbol{\omega} - \nabla(V + \tfrac{1}{2}\mathbf{u}^2). \tag{3.3}$$

The curl of (3.3) is
$$\partial_t \boldsymbol{\omega} = (\boldsymbol{\omega}\cdot\nabla)\mathbf{u} - (\mathbf{u}\cdot\nabla)\boldsymbol{\omega} \tag{3.4}$$

whence we obtain the Helmholtz theorem [15]
$$\frac{D}{Dt}\left(\frac{\boldsymbol{\omega}}{\rho}\right) = \left(\frac{\boldsymbol{\omega}}{\rho}\right)\cdot\nabla\mathbf{u}. \tag{3.5}$$

If $\boldsymbol{\omega}$ is initally zero, it remains zero. Therefore we can have
$$\boldsymbol{\omega} = 0 \tag{3.6}$$

in a realization of this expression of Newton's laws for a simple particle.

If we assume (3.6) we may write
$$\mathbf{u} = \nabla\phi, \tag{3.7}$$

where ϕ is the velocity potential. We introduce (3.7) into (3.4) and directly obtain
$$\partial_t \phi + \tfrac{1}{2}(\nabla\phi)^2 + V = \mathsf{F}, \tag{3.8}$$

where F is an arbitrary function of time. Since $\mathbf{u}$ is unaffected by the addition of a function of time to ϕ, we may choose a gauge so that F is zero or a constant. Thus, the Hamilton-Jacobi equation (3.8) is based on the assumption of irrotational flow of the representative points. This assumption is often hidden in contemporary discussions of (3.8), but its role in mechanics was not overlooked in the nineteenth century [1]. In the present context, where the particles are uncoupled, the assumption of irrotationality has no dramatic significance. The orbit of an individual particle is not affected by the relative motions of its neighbors. Then assumption (3.7) merely characterizes a particular choice of a set of initial conditions. This is no longer true when the particles interact, as in fluid dynamics or wave mechanics.

In classroom examples, V is often prescribed so as to permit separable solutions of (3.8). Such solutions preserve the conditions under which that equation was derived. But examples in which (2.7) breaks down in a finite time are abundant in cases with turning points, for instance. The simplest examples occur for $V=0$. Then (3.1) admits solutions in one dimension of the form

$$U(X,t) = V(X-Ut) \tag{3.9}$$

where $U = u_1$, $X = x_1$ and V is an arbitrary function. Suppose that the initial condition is

$$U(X,0) = -X^{\frac{1}{3}} \tag{3.10}$$

Then we find that

$$U^3 - Ut + X = 0. \tag{3.11}$$

The orbit is

$$x = X \equiv \alpha - \alpha^{\frac{1}{3}}t \tag{3.12}$$

and the Jacobian is

$$D = \frac{\partial X}{\partial \alpha} = 1 - \tfrac{1}{3}t/\alpha^{\frac{2}{3}}. \tag{3.13}$$

For any given α, D vanishes sooner or later. Such breakdown is typical of many problems and it is of interest to see what additional terms in the dynamics may help to avoid it.

4. GENERALIZED BAROTROPES

It is possible to rescue the Hamilton-Jacobi formalism from the breakdown discussed at the end of §3 by introducing forces that tend to keep the particles apart when the density is high. The most familiar example is the force associated with the pressure gradient.

Consider a barotropic fluid, that is, a fluid whose pressure is a function only of its density:

$$p = p(\rho). \tag{4.1}$$

The specific enthalpy is

$$h(\rho) = \int \frac{dp}{\rho} \tag{4.2}$$

and is like a potential. Then (3.1) is generalized to

$$\frac{D\mathbf{u}}{Dt} = -\nabla[V+h(\rho)]. \tag{4.3}$$

We still have (3.5) so that (3.7) describes possible solutions. Breakdown is still possible, if somewhat delayed, but the fact that we are dealing with fluids, provides a physical approach to this question: discontinuous solutions, or shocks, may be introduced.

If we want to avoid the breakdown completely, we may introduce nonlocal interactions among the particles. But why should we care to do this? It is not just that the Hamilton-Jacobi theory is lovely enough for us to try to save it, but rather that this is also a way to arrive at the Schrödinger equation from the fluid-dynamical viewpoint. The idea is that a nonlocal interaction may give the particles sufficient warning to avoid trouble, provided the interaction propagates fast enough. Suppose that in fact the interaction travels at infinite speed and that it is a functional of the density. Consider in particular the functional

$$\mathcal{W}[\rho] = \int K(\mathbf{x} - \boldsymbol{\xi}) g(\rho(\boldsymbol{\xi})) \, d_3\boldsymbol{\xi} , \qquad (4.4)$$

where g is an arbitrary function. Let

$$\int K(\mathbf{x}) \, d_3\mathbf{x} = 1. \qquad (4.5)$$

So the dimension of K is inverse volume. If no characteristic value of ρ is provided, it is dimensionally indicated that g should vary like ρ^n, where n is a pure number.

We are looking for a potential per unit mass, and unless suitable dimensional constants are provided, $\mathcal{W}$ will not have the right dimensions, namely energy per unit mass, or speed squared. We introduce c, a constant whose dimension is speed. Then,

$$W[\rho] = -c^2 \mathcal{W}/g(\rho) \qquad (4.6)$$

has the right dimensions to be a potential. I call a fluid with such a potential a generalized barotrope. If we add this potential to those we have already, we obtain the equation

$$\frac{D\mathbf{u}}{Dt} = -\nabla\{V(\mathbf{x},t) + h(\rho) + W[\rho]\}. \qquad (4.7)$$

We also have

$$\partial_t \rho + \nabla \cdot (\rho \mathbf{u}) = 0. \qquad (4.8)$$

Now (3.4) still holds and if we introduce once more the assumption

$$\mathbf{u} = \nabla \phi \qquad (4.9)$$

we obtain the Bernoulli integral

$$\partial_t\phi + \tfrac{1}{2}(\nabla\phi)^2 + V + h + W = 0. \qquad (4.10)$$

The motion is described by the two fields ρ and ϕ. We may combine them into the single complex function

$$\psi = \sqrt{\rho}\, e^{i\phi/\kappa} \qquad (4.11)$$

where κ is an arbitrary constant of dimensions length^2/time. With this Madelung transformation [4] we may combine (4.8) and (4.10) into the single equation

$$i\kappa\partial_t\psi = -\tfrac{1}{2}\kappa^2\nabla^2\psi + \{V + h + W + \tfrac{1}{2}\kappa^2\nabla^2\sqrt{\rho}/\sqrt{\rho}\}\psi. \qquad (4.12)$$

In short, the dynamics of any irrotational (generalized) barotrope may be described by a wave equation, even if the equation is not always so beautiful. If $W=0$, (4.12) is the wave equation for a perfect barotropic fluid and, for $h=0$, (4.12) is the wave equation for classical mechanics.

5. SCHRÖDINGER EQUATIONS

What kind of generalized barotropes, that is what choices for K, might be of physical interest? This can almost be read off from (4.12). Let

$$\lambda^2 = \int \mathbf{x}^2 K(\mathbf{x})\, d_3\mathbf{x}. \qquad (5.1)$$

Suppose that λ is small compared to the scale on which ρ varies. Then in (4.4) we may replace g by a Taylor series and we get, for symmetric K,

$$W = -1 - \tfrac{1}{2}\lambda^2(\nabla^2 g)/g + O(\lambda^4). \qquad (5.2)$$

If we choose

$$g = \sqrt{\rho} \qquad (5.3)$$

and

$$\kappa = c\lambda, \qquad (5.4)$$

(4.12) becomes a conventional looking Schrödinger equation:

$$i\kappa\partial_t\psi = -\tfrac{1}{2}\nabla^2\psi + \{V + h(\rho)\}\psi. \qquad (5.5)$$

(A constant term in W has been removed by a choice of gauge.) If

$$p = K\rho^\Gamma; \qquad (5.6)$$

then

$$h = \left(\tfrac{\Gamma}{\Gamma-1}\right) K\rho^{\Gamma-1}. \qquad (5.7)$$

For $\Gamma = 2$, (5.4) is the cubic Schrödinger equation.

To recapitulate, any irrotational (generalized) barotrope satisfies equation (4.12) but when λ, the range of K, becomes small, (4.12) reduces to a conventional nonlinear Schrödinger equation. Moreover, if we introduce (4.11) into (5.5), we get

$$\frac{D\mathbf{u}}{Dt} = -\nabla[V + W_q + h] \tag{5.8}$$

$$W_q = \text{const.} - \tfrac{1}{2}\lambda^2 c^2 \nabla^2 \sqrt{\rho}/\sqrt{\rho}. \tag{5.9}$$

When $\lambda \to 0$, that is in the WKB limit, the problem reduces to

$$\frac{D\mathbf{u}}{Dt} = -\nabla[V + h]. \tag{5.10}$$

When $\Gamma = 2$, this is the shallow-water equation. Bohm [6] named W_q the quantum-mechanical potential, and Feynman [8] called it a quantum-mechanical pressure, each in a certain context. As λ is decreased, W reduces to W_q in leading order and, finally, it becomes a constant in the limit where λ vanishes. Most of this arises from doing what comes not too unnaturally except for the choice of $g(\rho)$. That is a feature that does not have an obvious physical rationalization in this picture, but there are clues in the theory of Brownian movement [16].

6. ACOUSTICS

The Schrödinger equation has been painted here as an almost inevitable consequence of having a nonlocal but short-range force in a fluid. The nonlocal force was introduced as a device to keep Hamilton-Jacobi theory out of trouble. But if we put in both the ordinary fluid forces and the non-local forces, we can get singularities again. In one dimension, the non-local forces may make things awkward if they do not satisfy certain conditions. This may be seen in the acoustics of generalized barotropes.

Equations (4.7) and (4.8) have the solution $\mathbf{u} = 0$ with $\rho = \rho_0 = \text{constant}$. If we perturb this solution so that $|\mathbf{u}|$ and ρ_1 are small, where $\rho_1 = \rho - \rho_0$, we get, in linear theory,

and
$$\partial_t \rho_1 + \rho_0 \nabla \cdot \mathbf{u} = 0. \tag{6.1}$$

$$\partial_t \mathbf{u} = \nabla[h'(\rho_0)\rho_1 + nc^2\rho_1/\rho_0] + nc^2 \nabla\{K*\rho_1\}/\rho_0, \tag{6.2}$$

where

$$n = \frac{\partial \ln g}{\partial \ln \rho} \text{ for } \rho = \rho_0, \tag{6.3}$$

and

Then
$$K*\rho_1 = \int K(\mathbf{x}-\boldsymbol{\xi})\rho_1(\boldsymbol{\xi})d_3\boldsymbol{\xi}. \qquad (6.4)$$

$$\partial_t^2 \rho_1 - [h'(\rho_0)\rho_0 + nc^2]\nabla^2\rho_1 + nc^2\{K*\nabla^2\rho_1\} = 0. \qquad (6.5)$$

For plane wave solutions,
$$\rho_1 \propto \exp[i\omega t - i\mathbf{k}\cdot\mathbf{x}], \qquad (6.6)$$

we get the dispersion relation
$$\omega^2 = \{h'(\rho_0)\rho_0 + nc^2[1 - \hat{K}(k)]\}k^2 \qquad (6.7)$$

where $k = |\mathbf{k}|$ and
$$\hat{K}(k) = \int K(\mathbf{x}) e^{i\mathbf{k}\cdot\mathbf{x}} d_3\mathbf{x}. \qquad (6.8)$$

The conventional sound speed, a, is given by
$$a^2 = h'(\rho_0)\rho_0. \qquad (6.9)$$

When c = 0, (6.7) is the dispersion relation of ordinary acoustics. When a = 0, (6.7) is
$$\omega^2 = nc^2(1-\hat{K})k^2. \qquad (6.10)$$

For small k,
$$\hat{K} = 1 - \tfrac{1}{2}(\lambda k)^2 + \ldots \qquad (6.11)$$

and, for $n = \tfrac{1}{2}$, we get
$$\omega = \pm \tfrac{1}{2}(\lambda c)k^2, \qquad (6.12)$$

as in the linear Schrödinger theory.

For large k we have, as yet, no restriction. But for $\hat{K} > 1$, the fluid suffers instabilities. If we want to avoid this, we may introduce the condition
$$\hat{K} < 1. \qquad (6.13)$$

It would also seem desirable to have the phase speed $|\omega|/k$ bounded. Since, as yet, c is not specified, it might be chosen as an estimate of that bound, and we simply add the condition suggested by (6.10),
$$\hat{K} \geq 0. \qquad (6.14)$$

A number of simple kernels satisfying these conditions have the property
$$\lim_{k\to\infty} \hat{K} = \text{constant}. \qquad (6.15)$$

The motivation for all this comes from a subjective feeling that the potential, W_q in (5.9), that appears in Madelung's fluid-dynamical form of the usual Schrödinger equation is a strange-looking object. It makes more sense fluid-dynamically if it can be seen an approximation to a more natural looking potential. Of course, this is an idiosyncratic viewpoint, but I find it comforting that a somewhat sensible looking model can be written down. For example, we can take the following system:

$$\partial_t \mathbf{u} + (\mathbf{u}\cdot\nabla)\mathbf{u} = -\nabla\{V + h(\rho) + W[\rho]\} \tag{6.16}$$

$$\partial_t \rho + \nabla\cdot(\rho\mathbf{u}) = 0 \tag{6.17}$$

$$W = -c^2 \mathcal{W}/g \tag{6.18}$$

$$\tfrac{1}{2}\lambda\nabla^2\mathcal{W} - \mathcal{W} = -g. \tag{6.19}$$

For

$$g = \sqrt{\rho} \tag{6.20}$$

and λ^2 small, we get a (nonlinear) Schrödinger equation when $\mathbf{u}$ is irrotational. The system (6.16)-(6.19) does not seem especially peculiar to anyone who has worked with self-gravitating fluids. It gives a hint about the kinds of physical systems that give rise to Schrödinger equations. For example, in incompressible fluids, the pressure forces are nonlocal and, sure enough, the Schrödinger equation emerges from certain variants of the Bénard problem. But such Schrödinger equations are asymptotic limits. A model like that discussed in this section may used to examine the question of the robustness of such asymptotic limits.

In fact, my interest in these matters is more far-fetched. Gravitational theories, whether classical or relativistic, give rise to singularities in density. These resemble the infinities in the quantum mechanical probability density that arise in WKB theory near a turning point. The question that I am really trying to answer is whether it is possible to find a connection formula across such singularities. All I can say so far, is that in classical gravity, it looks as if the nonlocal effects need to be rather more nonlocal than in ordinary Schrödinger theory if this is to be possible.

7. VORTICITY

Madelung's transformation of Schrödinger's equation leads to fluid equations, but the flow is irrotational. In the case where W vanishes, the correspondence between the equation for an irrotational fluid and the usual Hamilton-Jacobi theory remains. It is natural to ask what happens when the flow is not irrotational and to see what correspondence exists between the equations for a particle and those for a fluid. This is a complicated problem, and I shall just give an introduction to it.

Consider the motion of a simple particle in an electromagnetic field. Let q be the ratio of its charge to its mass. The equation of motion is

$$\ddot{x}_i = qE_i + \frac{q}{c}\varepsilon_{ijk}\dot{x}_j B_k, \quad i = 1,2,3, \qquad (7.1)$$

where we assume that

$$\mathbf{B} = \nabla \times \mathbf{A}, \quad \mathbf{E} = -\nabla V - \frac{1}{c}\partial_t \mathbf{A}. \qquad (7.2)$$

As in the case without fields, we may rewrite (7.1) as

$$\frac{D\mathbf{u}}{Dt} = q\mathbf{E} + \frac{q}{c}\mathbf{u} \times \mathbf{B}. \qquad (7.3)$$

(To include the fluid case we simply replace V by $V + h(\rho)$.)
It is convenient to introduce

$$\mathbf{w} = \mathbf{u} + \frac{q}{c}\mathbf{A} \qquad (7.4)$$

and to let

$$\mathbf{\Omega} = \nabla \times \mathbf{w} = \boldsymbol{\omega} + \frac{q}{c}\mathbf{B}. \qquad (7.5)$$

We find

$$\partial_t \mathbf{w} + \nabla(qV + \tfrac{1}{2}\mathbf{u}^2) = \mathbf{u} \times \mathbf{\Omega} \qquad (7.6)$$

and hence that

$$\frac{D\mathbf{\Omega}}{Dt} = (\mathbf{\Omega}\cdot\nabla)\mathbf{u} - \mathbf{\Omega}(\nabla\cdot\mathbf{u}). \qquad (7.7)$$

Thus, if $\mathbf{\Omega}$ is initially zero, it remains so. The same is true if we include an enthalpy for the barotropic or the generalized barotropic case. But we are now interested in $\mathbf{\Omega} \neq 0$.

A number of conservation laws of fluid mechanics hold for this system of equations together with

$$\partial_t \rho + \nabla \cdot (\rho \mathbf{u}) = 0. \qquad (7.8)$$

Perhaps the most important of these is that the natural generalization of the Ertel theorem [1] to this case holds. That is, the commutator

$$\mathbf{C} = [\frac{D}{Dt}, \frac{1}{\rho}\mathbf{\Omega}\cdot\nabla] \qquad (7.9)$$

annihilates any component of any (here Cartesian) tensor.

Next consider the quantity

$$T = \oint \mathbf{w} \cdot d\mathbf{x}. \qquad (7.10)$$

where the integral is around a material contour, that is one that moves with the flow. Then, as in the derivation of Faraday's law for a moving medium [17], we get

$$\frac{DT}{Dt} = 0. \qquad (7.11)$$

This combination of Kelvin's theorem [15] with Faraday's law has been given in relativistic form [18].

Now consider the quantity

$$I = \int \mathbf{w} \cdot \mathbf{\Omega} \, dV \qquad (7.12)$$

where the integral is over a material volume. Then, if $\mathbf{\Omega}$ does not penetrate the boundary of this volume,

$$\frac{DI}{Dt} = 0. \qquad (7.13)$$

This combination of Woltjer's integral [19] with the helicity theorem of fluid mechanics [20] may be expressed especially nicely with differential forms [18]. Whether any more such results can be obtained I do not know, but I suspect it would not be possible (except perhaps for an analogue of enstrophy conservation) unless we go to higher spatial dimensions. (But see [21].)

These results apply to the description of the motion of a single particle. They are easily seen to hold for (generalized) barotropic fluids and, not surprisingly, also have relativisitic generalizations [18], [22]. The derivations are not given here because they require no special ingenuity beyond what is already involved in the standard fluid dynamical discussions. In short, the dynamics of a simple charged particle moving in an electromagnetic field has much in common with the hydromagnetics of a perfect fluid.

8. DYNAMICS OF A SPINNING PARTICLE

In an attempt to understand spin, Kramers [23] studied the classical dynamics of a charged particle endowed with a magnetic moment or spin. In fact, you can recover Kramers' result without having to endow the particle with a dipole moment by fiat if you start with the equations of a simple charged particle, but do not assume that the Hamilton-Jacobi flow is irrotational.

Euler [24] showed that the vector potential for a solenoidal vector is of the form $\sigma\nabla\chi$, where σ and χ are scalar fields. This representation has a gauge freedom and we can add a gradient, as Hankel observed [1]. Since **w** is a vector potential for **Ω** (see 7.5), we can represent **w** in this way, as Clebsch [25] did for the velocity field in fluid dynamics. However, if $\mathbf{w}\cdot\mathbf{\Omega}\neq 0$, this representation, which has only local validity, may give rise to coordinate singularities [26]. To avoid these (and for other reasons [27]) we may make the representation overcomplete and add another one-form [26]. We write, therefore,

$$\mathbf{w} = \nabla\phi + \sigma\nabla\chi + \xi\nabla\eta. \quad (8.1)$$

Now, following Lamb [25], we introduce

$$\Sigma = qV + \tfrac{1}{2}\mathbf{u}^2 + \partial_t\phi + \sigma\partial_t\chi + \xi\partial_t\eta \quad (8.2)$$

into (7.6) and we obtain

$$\nabla\Sigma = (\nabla\sigma)\partial_t\chi - (\partial_t\sigma)\nabla\chi + (\nabla\xi)\partial_t\eta - (\partial_t\xi)\nabla\eta + \mathbf{u}\times\mathbf{\Omega}. \quad (8.3)$$

But

$$\mathbf{u}\times\mathbf{\Omega} = \mathbf{u}\times(\nabla\times\mathbf{w}) = \mathbf{u}\times(\nabla\sigma\times\nabla\chi + \nabla\xi\times\nabla\eta)$$

$$= (\mathbf{u}\cdot\nabla\chi)\nabla\sigma - (\mathbf{u}\cdot\nabla\sigma)\nabla\chi + (\mathbf{a}\cdot\nabla\eta)\nabla\xi - (\mathbf{u}\cdot\nabla\xi)\nabla\eta,$$

so that

$$\nabla\Sigma = (\nabla\sigma)\frac{D\chi}{Dt} + (\nabla\xi)\frac{D\eta}{Dt} - (\nabla\chi)\frac{D\sigma}{Dt} - (\nabla\eta)\frac{D\xi}{Dt}. \quad (8.4)$$

If we suppose that

$$\Sigma = \Sigma(\sigma, \chi, \xi, \eta; t), \quad (8.5)$$

we are led to the relations

$$\frac{D\chi}{Dt} = \frac{\partial\Sigma}{\partial\sigma}, \quad \frac{D\sigma}{Dt} = -\frac{\partial\Sigma}{\partial\chi}$$

$$\frac{D\eta}{Dt} = \frac{\partial\Sigma}{\partial\xi}, \quad \frac{D\xi}{Dt} = -\frac{\partial\Sigma}{\partial\eta} \quad (8.6)$$

as a means of satisfying (8.4).

From (8.2) we obtain also

$$\frac{D\phi}{Dt} = \Sigma - (qV + \tfrac{1}{2}\mathbf{u}^2) - (\sigma\partial_t\chi + \xi\partial_t\eta) + \mathbf{u}\cdot\nabla\phi. \quad (8.7)$$

This can be rewritten as

$$\partial_t\phi + \tfrac{1}{2}(\nabla\phi - \tfrac{q}{m}\mathbf{A})^2 + qV =$$
$$\Sigma - \sigma\frac{\partial\Sigma}{\partial\sigma} - \xi\frac{\partial\Sigma}{\partial\xi} + \tfrac{1}{2}(\sigma\nabla\chi + \xi\nabla\eta)^2. \quad (8.8)$$

This is the generalization of the Hamilton-Jacobi equation for $\mathbf{\Omega}\neq 0$ with Σ as an arbitrary function. We can use (8.6) to rewrite this once more as

$$\partial_t\phi + \sigma\partial_t\chi + \xi\partial_t\eta + qV - \Sigma$$
$$-\tfrac{1}{2}(\nabla\phi + \sigma\nabla\chi + \xi\nabla\eta - \tfrac{q}{c}\mathbf{A})^2 = 0. \qquad (8.9)$$

This is the generalized Hamilton-Jacobi equation for the motion of charged particle in electromagnetic fields. For the case of a uniform magnetic field, we choose

$$\Sigma = -\tfrac{q}{c}\mathbf{s}\cdot\mathbf{B} \qquad (8.10)$$

where

$$\mathbf{s} = \left((\xi^2-\sigma^2)^{\tfrac{1}{2}}\sin\chi,\ (\xi^2-\sigma^2)^{\tfrac{1}{2}}\cos\chi,\ \sigma\right). \qquad (8.11)$$

Then (8.9) is the same as the ordinary Hamilton-Jacobi equation that Schiller [28] found for the Kramers model of a classical particle endowed with a dipole moment, in the case where the external magnetic field is uniform. But here we have not assumed a dipole moment and have studied only a simple particle. This result has nothing to do with the electromagnetic terms; I included them to make the comparison with the equations given by Schiller more complete. The point is that, in the generalized Hamilton-Jacobi theory for a simple structureless particle, the terms that describe the spin are automatically included. With their help it is possible to formulate a fluid dynamical version of nonlinear Pauli equations.

9. CONCLUSION

The formal equivalence between the dynamical descriptions of particles and fluids has long been known, but it has not been very popular. Madelung's [4] transformation on the Schrödinger equation made the connection even stronger and, once he made it, the Schrödinger equation itself entered the family of related dynamical descriptions. This connection is not peculiar to quantum mechanics. Given a density field and a suitable irrotational dynamics, we can concoct a (generally nonlinear) Schrödinger equation. If the problem is not quantum mechanical, the negative of the quantum mechanical potential (eqn. (5.9)) will appear in the wave equation. When the problem is quantum mechanical, the potential appears instead in the Hamilton-Jacobi-Bernoulli equation. These analogues help us to understand things, but do they lead anywhere? Pauli [5] found the fluid dynamical form of quantum mechanics to be "in keiner Weise vorteilhaft". It is not yet possible to disagree strongly, but I believe that when the analogy is fully developed it will have proved its worth. Already in the description of a spinning particle, it provides a correction to what I consider to be a widespread conceptual error.

We often read that spin is a purely quantum mechanical effect, yet this statement is rarely accompanied by an explanation of how spin disappears in the classical limit. In fact, the fluid dynamical picture or Hamilton - Jacobi picture, shows that spin does not disappear in the classical limit; it merely decouples from the orbital dynamics. For a simple particle, the representative points move independently, and any given one is indifferent to what the others do. In the Schrödinger fluid, the representative points of classical mechanics interact, and hence the vorticity field may well affect the motion of individual particles or fluid elements. While I described enough of the formalism to show this (I hope), I did not go on to a full treatment of the fluid description of the Pauli equation, and its nonlinear extensions. This is a complicated affair, that I shall postpone to a later installment. But I shall conclude by indicating briefly how this goes.

In his discussion of the fluid picture of quantum mechanics, Takabayasi [7] made the Madelung transfomation separately on the two components of the Pauli spinor. This led to a coupled pair of fluid equations. When we ask what this has to do with the Clebsch transformation, we are led to an interpretation of the Clebsch variables. Consider the velocity field

$$\mathbf{u} = \nabla \Phi + \sigma \nabla X. \qquad (9.1)$$

This is not the complete representation used in §8, but it will serve to give the idea. Consider also two interpenetrating fluids with densities ρ_1 and ρ_2. Suppose that both fluids are irrotational:

$$\mathbf{u}_1 = \nabla \phi_1, \quad \mathbf{u}_2 = \nabla \phi_2. \qquad (9.2)$$

The composite fluid has the properties

$$\rho = \rho_1 + \rho_2 \qquad (9.3)$$

and

$$\rho \mathbf{u} = \rho_1 \mathbf{u}_1 + \rho_2 \mathbf{u}_2. \qquad (9.4)$$

Then, if we let

$$\Phi = \tfrac{1}{2}(\Phi_1 + \Phi_2) \qquad (9.5a)$$

$$X = \tfrac{1}{2}(\Phi_1 - \Phi_2) \qquad (9.5b)$$

$$\sigma = \tfrac{1}{2}(\rho_1 + \rho_2)/\rho, \qquad (9.5c)$$

we see that (9.2) - (9.4) give (9.1). If we write a suitable set of two-fluid equations, we can get variously Pauli and nonlinear Pauli equations. The trick is to put in the appropriate couplings between the component fluids. With the right choice of couplings, you can also get ordinary rotational fluid dynamics or the Landau equations for a superfluid.

This may still leave us far from countering Pauli's negativism, but I remain optimistic. I am also grateful to many for helpful advice and attention. Explicitly, I want to thank Ted Scharlemann for criticizing this account and Mark Ablowitz for correcting an error. I am also indebted to the N.S.F. for support, most recently under grant NSF PHY 80-23721.

REFERENCES

1. Truesdell, C., **Kinematics of Vorticity**, Indiana University Press, (1954).

2. Rund, H., "Clebsch Potentials and Variational Principles in the Theory of Dynamical Systems", **Archive for Rational Mechanics and Analysis, 65,** 305-334 (1977).

3. Baumeister, R., "Generalized Hamilton-Jacobi Theories", **J. Math. Phys., 19,** 2377-2387 (1978).

4. Madelung, E., "Quantentheorie in Hydrodynamischen form", **Zts. F. Phys., 40,** 322-326 (1926).

5. Pauli, W., "Allgemeine Grundlagen der Quantentheorie des Atombaues", **Muller - Pauillets Lehrbuch, 2,** part 2, 11th ed, Braunschweig, pp. 1709-1847 (1929).

6. Bohm, D., "A Suggested Interpretation of the Quantum Theory in Terms of Hidden Variables. I", **Phys. Rev., 85,** 166-179 (1952).

7. Takabayasi, T., "On the Formulation of Quantum Mechanics Associated With Classical Pictures", **Prog. Theor. Phys., 8,** 143-182 (1952).

8. Feynman, R.P., **Statistical Mechanics,** W.A. Benjamin, Reading, MA, pp. 303ff (1972).

9. Broer, L. J. F.,"Generalized Fluid Dynamics and Quantum Mechanics", **Physica, 76,** 364-372 (1974).

10. Bialynicki-Birula, I. & Mycielski, J., "Nonlinear Wave Mechanics", **Ann. Phys. 100,** 62-93 (1976).

11. Shukla, G.C. & Barbaro, M., "Multipole polarizabilities and shielding factor of the hydrogen atom from the hydrodynamic analogy to quantum mechanics", **Phys. Rev. 15A**, 23-26 (1977).

12. Delion, D. S., Gridnev, K. A., Hefter, E. F. & Semjonov, V. M., "The Non-Linear Schrodinger Equation and Anomalous Backward Scattering", **J. Phys. G. Nucl. Phys.**, **4**, 125-132 (1978).

13. Spiegel, E.A., "Fluid Dynamical Form of the Linear and Nonlinear Schrodinger Equations", **Physica, 1-D**, 236-240 (1980).

14. Roberts, P.H., **An Introduction to Magnetohydrodynamics**, American Elsevier Pub. Co. Inc., NY, pp. 28-31 (1967).

15. Serrin, J., "Mathematical Principles of Classical Fluid Dynamics", **Hdbch. d. Phys. 28**, part 1, S. Flugge and C. Truesdell, eds. Springer-Verlag, Berlin (1959).

16. Nelson, E., **Dynamical Theories of Brownian Motion**, Mathematical Notes, Princeton University Press (1967).

17. Landau, L.D. & Lifshitz, E.M., **Electrodynamics of Continuous Media**, Addison-Wesley, Reading MA, p. 207 (1960).

18. Carter, B., "Perfect Fluid and Magnetic Field Conservation in the Theory of Black Hole Accretion Rings", **Active Galactic Nuclei**, C. Hazard and S. Mitton, eds., Cambridge University Press, pp. 273-300 (1979).

19. Woltjer, L., "A theorem on force free magnetic fields", **Proc. Nat. Acad. Sci.**, **44**, 489-91 (1958).

20. Moffatt, H. K., "The degree of knottedness of tangled vortex lines, **J. Fluid Mechanics**, **35**, 117-129 (1969).

21. Hide, R., "On the Magnetic Flux Linkage of an Electrically Conducting Fluid", **J. Geophys. Astrophys. Fluid Dynamics**, **12**, 171-176 (1979).

22. Schmid, L.A., "Larmor and Helmholtz Theorems for Relativistic Charged Fluid Flow", **Il Nuovo Cemento, 52B**, 288-312 (1967).

23. Kramers, H. A., **Quantum Mechanics**, North-Holland Co., Amsterdam, p. 233 (1957).

24. Phillips, H. B., **Vector Analysis**, John Wiley & Sons, Inc., NY, pp 104-105 (1949).

25. Lamb, H., **Hydrodynamics**, Dover Publications, N.Y., pp. 248 - 249 (1945).

26. Bretherton, F. P., "A Note on Hamilton's Principle for Perfect Fluids", **J. Fluid Mech.**, **44**, 19-31 (1970).

27. Seliger, R.L., & Whitham, G.B., **Proc. Roy. Soc. A**, **305**, 1-25 (1968).

28. Schiller, R., "Quasi-Classical Theory of the Spinning Electron", **Phys. Rev. 125**, 1116-1123 (1962).

DUAL HAMILTONIAN FORMULATIONS
AND COMPLETELY INTEGRABLE SYSTEMS*

K. M. Case
Center for the Study of Nonlinear Dynamics
La Jolla Institute
P. O. Box 1434
La Jolla, CA 92038

and

The Rockefeller University**
New York, NY 10021

ABSTRACT

It is shown that almost all the special properties of completely integrable systems fall out immediately when it is possible to write the equations in Hamiltonian form two ways. In particular, the construction of constants, the proof that they are in involution, and the forms for Lax pairs becomes straightforward.

I. INTRODUCTION

We are here interested in nonlinear evolution equations of the form

$$\partial_t u = K(u) . \qquad (1)$$

It is assumed that this is a completely integrable Hamiltonian system. By Hamiltonian it is meant that between any two functionals of u ($F_i[u]$, $F_j[u]$) it is possible to define a Poisson Bracket with the properties

$$[F_i, F_j] = - [F_j, F_i] \qquad \text{(antisymmetry)} \qquad (2)$$

and

$$\left[[F_i, F_j], F_k\right] + \left[[F_j, F_k], F_i\right] + \left[[F_k, F_i], F_j\right] = 0 \qquad (3)$$

(the Jacobi identity).

Further, there is a functional H (the Hamiltonian) such that Eq.(1) is

---

*Supported in part by the National Science Foundation under grant MCS 80-17781 and in part by the Office of Naval Research under Contract No. N00 14-79-0537.

**Permanent address.

0094-243X/82/880163-26$3.00 Copyright 1982 American Institute of Physics

$$\partial_t u = [u, H] \tag{4}$$

The definition of completely integrable will te taken to mean (for the systems of infinite numbers of degree of freedom considered here) the existence of an infinite number of constants of motion I_n, $n = 1, 2, ---$.

If it is not known whether a given system is completely integrable, the procedure to be discussed can demonstrate that it is. However, the converse is not true. If the procedure does not work the system can still be completely integrable.

The work to be discussed is based on two observations.

Define:
$$Q_n = \delta I_n / \delta u \tag{5}$$

(These will be called the conserved gradients).

(i) The conserved gradients provide solutions of the linearized evolution equation.[1]

(ii) If there exists a second Hamiltonian formulation it is possible to construct a recursion relation for the conserved gradients.[2] This is of the form

$$L_u Q_{n+1} = M_u Q_n \tag{6}$$

where L_u, M_u are certain symplectic operators defining Poisson Brackets.

The main result obtained is that almost all the special properties of completely integrable systems are contained in these two observations. For example:

1) The construction of constants of motion becomes obvious.

2) The fact that these constants are in involution (i.e. $[I_n, I_m] = 0$) is trivial to prove.

3) The construction of a Lax pair becomes straight-forward. (This is not the usual pair - but is closely related).

Our program is as follows:

In Section II the KdeV equation is treated as a paradigm. The general situation is described in Sec. III. The Sec. IV treats in detail all the nonlinear evolution equations associated with the Zakharov-Shabat[3] eigenvalue problem. Lastly, we treat an equation which is not related to that eigenvalue problem.

II. THE KORTEWEG-DE VRIES EQUATION

This is taken in the form

$$u_t = -\partial_x \{u^2 + 2 u_{xx}\} \tag{7}$$

Some constants are readily found. For example:

$$I_1 = \int_{-\infty}^{\infty} u\, dx \quad \text{and} \quad I_2 = \int_{-\infty}^{\infty} \frac{u^2}{2} dx.$$

Now note that Eq. (7) can be written in Hamiltonian form.

$$\text{Let } [F_i, F_j] = -\int_{-\infty}^{\infty} \frac{\delta F_i}{\delta u} \partial_x \frac{\delta F_j}{\delta u} dx \tag{8}$$

and

$$H = \int_{-\infty}^{\infty} \left\{ \frac{u^3}{3} - (u_x)^2 \right\} dx. \tag{9}$$

It is readily checked that Eq. (7) is just

$$u_t = [u, H]. \tag{10}$$

Of course it follows that $H = I_3$ is also a constant for Eq. (7). Look for a second Hamiltonian form for Eq. (7). Suppose the new Hamiltonian (H') is

$$H' = I_2. \tag{11}$$

A new Poisson Bracket will be needed. Suppose this to be of the form

$$[F_i, F_j]' = \int_{-\infty}^{\infty} \frac{\delta F_i}{\delta u} M_u \frac{\delta F_j}{\delta u} dx$$

where the operator M_u is to be determined. There are three conditions:

a) Eq. (7) is to result.

$$\therefore M_u(u) = -\partial_x (u^2 + 2 u_{xx}).$$

b) The Poisson Bracket is to be antisymmetric.

c) The Jacobi identity is to be satisfied.

By inspection it is clear that if there is to be such an M_u it must be of the form

$$M_u = -2 \partial_x^3 + c_1 u \partial_x + c_2 u_x 1$$

a) implies $c_1 + c_2 = -2$

b) implies $2 c_2 = c_1$

Therefore if there is an M_u it must be

$$M_u = -2 \partial_x^3 - \frac{4}{3} u \partial_x - \frac{2}{3} u_x 1 \qquad (12)$$

It remains only to be seen whether the Jacobi identity is satisfied. A straightforward calculation shows it is.

Quantities Q_n can now be calculated recursively from the relation

$$-\partial_x Q_{n+1} = M_u Q_n \qquad (13)$$

We note that except for unimportant numerical factors $\delta I_1/\delta u$, $\delta I_2/\delta u$, and $\delta I_3/\delta u$ obey this relation. In general, as will be shown later, if the Q_n are variational derivatives, i.e. $Q_n = \delta I_n/\delta u$ then the I_n are constants of motion and indeed are in involution.

We postpone the proof that the Q_n defined by Eq. (13) are variational derivatives until appropriate machinery is developed.

Explicitly we can write out Eq. (13). It is

$$\partial_x Q_{n+1} = 2 \partial_x^3 Q_n + \frac{4}{3} u \partial_x Q_n + \frac{2}{3} u_x Q_n \qquad (14)$$

Let us introduce a generating function Ψ for the Q_n in the form

$$\Psi = \sum_n \frac{Q_n}{(2\lambda)^n} . \qquad (15)$$

Multiplying Eq. (14) by $(2\lambda)^{-n}$ and summing yields the linear eigenvalue problem

$$\lambda \Psi_x = \Psi_{xxx} + \frac{2}{3} u \Psi_x + \frac{1}{3} u_x \Psi \qquad (16)$$

Alternatively we can write this as

$$L\Psi = \lambda \Psi \qquad (17)$$

where

$$L = \partial_x^2 + \frac{2}{3} u \mathbf{1} - \frac{1}{3} \partial_x^{-1} u_x \qquad (18)$$

and ∂_x^{-1} is to be the antisymmetric operator

$$\partial_x^{-1} = \frac{1}{2} \left\{ \int_{-\infty}^{x} \ldots dx' - \int_{x}^{\infty} \ldots dx' \right\} \qquad (19)$$

We would also like an equation telling how Ψ depends on time. Use the result[1] that

$$\delta u = [u, I_n] \equiv -\partial_x Q_n \qquad (20)$$

satisfies the linear equation

$$\delta u_t = -\partial_x \{ 2u\delta u + 2\delta u_{xx} \} . \qquad (21)$$

Combining Eqs. (20) and (21) yields

$$\partial_x \partial_t Q_n = -\partial_x \{ 2u \partial_x Q_n + 2(Q_n)_{xxx} \} . \qquad (22)$$

Integrating, we obtain

$$\partial_t Q_n = -2u \partial_x Q_n - 2(Q_n)_{xxx} . \qquad (23)$$

If we multiply this equation by $(2\lambda)^{-n}$ and sum we finally obtain

$$\partial_t \Psi = -2u \Psi_x - \Psi_{xxx} \equiv B\Psi . \qquad (24)$$

The compatibility of Eqs. (17) and (24) requires

$$\frac{\partial L}{\partial t} = [B, L], \qquad (25)$$

but this is just the condition that u satisfies Eq. (7). Thus B, L form a Lax[4] pair. Note, however, that:

(i) This is not the usual pair.

(ii) The eigenvalue problem we encounter is somewhat unfamiliar.

These differences are readily taken care of by the following remark: If ψ satisfies

$$(\partial_x^2 + \frac{u}{6})\psi = \frac{\lambda}{4}\psi, \qquad (26)$$

then

$$\Psi = \psi^2 \qquad (27)$$

satisfies Eq. (16).

The time dependence of ψ is obtained by substituting Eq. (27) into Eq. (24). The result is

$$\partial_t \psi = -8\partial_x^3 \psi - 2u\partial_x \psi - u_x \psi \qquad (28)$$

or, alternatively,

$$\partial_t \psi = -[2\lambda + \frac{4u}{3}]\partial_x \psi + \frac{u_x}{3}\psi. \qquad (29)$$

We are now in a position to prove the theorem:
The Q_n given by Eq. (13) are all variational derivatives.

Proof:

We do this by showing that the generating function Ψ is of the form

$$\Psi(x) = \frac{\delta \Lambda}{\delta u(x)}. \qquad (30)$$

The Laurent series for Ψ (in λ) then shows all Q_n are derivatives. Motivation for the method of proof arises first from the fact that $\Psi = \psi^2$ [Eq. (27)] and the usual variational principle for bound states in quantum mechanics. Thus, if $\lambda = \lambda_i$, $\psi = \psi_i$ in Eq. (26) corresponding to a bound state then we know

$$\frac{\lambda_i}{4} = \frac{\int_{-\infty}^{\infty}\{-(\partial_x\psi)^2 + \frac{u}{6}\psi^2\}dx}{\int_{-\infty}^{\infty}\psi^2 dx} \tag{31}$$

is stationary with respect to small variations in ψ. Accordingly, when we look for the change in λ_i corresponding to a small change in u, we merely need to look at the explicit change in u, i.e.,

$$\frac{\delta\lambda_i}{4} = \frac{1}{6}\frac{\int_{-\infty}^{\infty}\delta u\psi^2 dx}{\int_{-\infty}^{\infty}\psi^2 dx}$$

$$\therefore \frac{\delta\lambda_i}{\delta u(x)} = \frac{2}{3}\frac{\psi^2(x)}{\int_{-\infty}^{\infty}\psi^2(x')dx'}. \tag{32}$$

(This is just first-order perturbation theory.) We intend to extend this to the continuous spectrum. Consider Eq. (26) as a scattering problem. With $v = -\frac{u}{6}$, $k^2 = \frac{-\lambda}{4}$, this is:

$$(\partial_x^2 + k^2)\psi = v\psi. \tag{33}$$

The general solution of this equation with k real can be written as

$$\psi^{(i)} = \phi^{(i)} + Gv\psi^{(i)}, \tag{34}$$

where $(\partial_x^2 + k^2)\phi^{(i)} = 0$,

$$Gv\psi^{(i)} = \frac{1}{2k}\int_{-\infty}^{\infty}\text{sgn}(x-x')\sin k(x-x')v\psi(x)dx'. \tag{35}$$

Variational principles for problems described by equations similar to Eq. (33) have been found in a somewhat different context. We adopt the methods for the present purpose. Define an operator R by

$$R = v + vGR. \tag{36}$$

It is readily seen that

$$v\psi^{(i)} = R\phi^{(i)} \tag{37}$$

and

$$\int_{-\infty}^{\infty} \chi R\eta dx = \int_{-\infty}^{\infty} \eta R\chi. \tag{38}$$

Accordingly,

$$R_{ij} = \int_{-\infty}^{\infty} \phi^{(i)} R\phi^{(j)} dx$$

$$= \int_{-\infty}^{\infty} \phi^{(i)} v\psi^{(j)} dx \tag{39}$$

$$= \int_{-\infty}^{\infty} \psi^{(i)} v\phi^{(j)} dx. \tag{40}$$

A third expression for R_{ij} is obtained by multiplying the Eq. (34) for $\psi^{(j)}$ by $\psi^{(i)} v$ and integrating. The result is:

$$R_{ij} = \int_{-\infty}^{\infty} \psi_i [v - vGv] \psi_j dx. \tag{41}$$

Finally, from the three expressions of Eqs. (38) — (41) we have

$$R_{ij} = \frac{\left(\int_{-\infty}^{\infty} \phi^{(i)} v\psi^{(j)} dx\right)\left(\int_{-\infty}^{\infty} \psi^{(i)} v\phi^{(j)} dx'\right)}{\int_{-\infty}^{\infty} \psi_i [v - vGv] \psi_j dx}. \tag{42}$$

It is easily shown that this expression for R_{ij} is stationary with respect to small variations in $\psi^{(i)}$ and $\psi^{(j)}$. Accordingly, to get the change when v is slightly changed, we need only to calculate the change in the explicit occurrence of v. It follows that

$$\frac{\delta R_{ij}}{\delta v(x)} = \psi^{(i)}(x)\psi(x)^{(j)}. \tag{43}$$

Note that Ψ is the solution of a third-order differential equation — which has only three linearly independent solutions.

Suppose that $\psi^{(1)}$, $\psi^{(2)}$ are two linearly independent solutions of Eq. (33). Then the Wronskian

$$W[\psi^{(1)}, \psi^{(2)}] \equiv \psi^{(1)}\psi_x^{(2)} - \psi^{(2)}\psi_x^{(1)} \neq 0. \tag{44}$$

Two solutions of Eq. (29) are

$$\Psi^{(1)} = [\psi^{(1)}]^2, \tag{45}$$

and

$$\Psi^{(2)} = [\psi^{(2)}]^2 \tag{45}$$

Then since $\psi^{(1)} + \psi^{(2)}$ is also a solution of Eq. (33) $[\psi^{(1)} + \psi^{(2)}]^2$ is a solution of Eq. (39). Therefore as a third solution we can choose

$$\Psi^{(3)} = \psi^{(1)}\psi^{(2)}. \tag{46}$$

By direct calculation we find the Wronskian

$$W[\Psi^{(1)}, \Psi^{(2)}, \Psi^{(3)}] = \begin{vmatrix} \Psi^{(1)} & \Psi^{(2)} & \Psi^{(3)} \\ \partial_x \Psi^{(1)} & \partial_x \Psi^{(2)} & \partial_x \Psi^{(3)} \\ \partial_x^2 \Psi^{(1)} & \partial_x^2 \Psi^{(2)} & \partial_x^2 \Psi^{(3)} \end{vmatrix}$$

$$= -2W^3[\psi^{(1)}, \psi^{(2)}]$$

$$\neq 0, \tag{47}$$

in virtue of Eq. (43). Thus the $\Psi^{(1)}$, $\Psi^{(2)}$, $\Psi^{(3)}$ are linearly independent and the general solution of Eq. (29) can be written

$$\Psi = a\Psi^{(1)} + b\Psi^{(2)} + c\Psi^{(3)}. \tag{48}$$

Using Eq. (42) we then see

$$\Psi = \frac{\delta}{\delta v}\{aR_{11} + bR_{22} + cR_{12}\}. \tag{49}$$

Hence Ψ is indeed a gradient and our Q_n are also such.

III. THE GENERAL CASE

We assume the system described by Eq. (1) is Hamiltonian. Thus there is a bilinear form $<\phi_1,\phi_2>$, e.g.,

$$<\phi_1,\phi_2> = \int_{-\infty}^{\infty} \phi_1\phi_2 dx, \tag{50}$$

and a symplectic operator L_u such that

$$[F_i,F_j] = \left\langle \frac{\delta F_i}{\delta u}, L_u \frac{\delta F_j}{\delta u} \right\rangle \tag{51}$$

is antisymmetric and satisfies the Jacobi identity. In addition there is a Hamiltonian such that Eq. (1) is equivalent to Eq. (4). Let there be constants I_n with gradients $Q_n \equiv \delta I_n/\delta u$.

First we obtain a linear time-dependence equation. Using the theorem of Ref. 1 we note that the equation

$$\partial_t(\delta u) = K'_u(\delta u), \tag{52}$$

where

$$K'_u(\delta u) = \lim_{\varepsilon \to 0} \frac{K(u + \varepsilon\delta u) - K(u)}{\varepsilon} \tag{53}$$

is satisfied by

$$\delta u = [u, I_n] = L_u \frac{\delta I_n}{\delta u} = L_u Q_n. \tag{54}$$

Thus:

$$\partial_t L_u Q_n = K'_u(L_u Q_n). \tag{55}$$

With

$$\Psi = \sum_n \frac{Q_n}{\lambda^n} \tag{56}$$

we obtain <u>a</u> time-dependent linear Lax equation for Ψ,

$$\partial_t(L_u\Psi) = K'(L_u\Psi). \tag{57}$$

If there is another Hamiltonian formulation we can construct a linear eigenvalue problem. Thus if there is another symplectic operator M_u and Hamiltonian H' so that

$$u_t = [u, H] = [u, H']' \qquad (58)$$

where

$$[F_i, F_j]' = \left\langle \frac{\delta F_i}{\delta u}, M_u \frac{\delta F_j}{\delta u} \right\rangle, \qquad (59)$$

we can define a sequence of functions by the recursion relation[2]

$$L_u Q_{n+1} = M_u Q_n. \qquad (60)$$

For the generating function Ψ we then obtain the linear eigenvalue problem

$$\lambda L_u \Psi = M_u \Psi. \qquad (61)$$

If the Q_n are gradients, ($Q_n = I_n/\delta u$), then the I_n are constants and are in involution, i.e., $[I_n, I_m] = 0$.

Proof[2]:

Without loss of generality we can assume $m > n$. Then

$$[I_n, I_m] = \langle Q_n, LQ_m \rangle = \langle Q_n, MQ_{m-1} \rangle$$
$$= -\langle MQ_n, Q_{m-1} \rangle = -\langle LQ_{n+1}, Q_{m-1} \rangle$$
$$= +\langle Q_{n+1}, LQ_{m-1} \rangle.$$

Repeating this process we finally obtain either

$$[I_n, I_m] = [I_i, I_i] \equiv 0 \qquad (62)$$

or

$$[I_n, I_m] = [I_{i+1}, I_i]$$
$$= [I_i, I_{i+1}]$$
$$= -[I_{i+1}, I_i]$$
$$= 0. \qquad (63)$$

An important consequence is that any of the constants can be used as a Hamiltonian. All the constants are then such for the new system of equations. Thus from a completely integrable Hamiltonian system we get an infinite number of such systems.

The Lax pairs for each member of the hierarchy can be succinctly characterized.

1) The eigenvalue problem is the same for all members of the hierarchy.

2) The equations for time dependence differ. However, they are just the evolution equation linearized.

We can now describe the general equation of the KdeV class. The Poisson bracket is given by Eq. (8). Conserved gradients are given by

$$Q_n = (L)^n Q_0 \tag{64}$$

where L is given in Eq. (18). The general equation is

$$\partial_t u = \partial_x F(L) u, \tag{65}$$

where F is any entire function. This has been given previously[6] in the form

$$\partial_t u = F(L^+) \partial_x u. \tag{66}$$

The connection between Eqs. (64) and (65) is obtained by noting that

$$\partial_x L = L^+ \partial_x \tag{67}$$

IV. EQUATIONS ASSOCIATED WITH THE Z-S EIGENVALUE PROBLEM

In Ref. 6 a detailed discussion of evolution equations related to the Zakharov-Shabat[3] eigenvalue problem is given. This consists of the two first-order equations with eigenvalue ζ and potentials $q(x, t)$, $r(x, t)$.

$$\begin{aligned} \psi_{1x} + i\zeta\psi_1 &= q\psi_2 \\ \psi_{2x} - i\zeta\psi_2 &= r\psi_1 \end{aligned} \tag{68}$$

Here we will show how these evolution equations fit into the dual Hamiltonian formalism.

In general the evolution equations will be for two different and non-trivial functions, q and r. However, there is considerable simplification if there is really only one function. The special cases are:

1) Either q or r is a constant; for example r = -1. Then we have the general class of KdeV equations just discussed.

2) r = ±q. This is the class of equations which includes the Sine and Sinh Gordon equations as well as the modified KdeV equations. They are discussed elsewhere in these Proceedings by A. M. Roos.

Here we consider the general case. (The prototype is the non-linear Schrödinger equation.)
Consider the Hamiltonian system:

Poisson Bracket

$$[F_i, F_j] = \int_{-\infty}^{\infty} \left\{ \left(\frac{\delta F_i}{\delta q}, \frac{\delta F_i}{\delta r} \right) L \begin{pmatrix} \delta F_j/\delta q \\ \delta F_j/\delta r \end{pmatrix} \right\} dx \qquad (69)$$

$$L = c \begin{pmatrix} 0 & 1 \\ -1 & 0 \end{pmatrix} \qquad (70)$$

Hamiltonian

$$H = a \int_{-\infty}^{\infty} (q_x r_x + q^2 r^2) dx. \qquad (71)$$

The equations of motion are then:

$$\dot{q} = -ac\{q_{xx} - 2q^2 r\},$$
$$\dot{r} = ac\{r_{xx} - 2r^2 q\}. \qquad (72)$$

Note that if

$$-ac = i, \quad r = -\sigma q^*, \quad \sigma = \pm 1$$

these are

$$\dot{q} = i\{q_{xx} + 2\sigma q^* q\, q\}$$
$$\dot{q}^* = -i\{q^*_{xx} + 2\sigma q^* q\, q^*\}, \qquad (73)$$

which is just the nonlinear Schrödinger equation.

To construct a second Hamiltonian form we note that in our other examples we have always related H' to the __momentum__ integral P. This must be such that

$$\begin{pmatrix} \delta q \\ \delta r \end{pmatrix} = \left[\begin{pmatrix} q \\ r \end{pmatrix}, P \right] = \begin{pmatrix} -q_x \\ -r_x \end{pmatrix}. \tag{74}$$

(From the spatial translation invariance of H it is clear that P exists.) It is readily guessed that

$$P = -\frac{1}{c} \int_{-\infty}^{\infty} \frac{(r\partial_x q - q\partial_x r)}{2} dx. \tag{75}$$

Then

$$\frac{\delta P}{\delta q} = \frac{1}{c} \partial_x r, \quad \frac{\delta P}{\delta r} = -\frac{1}{c} \partial_x q. \tag{76}$$

For a dual Hamiltonian formulation we need a symplectic operator M such that

$$\begin{pmatrix} -ac[q_{xx} - 2q^2 r] \\ +ac[r_{xx} - 2r^2 q] \end{pmatrix} = M \begin{pmatrix} \delta P/\delta q \\ \delta P/\delta r \end{pmatrix} \equiv M \begin{pmatrix} \frac{1}{c} \partial_x r \\ -\frac{1}{c} \partial_x q \end{pmatrix} \tag{77}$$

By trial we suggest

$$M = d \begin{pmatrix} 2q\partial_x^{-1} q & \partial_x - 2q\partial_x^{-1} r \\ \partial_x - 2r\partial_x^{-1} q & 2r\partial_x^{-1} r \end{pmatrix}. \tag{78}$$

Inserting, we find Eq. (77) is satisfied with $d = ac^2$.

The antisymmetry of M is obvious. That the Jacobi identity is satisfied follows from straightforward, if tedious, calculation. Conserved gradients

$$Q_n = \begin{pmatrix} \delta I_n/\delta q \\ \delta I_n/\delta r \end{pmatrix} \tag{79}$$

are then calculated recursively from the relations

$$LQ_{n+1} = MQ_n,$$

i.e.,

$$Q_{n+1} = \mathcal{L} Q_n \tag{80}$$

where $\mathcal{L} = L^{-1}M$. Since $L^2 = -c^2 1$,

$$L^{-1} = -\frac{1}{c^2} L,$$

and thus

$$L = -a \begin{pmatrix} \partial_x - 2r\partial_x^{-1} q & 2r\partial_x^{-1} r \\ -2q\partial_x^{-1} q & -\partial_x + 2q\partial_x^{-1} r \end{pmatrix}. \quad (81)$$

Note: We have partially begged the question by writing Eq. (79). The Q_n are defined by Eq. (80).

It is still to be shown that they are gradients. However:

1) If we call $P = I_2$, $H = I_3$

$$Q_2 = \begin{pmatrix} \delta I_2/\delta q \\ \delta I_3/\delta q \end{pmatrix}.$$

$$Q_3 = \begin{pmatrix} \delta I_3/\delta q \\ \delta I_3/\delta r \end{pmatrix}. \quad (82)$$

2) We readily calculate that

$$I_1 = -\frac{1}{ac} \int_{-\infty}^{\infty} rq \, dx = \text{constant}, \quad (83)$$

i.e.,

$$[I_1, H] = 0$$

and

$$LQ_1 = Q_2,$$

i.e.,

$$Q_1 = L^{-1} Q_2. \quad (84)$$

3) Below we prove the gradient property generally.

We conclude that a general conserved gradient is given by

$$Q_n = L^{n-1} Q_1. \quad (85)$$

The I_n are then in involution. We conclude further that:

a) The general equation of the form considered here is

$$\begin{pmatrix} q_t \\ r_t \end{pmatrix} = \begin{pmatrix} 0 & 1 \\ -1 & 0 \end{pmatrix} f(L) \begin{pmatrix} \partial_x r \\ -\partial_x q \end{pmatrix} \tag{86}$$

or equivalently

$$\begin{pmatrix} q_t \\ r_t \end{pmatrix} = \begin{pmatrix} 0 & 1 \\ -1 & 0 \end{pmatrix} \bar{f}(L) \begin{pmatrix} r \\ q \end{pmatrix} \tag{87}$$

with f or $\bar{f}$ arbitrary entire functions.

b) The linear eigenvalue problem associated with this equation is:

$$\lambda\Psi = L\Psi. \tag{88}$$

c) The linear time-dependent problem is obtained by linearizing Eq. (87).

It is somewhat interesting to look at the first four constants generated by our recursion relation. These are:

$$I_1 \sim \int_{-\infty}^{\infty} qr\, dx$$

$$I_2 \sim \int_{-\infty}^{\infty} (q\partial_x r - r\partial_x q)\, dx$$

$$I_3 \sim \int_{-\infty}^{\infty} (q_x r_x + q^2 r^2)\, dx$$

$$I_4 \sim \int_{-\infty}^{\infty} \{q\partial_x^3 r - 3q^2 rr_x\}\, dx$$

$$I_5 \sim \int_{-\infty}^{\infty} \{(\partial_x^2 r)(\partial_x^2 q) + 2r^3 q^3 + 2(\partial_x r^2)(\partial_x q^2)$$

$$+ r_x^2 q^2 + q_x^2 r^2\}\, dx.$$

Consider what happens in the two special cases:

1) <u>r = constant</u> (KdeV)
   We see I_2, I_4, ... I_{2n}... are all identically zero.
   I_1, I_3, I_5 are just the first three constants previously given.

2) $r = \pm q$ (Sine-Gordon et al.)
   Here again $I_{2n} = 0$, while the I_{2n+1} are again the constants given elsewhere.

We now must prove the Q_n are gradients. To do this we first make the connection to the Z–S problem — Eqs. (68). Rewrite Eq. (88) as

$$\lambda L\Psi = M\Psi \tag{89}$$

and let $\lambda = -2i\zeta ac$. Written out, the equation is then

$$-zi\zeta\Psi_2 = (\partial_x - 2q\partial_x^{-1} r)\Psi_2 + 2q\partial_x^{-1} q\Psi_1$$
$$zi\zeta\Psi_1 = (\partial_x - 2r\partial_x^{-1} q)\Psi_1 + 2r\partial_x^{-1} r\Psi_2. \tag{90}$$

It is convenient to remove the awkward operator ∂_x^{-1}. To do this, introduce Ψ_3 by

$$\Psi_3 = \partial_x^{-1}(q\Psi_1 - r\Psi_2). \tag{91}$$

Then Eqs. (89) - (91) can be rewritten as

$$-\partial_x \Psi_2 - 2i\zeta\Psi_2 = 2q\Psi_3 \tag{92}$$

$$\partial_x \Psi_1 - 2i\zeta\Psi_1 = 2r\Psi_3 \tag{93}$$

$$\partial_x \Psi_3 = q\Psi_1 - r\Psi_2. \tag{94}$$

The connection with the Zakharov-Shabat problem is made by noting that if ψ satisfies Eqs. (68), then if

$$\Psi_1 = (\psi_2)^2, \quad \Psi_2 = -(\psi_1)^2, \quad \Psi_3 = \psi_1 \psi_2,$$

Eqs. (91) - (93) are satisfied.
We now turn to the proof of the

Theorem:

If Ψ_1, Ψ_2 satisfy Eqs. (90), or alternatively Ψ_1, Ψ_2, Ψ_3 satisfy Eqs. (92) - (94), then

$$\begin{pmatrix} \Psi_1 \\ \Psi_2 \end{pmatrix} = \begin{pmatrix} \delta\Lambda/\delta q \\ \delta\Lambda/\delta r \end{pmatrix} \tag{95}$$

for some Λ.

Proof: This is much like that given earlier for the K-deV problem. It is again based on a variational principle.

We consider several solutions $\psi^{(i)}$ of Eqs. (67), i.e.,

$$\psi_{1x}^{(i)} + i\zeta\psi_1^{(i)} = q\psi_2^{(i)}, \tag{96}$$

$$\psi_{2x}^{(i)} - i\zeta\psi_2^{(i)} = r\psi_1^{(i)}. \tag{97}$$

It is also convenient to consider adjoint functions $\psi^{+(j)}$ which satisfy

$$\psi_{1x}^{+(j)} - i\zeta\psi_1^{+(j)} = -q\psi_2^{+(j)}, \tag{98}$$

$$\psi_{2x}^{+(j)} + i\zeta\psi_2^{+(j)} = -r\psi_1^{+(j)}. \tag{99}$$

Remark:

Solutions to these adjoint equations are readily constructed in terms of solutions of the original equations. Thus if $\psi^{(j)}$ satisfies Eqs. (96), (97), then

$$\psi^{+(j)} = \begin{pmatrix} \psi_2^{(j)} \\ -\psi_1^{(j)} \end{pmatrix} \tag{100}$$

satisfies Eqs. (98) — (99). Reformulate the equations as integral equations. Thus

$$\psi^{(i)} = \phi^{(i)} + Gv\psi^{(i)} \tag{101}$$

where
$$\phi^{(i)} = \begin{pmatrix} A^{(i)} e^{-i\zeta\psi} \\ B^{(i)} e^{i\zeta\psi} \end{pmatrix} \quad (102)$$

$$V = \begin{pmatrix} 0 & q \\ r & 0 \end{pmatrix} \quad (103)$$

and

$$G\Phi = \frac{1}{2}\int_{-\infty}^{\infty} \text{sgn}(x-x') \begin{pmatrix} e^{-i\zeta(x-x')} & 0 \\ 0 & e^{i\zeta(x-x')} \end{pmatrix} \Phi(x')dx'. \quad (104)$$

For the adjoint functions we have

$$\psi^{+(j)} = \phi^{+(j)} + G^+ V^+ \psi^{+(j)}, \quad (105)$$

$$\phi^{+(j)} = \begin{pmatrix} B^{(j)} e^{i\zeta x} \\ -A^{(j)} e^{-i\zeta x} \end{pmatrix} \quad (106)$$

$$V^+ = \begin{pmatrix} 0 & r \\ q & 0 \end{pmatrix} \quad (107)$$

$$G^+\Phi = -\frac{1}{2}\int_{-\infty}^{\infty} \text{sgn}(x-x') \begin{pmatrix} e^{i\zeta(x-x')} & 0 \\ 0 & e^{-i\zeta(x-x)} \end{pmatrix} \Phi(x')dx_x' \quad (108)$$

Some properties:
  Define:
$$\int_{-\infty}^{\infty} (\phi, \psi)dx = \int_{-\infty}^{\infty} (\phi_1 \psi_1 + \phi_2 \psi_2)dx \quad (109)$$

  Then:
$$\int_{-\infty}^{\infty} (\phi, v\psi)dx = \int_{-\infty}^{\infty} (v^+ \phi, \psi)d \quad (110)$$

  and

$$\int_{-\infty}^{\infty} (\phi, G\psi)dx = \int_{-\infty}^{\infty} (G^+ \phi, \psi)dx. \tag{111}$$

Introduce operators R, R^+ by virtue of the integral equations

$$R = v + vGR, \tag{112}$$

$$R^+ = v^+ + v^+ G^+ R^+. \tag{113}$$

Using Eqs. (110) and (111) we see that

$$\int_{-\infty}^{\infty} (\phi, R\psi)dx = \int_{-\infty}^{\infty} (R^+ \phi, \psi)dx. \tag{114}$$

In virtue of Eqs. (112) and (113) we have

$$\int_{-\infty}^{\infty} (\phi, R\psi)dx = \int_{-\infty}^{\infty} (R^+ \phi, \psi)dx. \tag{115}$$

From the integral Eqs. (112), (113) we show that

$$R\phi^{(i)} = v\psi^{(i)}, \tag{116}$$

$$R^+ \phi^{+(j)} = v^+ \psi^{+(j)}. \tag{117}$$

Introduce the "matrix element" R_{ji} by

$$R_{ji} = \int_{-\infty}^{\infty} (\phi^{+(j)}, R\phi^{(i)})dx. \tag{118}$$

Let us obtain some alternate expressions for R_{ji}. Using Eq. (115)

$$R_{ji} = \int_{-\infty}^{\infty} (\phi^{+(j)}, v\psi^{(i)})dx. \tag{119}$$

Using Eqs. (115) and (117)

$$R_{ji} = \int_{-\infty}^{\infty} (\psi^{+(j)}, v\psi^{(i)})dx. \tag{120}$$

Finally multiplying Eq. (101) by $\psi^{+(j)} v$ and integrating we have

$$R_{ji} = \int_{-\infty}^{\infty} (\psi^{+(j)}, [v - vGv]\psi^{(i)}) dx. \tag{121}$$

From the three expressions of Eqs. (119), (120), and (121) we still have a fourth form

$$R_{ji} = \frac{\left\{\int_{-\infty}^{\infty} (\phi^{+(j)}, v\psi^{(i)}) dx\right\}\left\{\int_{-\infty}^{\infty} (\psi^{+(j)}, v\phi^{(i)}) dx\right\}}{\int_{-\infty}^{\infty} (\psi^{+(j)}, [v - vGv]\psi^{(i)}) dx}. \tag{122}$$

The important property of this form (which is readily checked) is that it is stationary with respect to variations of $\psi^{(i)}$ and $\psi^{+(j)}$. I.e.,

$$\frac{\delta R_{ji}}{\delta \psi^{(i)}} = \frac{\delta R_{ji}}{\delta \psi^{+(j)}} = 0. \tag{123}$$

Hence to calculate the first-order change in R_{ji} when v varies we need only consider the explicit occurrence of v. Thus one finds

$$\delta R_{ji} = \int_{-\infty}^{\infty} (\psi^{+(j)}, \delta v \psi^{(i)}) d_x$$

$$= \int_{-\infty}^{\infty} \{\psi_1^{+(j)} \psi_2^{(i)} \delta q + \psi_2^{+(j)} \psi_1^{(i)} \delta r\} dx, \tag{124}$$

and so

$$\frac{\delta R_{ji}}{\delta q} = \psi_1^{+(j)} \psi_2^{(i)},$$

$$\frac{\delta R_{ji}}{\delta r} = \psi_2^{+(j)} \psi_1^{(i)}. \tag{125}$$

Using the relation of Eq. (100) between ψ and ψ^+ we then have

$$\frac{\delta R_{ji}}{\delta q} = \psi_2^{(j)} \psi_2^{(i)}$$

$$\frac{\delta R_{ji}}{\delta r} = -\psi_1^{(j)} \psi_1^{(i)} \tag{126}$$

Now suppose two linearly independent solutions $\psi^{(1)}, \psi^{(2)}$ are given. Then

$$W[\psi^{(1)}, \psi^{(2)}] = \psi_1^{(1)} \psi_2^{(2)} - \psi_1^{(2)} \psi_2^{(1)} \neq 0. \qquad (127)$$

Three solutions of Eqs. (92) – (93) are readily constructed

$$\Psi^{(1)} = \begin{pmatrix} (\psi_2^{(1)})^2 \\ -(\psi_1^{(1)})^2 \\ \psi_1^{(1)} \psi_2^{(1)} \end{pmatrix} \qquad (128)$$

$$\Psi^{(2)} = \begin{pmatrix} (\psi_2^{(2)})^2 \\ -(\psi_1^{(2)})^2 \\ \psi_1^{(2)} \psi_2^{(2)} \end{pmatrix} \qquad (129)$$

A third solution is obtained from noting that if $\psi^{(1)}, \psi^{(2)}$ are solutions of Eq.s (68) so is $\psi^{(1)} + \psi^{(2)}$. The Ψ obtained from the squares of this solution is then a solution. Subtracting out parts $\Psi^{(1)}, \Psi^{(2)}$ gives a third solution,

$$\Psi^{(3)} = \begin{pmatrix} 2\psi_2^{(1)} \psi_2^{(2)} \\ -2\psi_1^{(1)} \psi_1^{(2)} \\ \psi_1^{(2)} \psi_2^{(1)} + \psi_1^{(1)} \psi_2^{(2)} \end{pmatrix} \qquad (130)$$

To show the $\Psi^{(1)}, \Psi^{(2)}, \Psi^{(3)}$ are linearly independent we need to calculate their Wronskian.

$$W[\Psi^{(1)}, \Psi^{(2)}, \Psi^{(3)}] = \begin{vmatrix} \Psi_1^{(1)} & \Psi_1^{(2)} & \Psi_1^{(3)} \\ \Psi_2^{(1)} & \Psi_2^{(2)} & \Psi_2^{(3)} \\ \Psi_3^{(1)} & \Psi_3^{(2)} & \Psi_3^{(3)} \end{vmatrix} \qquad (131)$$

One finds

$$W = W^3(\psi_1\psi_2) \neq 0 \tag{132}$$

Any solution of Eqs. (92)-(95) can then be written as

$$\Psi = a\Psi^{(1)} + b\Psi^{(2)} + c\Psi^{(3)} \tag{133}$$

$$\therefore \quad \Psi_1 = a(\psi_2^{(1)})^2 + b(\psi_2^{(2)})^2 + c\psi_2^{(1)}\psi_2^{(2)}$$

$$\Psi_2 = -a(\psi_1^{(1)})^2 + b(\psi_1^{(2)})^2 - c\psi_1^{(1)}\psi_1^{(2)} \tag{134}$$

Thus from Eqs. (125) we have

$$\begin{pmatrix}\Psi_1 \\ \Psi_2\end{pmatrix} = \begin{pmatrix}\delta\Lambda/\delta q \\ \delta\Lambda/\delta r\end{pmatrix} \tag{135}$$

where

$$\Lambda = aR_{11} + bR_{22} + cR_{12} \tag{136}$$

IV. THE HARRY DYM EQUATION

So far all our examples of dual Hamiltonian formalisms has been evolution equations related to the Zakharov-Shabat eigenvalue problem. Here we consider one which is not. It is the equation

$$u_t = \partial_x^3 u^{1/2}$$

Some constants are readily calculated. These are:

$$I_1 = \int_{-\infty}^{\infty} u\, dx, \quad I_2 = 2\int_{-\infty}^{\infty} u^{1/2}\, dx$$

$$I_3 = \int_{-\infty}^{\infty} \frac{u^{5/2}(u_x)^2}{8}\, dx$$

A First Hamiltonian Form

Let $[F_i, F_j] = \int_{-\infty}^{\infty} \frac{\delta F_i}{\delta u} L \frac{\delta F_j}{\delta u}\, dx$

where $L = 2u\, \partial_x + u_x\, 1$,

and $H = I_3$.

Second Formulation

$$[F_i, F_j]' = \int_{-\infty}^{\infty} \frac{\delta F_i}{\delta u} M \frac{\delta F_j}{\delta u}\, dx,$$

$$M = \partial_x^3,$$

and $H' = I_2$

A Lax pair is then:

1) The eigenvalue problem is

$$\lambda(2u\, \partial_x + u_x)\Psi = \partial_x^3 \Psi$$

2) The time dependent problem is

$$\partial_t \Psi = -\frac{1}{2} u^{-3/2} \partial_x^3 \Psi .$$

A reduced pair is obtained by writing

$$\Psi = \psi^2 .$$

Then we find $\partial_x^2 \psi = \frac{\lambda u}{2} \psi$,

and $\partial_t \psi = -\frac{1}{2} u^{-3/2} \partial_x^3 \psi - \frac{3\lambda}{4} u^{-3/2} \partial_x \psi$

or $\partial_t \psi = -\lambda u^{-1/2} \partial_x \psi - \frac{\lambda}{4} u^{-3/2} u_x \psi .$

REFERENCES

(1) K. M. Case, Phys. Rev. Letters $\underline{40}$, (1978), p. 351-354.

(2) F. Magri, Journal Math-Phys. $\underline{19}$, (1978), p. 1156-1162.

(3) V. E. Zakharov and A. B. Shabat, Sov. Phys. JETP, $\underline{34}$, (1972), p. 62.

(4) P. D. Lax, C. D. A.M., $\underline{21}$, (1968), p. 467.

(5) J. Schwinger, Unpublished lecture notes. Also mentioned in "Collision Theory", M. L. Goldberger and K. M. Watson, John Wiley & Sons, New York (1964), p. 320.

(6) M. J. Ablowitz, D. J. Kaup, A. C. Newell, and H. Segur, Studies in Applied Mathematics, L. III, (1976), p. 249-336.

ON COMMON ORIGINS OF INTEGRABLE HAMILTONIAN SYSTEMS

H. Flaschka

Department of Mathematics
University of Arizona
Tucson, AZ 85721

This is a (polished) summary of the lecture delivered at the Workshop. A somewhat expanded survey will appear soon [1]. Various details are described in forthcoming papers [2,3,4]; some other work referred to will be written up eventually.

A Hamiltonian system is <u>algebraically integrable</u>, roughly speaking, if it is integrable (has a maximal number of constants of motion in involution) and if the solution is meromorphic as function of complex time t (a more precise definition may be found in [5]). These systems appear (with just a few exceptions that are not yet understood) to be obtained by restricting a "soliton" partial differential equation to a finite-dimensional invariant subset of its phase space: the rational, soliton, or quasiperiodic solutions.

<u>Example 1</u> In the AKNS approach to soliton equations [6,7], there is an <u>eigenvalue problem</u>

(1)
$$V_x = (\zeta R + P)V,$$

(R = constant diagonal matrix, ζ = eigenvalue, $P = (p_{ij})$ = matrix of potentials) and a <u>time evolution</u>

(2)
$$V_t = Q(\zeta)V,$$

$Q = Q_0 \zeta^N + \ldots + Q_N$ being a matrix polynomial in ζ. The integrability condition of (1) and (2) is

(3)
$$P_t - Q_x + [\zeta R + P, Q] = 0.$$

This can represent many familiar soliton equations.

If P is independent of time t, one has

(4)
$$Q(\zeta)_x = [\zeta R + P, Q(\zeta)],$$

which is the general <u>finite-dimensional</u> integrable system called <u>spinning top equation</u> in [8]. The equations for 3-dimensional rotating ridgid bodies are included in, and give the name to, this class.

Solutions of the o.d.e. (4) turn out to be initial conditions, for p.d.e.'s of type (3), that turn into rational, soliton, or quasiperiodic solutions. This is a simple instance of how restricting a p.d.e. ((3), here) to a finite-dimensional manifold (by setting t-derivatives equal to zero, here) produces a finite-dimensional integrable system ((4), here).

The theory of (4) transfers relatively easily to (3) in many respects. Compare the outlines of the Lie-algebra theory of the p.d.e. (3) in [9] with the Lie-algebra theory of the o.d.e. (4) in [8]. The major difference is in the Hamiltonian formalisms for infinitely many as opposed to finitely many degrees of freedom. A unified theory, even a formal one, is in principle impossible because solutions of (4) are always analytic, hence determined by their derivatives at a single point; this is not true of the merely differentiable solutions of (3). [4] describes the extent of the analogy. (End of example).

There are other, far less obvious, ways of making a p.d.e. finite-dimensional. The question, in each case, is: what is the interplay between the Lie-algebra/Riemann surface/Lax pair approaches to the p.d.e. and the o.d.e.? How can one solve the o.d.e., knowing the inverse scattering theory of the p.d.e.?

Example 2: If $q(x)$ is a finite-gap potential of Hill's equation $\ddot{y} + qy = Ey$, then q may be written

(5)
$$q = \sum_{j=0}^{N} E_j \phi_j^2 + \dot{\phi}_j^2 ,$$

where the E_i are certain eigenvalues and ϕ_i the corresponding eigenfunctions:

(6)
$$\ddot{\phi}_i + q\phi_i = E_i \phi_i, \quad i = 0,\ldots,N.$$

Substitute (5) into (6) to get the system of nonlinear o.d.e.'s

(7)
$$\ddot{\phi}_i + (\sum_{j=0}^{N} E_j \phi_j^2 + \dot{\phi}_j^2)\phi_i = E_i \phi_i, \quad i=0,\ldots,N.$$

(7) is the <u>Neumann system</u>. It has a Lax-pair representation, and is related to various geometric problems in a most surprising way (see [10] for a description of this circle of ideas). The papers [2,3] try to make the connection between Hill's equation and Neumann's system systematic, and show how to derive Neumann-type systems for eigenvalue problems more general than Hill's equation.

Example 3: The Toda-lattice equations
$$\ddot{q}_n = e^{-(q_n - q_{n-1})} - e^{-(q_{n+1} - q_n)}$$

are related to successive Bäcklund -type transformations of the AKNS
eigenvalue problem. The formal theory is hinted at in [11], and is
developed in [12]. There is, however, no coherent analytic
connection as yet. For example, I noticed that the finite Toda
lattice [13] describes N-soliton potentials of the AKNS system, that
the Kac-van Moerbecke system [14] describes N-soliton potentials of
the derivative nonlinear Schrödinger equation [15], etc. The
proposal is to relate all the Toda lattices associated with
semisimple Lie algebras (finite [16] or periodic [8]) to AKNS
eigenvalue problems, in a systematic way, and to generalize this to
other eigenvalue problems.

Example 4: Rational solutions

$$q(x,t) = -2 \sum_{1}^{N} \frac{1}{x-x_j(t)}$$

of the Korteweg-de Vries equation $q_t - 6qq_x + q_{xxx} = 0$ lead to the
following system for the poles $x_j(t)$:

(8)
$$\dot{x}_j = 2 \sum_{k(\neq j)} (x_j - x_k)^{-3}$$

(see [17], for example). (8) has its own Lax pair again, [13] and
there is a very interesting Lie-algebra approach [18] that has
generalized (8) (surveyed in [19]).

There are pole-motion equations for all the other soliton
p.d.e.'s. It is trivial, in any given case, to deduce the analog of
system (8). With some guessing, one can find Lax pairs and whatever
else one wants, in any given case. The problem, again, is to derive
everything directly from the inverse-scattering theory of the p.d.e.
(End of example.)

There are, it seems to me, many individual problems of
mathematical interest involved in this program. Perhaps eventually
a classification of algebraically integrable systems may emerge from
the now beginning classification of soliton p.d.e.'s.

Acknowledgements: I thank the Kyoto University RIMS for
support during the academic year 1980-81, and NSF and DOA for
support thereafter. Alan Newell, Tudor Ratiu, and Randy Schilling -
my collaborators on this project - contributed many of the ideas. I
am also grateful to Profs. E. Date, M. Jimbo, M. Kashiwara, T. Miwa,
and M. Sato for helpful explanations.

References

1. H. Flaschka, On the common origins of integrable finite-dimensional systems, to appear in the proceedings of a 1981 Kyoto RIMS conference.

2. H. Flaschka and R. Schilling, Neumann systems for finite-gap potentials, University of Arizona preprint (1981).

3. R. Schilling, Particle systems for 1st order matrix eigenvalue problems, in preparation.

4. H. Flaschka, A. C. Newell, T. Ratiu, Estabrook-Wahlquist prolongation and Kac-Moody algebras, in preparation.

5. M. Adler and P. van Moerbecke, A criterion for algebraic complete integrability and Kowalewski's method, Brandeis University preprint (1981).

6. M. Ablowitz, D. J. Kaup, A. C. Newell, H. Segur, Stud. Appl. Math. 53 (1974) 249.

7. A. C. Newell, Proc. Roy. Soc. London A 365 (1979) 283.

8. M. Adler, P. van Moerbecke, Adv. Math. 38 (1980) 267.

9. G. Wilson, Ergodic Theory and Dynamical Systems, 1 (1981).

10. J. Moser, Proc. of Chern Symposium (1981), 147.

11. D. Levi, R. Benguria, Bäcklund transformations and nonlinear differential difference equations, Univ. of Rome preprint, 1980.

12. M. Jimbo, T. Miwa, Physica D 2 (1981) 407.

13. J. Moser, Adv. Math. 16 (1975) 1.

14. M. Kac, P. van Moerbecke, Adv. Math. 16 (1975) 160.

15. D. J. Kaup, A. C. Newell, J. Math. Phys. 19 (1978) 798.

16. B. Kostant, Adv. Math 34 (1979) 195.

17. H. Airault, H. P. McKean, Jr., J. Moser, Comm. Pure Appl. Math. 30 (1977) 95.

18. D. Kazhdan, B. Kostant, S. Sternberg, Comm. Pure Appl. Math. 31 (1978) 481.

19. M. A. Ol'shanetsky, A. M. Perelomov, Physics Reports 71 (1981) 313.

THE METHOD OF ESTABROOK AND WAHLQUIST

D. J. Kaup[†]

Institute for Nonlinear Studies
Clarkson College of Technology
Potsdam, NY 13676

ABSTRACT

The problem of determining when a nonlinear partial differential equation is integrable is discussed, and it is shown that the method of Estabrook and Wahlquist may be used to determine such.

I. INTRODUCTION

So far, at least to a very large extent, solving nonlinear partial differential equations (NLPDE) has been mostly a one-way street. By this I mean that once you are given the "Lax pair" [1] for a NLPDE, then one can inevitably find a method of solution. In fact, one of the very first methods [2] for solving NLPDE's was to assume a general form for the Lax pair, and then to see what kind of a NLPDE popped out. Frequently it was indeed one of physical interest [3].

However, such crude techniques are no longer successful, possibly because almost all of the simple cases have already been treated and classified, leaving only the complex and complicated NLPDE's to be found. And the solving of these systems, if at all, is going to require much more sophisticated techniques than those which were used at first. Thus we are being forced by the circumstances to tackle the major problem of: "Given a NLPDE, how does one know when the NLPDE is integrable (exactly solvable), and if so, how does one then construct the appropriate Lax pair?" This is an area where we are actually behind, and even backward, when compared to the sophistication of other techniques. I might add that in part this is probably due to the relative ease with which other results could have been had. However, out of necessity it may now be necessary to finally develop these more sophisticated techniques.

Before discussing the method introduced by Estabrook and Wahlquist, I shall mention and comment on some of the other methods by which one could currently ascertain whether a NLPDE might be integrable. The oldest method consists simply of numerically colliding two solitary wave pulses together. This was how Kruskal

---

[†]Research supported by the Office of Naval Research and the National Science Foundation.

and Zabusky [4] first defined the term "soliton". Solitons were defined to be those pulses which when collided together, passed through each other with no more than a phase shift resulting from the collision. Although this definition is still valid, attempting to determine when a NLPDE is integrable by this test does have some drawbacks. First, because of the nature of numerical calculations, it is only a negative test. By this I mean that if you are numerically testing a real soliton equation, you can never be absolutely sure that the pulses are indeed solitons, because of round-off errors. Some NLPDE's have been numerically observed to be very close to soliton systems by this test, but have eventually proven to be nonintegrable. But on the other hand, once one can establish that more than just a simple phase shift occurs upon colliding two solitary pulses together, then one is considered to have established that the corresponding NLPDE is not an integrable system. Consequently, this is only a negative test.

A second method is to attempt to construct a closed-form solution corresponding to N colliding pulses. The existence of a closed-form solutions for 2 or 3 or even a finite number of colliding pulses cannot be considered to be a valid test, because counterexamples now exist. For example, for the random phase three-wave interaction, one can construct [5] closed-form solutions consisting of 2-,3-, and 4-shocks, all colliding. However, except for this, these equations appear to be nonintegrable. So this method seems to be a valid positive test only if such solutions exist for an arbitrary number of colliding pulses. One systematic method for constructing and testing such solutions has been developed and used by Hirota [6].

A third test is whether or not a Bäcklund transformation [7] (BT) can be found. Using a BT, one can actually construct the N-soliton solution discussed above, so the existence of such a transformation can be considered to be a positive test for integrability. Note that one cannot consider the existence of a BT to imply the existence of solitons in a NLPDE, since even the Liouville equation has a BT [8], but the existence of a BT does imply integrability. Thus a BT also gives a positive test for integrability, and seems to be more basic than an N-soliton solution. But at the same time, it seems to be more elusive to find when it exists. To my knowledge, the most systematic and direct method for searching for a BT is that which was introduced by Clairin [9,10] in 1902. Otherwise, all BT's seem to found by luck and by guess.

A fourth and more recent method has been introduced by Ablowitz et al [11], and it relys on finding a connection between a similarity solution of a NLPDE and a solution of one of the Painlevé equations. This has been put forward as a test of integrability after observing that in a large number of cases such a connection is always associated with integrability, and is absent when the NLPDE is nonintegrable. However at the same time, it has been remarked [12] that you have to be very careful in using this test, in that the test seems to be very sensitive to how one selects the form of the asymptotic variable in the similarity solution. So until a more systematic and consistent procedure is developed for this method, as with other methods already mentioned, one should use this test with caution. In any case there seems to be no connection between this test and the problem of constructing the Lax pair. Whereas given a BT, one does have some information on the structure of the appropriate Lax pair [7,8].

Also in the same class, I should mention the recent work of Fokas [13], whose approach is to consider the symmetries of a given NLPDE. He notes that if a Lax pair exists, then the NLPDE has an infinite number of symmetries. He then proposes that a necessary condition for a nth order NLPDE to possess a Lax pair, is for the NLPDE to admit a "Lie-Bäcklund operator" of order $2n-1$. He illustrates this by showing that the most general case for a second-order equation is the Burger's equation, while for a third-order equation, it is a new NLPDE, which contains the KdV and modified KdV as special cases. And since this new equation can be related to the MKdV by a Bäcklund transformation [13], it is clear that the appropriate eigenvalue problem for his more general NLPDE is the same as for the MKdV. But still, there is no procedure proposed for going from a "Lie-Bäcklund operator" to the Lax pair.

Outside of the method developed by Estabrook and Wahlquist, the above five methods are the only other methods known to me which are useful for testing a NLPDE for integrability. What is most required and needed is a method which allows one to systematically progress from a chosen NLPDE, on through to the actual construction of a Lax pair, with a minimum amount of "guess" and "luck" required. In my judgement, of all these methods, the Estabrook-Wahlquist (EW) method does come the closest to satisfying this requirement. But, even it has drawbacks, some of which we shall later point out. But first, let us describe the EW method and its roots.

An excellent reference for the EW method (and incidently also for BT's) is Ref. 7, which is the conference proceedings of the 1979 NSF Workshop on Contact Transformations. At this workshop, Estabrook, Wahlquist, and Corones each presented detailed expositions on the EW method, each exposition of which is recommended for basic reading. The roots of this method goes back into the 19th century where Bäcklund was studying the transformation which now bears his name, and was applying it to various geometrical surfaces. The history of this was outlined in part by George Lamb [7] at that conference. Quite interestingly, Lamb also pointed out that it was Clairin [9,10] who had introduced a different form of Bäcklund's theory, which was more direct, in that one could take an arbitrarily chosen NLPDE, and systematically search for a BT. However, research in this area was interrupted by World War I, after which the subject seems to have been dropped as a research topic. And it has been only after the development of the inverse scattering transform (IST), that an interest in this topic was rekindled.

The gist of the EW method is the same as that which was introduced by Clairin in 1902, where he sought to find a general method for finding BT's. And whereas the Clairin method directed the attack onto the BT, the EW method goes beyond that and instead seeks to find the Lax pair directly, thereby bypassing the BT step. Of course, once a Lax pair is found, it is usually a trivial matter to construct a BT directly from the Lax pair, whereas the opposite is considerably more difficult. So, the EW approach is to aim directly for that which is required for complete integrability: namely the Lax pair.

The original approach [14,15] of Wahlquist and Estabrook also made use of the extreme power of differential forms, and Refs. 7 and 15 contain valuable expositions by them on the basic operations of differential forms and their interpretation. However, we do point out that differential forms are not necessary. They are only convenient and allow a quite compact notation. One could just as well translate all of their results into more classical terms, as was pointed out and done by J. Corones [16]. However this classical notation can get to be awkward. And, if one is to follow the current developments in this field, a fluent understanding of differential forms does then become a necessity.

Since those early years, there has been an explosion in the literature on the EW method. A lot of it currently lies beyond my interest, which fundamentally is only to refine a scheme for determining when a NLPDE is integrable. Examples of such can be found in another article by Corones [17] and an entire collection of papers [18-22] by H.C. Morris detailing the EW "prolongation structure" for most of the then known integrable NLPDE's, including even the higher dimensional cases.

Certainly one also has to mention the works of Belinsky and Zakharov [23], in showing how a very large class of solutions of the Einstein equations can be obtained from an IST. The relation of this Belinsky-Zakharov method to other known transform methods of general relativity has been detailed by Cosgrove [24]. Other than this, we shall choose to say nothing further, but instead shall discuss simple examples to illustrate the possible applications of the EW method.

II. APPLICATIONS

We are considering the problem of given a NLPDE, how do we construct a method of solution. We do this by constructing the Lax pair, and to construct the Lax pair, we use the EW method.

In using the EW method, there are only four basic steps that one must follow. These four steps have been detailed and outlined in a recent publication [25] of mine wherein I have used many examples to illustrate these steps. We shall now illustrate the method by applying it to an integrated form of the KdV equation, which is

$$q_t + q_{xxx} + \frac{3}{2}q_x^2 = 0, \qquad (1)$$

where subscripts refer to partial differentiation. If one differentiates Eq. (1) with respect to x and then sets $q_x = u$, then the usual form of the KdV equation results. We shall use instead the above form, because it will require an eigenvalue problem nonlinear in the potential, q, but still will illustrate all of the usual features of the EW method.

The first step [25] in the EW method is: <u>Choose an appropriately general form for the Lax pair</u>. The form which is adequate for (1) is:

$$Q_x^k = A^k(q, Q^k), \qquad (2a)$$

$$Q_t^k = B^k(q, q_x, q_{xx}, Q^k), \qquad (2b)$$

where Q^k is a (in general) vector pseudopotential and A^k and B^k are, at this point, arbitrary functions of their variables. But before we are finished, we shall have A^k and B^k explicitly specified. The criteria used to choose what variables A and B will depend on is discussed in Ref. [25], but basically one simply wants to use the simplest combination such that when we cross-differentiate Eq. (2), Eq. (1) will appear as an integrability condition.

After Eq. (2) has been chosen, then one has only to perform the simple calculus and algebra steps to determine the integrability of (2). We shall now detail these in this example. From cross-differentiation of (2), we obtain

$$Q_{xt}^k - Q_{tx}^k = 0$$

$$= A_{,q}^k q_t + [A,B]^k - B_{,q}^k q_x$$

$$\quad - B_{,q_x}^k q_{xx} - B_{,q_{xx}}^k q_{xxx}, \qquad (3)$$

where the bracket is defined by

$$[A,B]^k \equiv \sum_\ell \{B^\ell \frac{\partial A^k}{\partial Q^\ell} - A^\ell \frac{\partial B^k}{\partial Q^\ell}\}, \qquad (4)$$

and is the Lie derivative between the vectors A^k and B^k. Next one injects (1) into (3), and requires A and B to be local functions of their variables. This requirement is sufficient to totally determine the dependence of A and B on the variable q and its derivatives.

Simply equating coefficients yields

$$A^k = x_0^k(Q) + q x_1^k(Q) + \tfrac{1}{2} q^2 x_2^k(Q), \qquad (5a)$$

$$B^k = x_3^k(Q) - (q_{xx} + q_x q) x_1^k(Q)$$

$$+ \tfrac{1}{2}(q_x^2 - 2q_{xx}q - q_x q^2) x_2^k(Q) - q[0,0,1]^k$$

$$- \tfrac{1}{2}q^2[1,0,1]^k - (q_x + \tfrac{1}{2}q^2)[0,1]^k, \qquad (5b)$$

and where we are introducing the shorthand notation of

$$[n,m] \equiv [x_n, x_m], \qquad (6a)$$

$$[n,m,\ell] \equiv [x_n, [x_m, x_\ell]], \qquad (6b)$$

with the bracket on the rhs being the Lie derivative as defined by (4). In (6) we are now dropping the superscripts from the brackets, which are simply understood to be there, if needed. When our pseudopotential, Q, is a scalar and not a vector, then superscripts are not needed. But should it turn out that a vector form is required, it is a very simple matter to replace these superscripts.

However, (5) is not the only result of this operation. In addition to (5), there is also a complex of relations involving the brackets which are required to be satisfied. In our case these are:

$$[0,2] = x_1, \qquad (7a)$$

$$[1,2] = x_2, \qquad (7b)$$

$$[0,3] = 0, \qquad (7c)$$

$$[1,3] = [0,0,0,1], \qquad (7d)$$

$$[2,3] = 2[1,0,0,1] + [0,1,0,1] + [0,0,1], \qquad (7e)$$

$$0 = [2,0,0,1] + [1,1,0,1] + [1,0,1], \qquad (7f)$$

$$0 = [2,1,0,1] + [2,0,1], \qquad (7g)$$

In (7), the four-index bracket is defined similarly as is done in (6).

Once (7) has been obtained, we have completed Step 1. What (7) represents is an (possibly) incomplete Lie algebra, and as it stands, it is also unclosed. If this (possible) Lie algebra is completed and closed in a nontrivial fashion, then the NLPDE will be integrable, because we may then explicitly construct the necessary Lax pair. But on the other hand, if one can show that no such nontrivial closure exists, then the NLPDE is nonintegrable (modulo the assumed forms of A and B in Step 1, see Ref. 25). Consequently, the problem of the solvability of the NLPDE has now been reduced to a problem in Lie algebra.

This leads us now to Step 2, which is: <u>close the system in (7) into a Lie algebra, or else show that no such closures can exist.</u> This is easier said than done in many cases. This example that we are using has been chosen to be of an integrable form, so such a closure is known to exist. Some of the tricks and considerations that one could use at this step has been outlined in Ref. 25. One should note that the brackets of the final closed Lie algebra, if one is found, must satisfy all of the Jacobi indentities, which are

$$[i,j,k] + [j,k,i] + [k,i,j] = 0 \qquad (8)$$

These are necessary and sufficient conditions for a Lie algebra to exist. And, they impose some very stringent conditions on an arbitrary set of brackets, like those in (7). One will best appreciate these statements after a little experience. But we may now (with remarkable hindsight!) make Step 2 appear to be very simple by choosing:

$$x_3 = -4\lambda x_0, \qquad (9a)$$

$$x_0 = x_4 - 4\lambda x_2, \qquad (9b)$$

where x_4 is some other independent function of Q, and taking

$$[1,2] = x_2, \tag{10a}$$

$$[1,4] = 4\lambda x_2 - x_4, \tag{10b}$$

$$[2,4] = -x_1. \tag{10c}$$

In (9) and (10), λ is an arbitrary scalar constant.

One may show that (9) and (10) is a solution of (7), and that the single Jacobi identity for (10) is satisfied. This then completes Step 2.

We next consider Step 3, which is: <u>construct a representation of the x_i's, and then A and B.</u> Once we know that the x_i's are integrable, then we may proceed to solve (10) using the definition of the bracket as given by (4). A unique solution will never exist, since (2) is invariant under a very wide class of transformations on the pseudopotentials. And the general solution of (4) and (10) must reflect this invariance. However we may utilize this invariance to transform some of the x_i's to a chosen (canonical) and arbitrary form, provided that that form is consistent with (10). Thus we shall choose

$$x_2(Q) = 1, \tag{11a}$$

which when injected into (10) gives

$$x_1(Q) = Q, \tag{11b}$$

$$x_4(Q) = \tfrac{1}{2}Q^2 + 2\lambda, \tag{11c}$$

and from (9),

$$x_0(Q) = \tfrac{1}{2}Q^2 - 2\lambda, \tag{11d}$$

$$x_3(Q) = -2\lambda Q^2 + 8\lambda^2. \tag{11e}$$

We may now construct the Lax pair from (2), (5), and (11) which gives

$$Q_x = \frac{1}{2}Q^2 + Qq + \frac{1}{2}q^2 - 2\lambda, \qquad (12a)$$

$$Q_t = -\frac{1}{2}Q^2(q_x+4\lambda) - Q(q_{xx}+q_x q+4\lambda q)$$

$$+ \frac{1}{2}(q_x^2 - 2q_{xx}q - q_x q^2) - 2\lambda(q_x+q^2) + 8\lambda^2. \qquad (12b)$$

One may readily verify that cross-differentiation of (12) does indeed give (1) as being the integrability condition for (12). This completes Step 3.

The last step, Step 4, is simply: <u>prove that the resulting Lax pair may be used to construct a method of solution.</u> Although one may have gotten this far, there is no guarantee that the system is yet integrable, until one can give an explicit method of solution, based on these operators. In fact, pairs of functions similar to (12) may be found [25], but these pairs do not give rise to any method of solution. Furthermore, one could obtain other pairs which are less trivial, but with which no method of solution can be associated [5]. So this step is not necessarily trivial.

However, in our case, we only need to linearize the Riccatti Eq. in (12a). Let

$$Q = -2\frac{\psi_x}{\psi} - q, \qquad (13)$$

then (12a) transforms into

$$\psi_{xx} + \psi(\frac{1}{2}q_x+\lambda) = 0. \qquad (14)$$

The method of solution is now obvious.

To illustrate the opposite, we shall now consider another example. This is one which was suggested to me by Dave Benney [26]. As is well known, when one has an almost monochromatic wave and a weak dispersion, the generic NLPDE is the "nonlinear Schrödinger equation"

$$iq_t = -q_{xx} + \alpha(q^*q)q, \qquad (15)$$

where α is a real constant. However this supposes that the dispersion, ω_{kk}, evaluated at the central wave-vector, does not vanish. One can easily have a case where such is not so, with $\omega(k)$ having an inflection point ($\omega_{kk}=0$) at the value of the wave-vector of interest. In this case, the nonlinear Schrödinger equation is not the appropriate equation. Instead, one has to go on to the next order, obtaining instead

$$q_t = q_{xxx} + i\alpha(q^*q)q, \qquad (16)$$

where q is still complex. Now, one has a sixth-order system, with the dispersion being KdV-like, coupled to a nonlinear Schrödinger nonlinearity. The question now is: "Is this equation integrable?" We shall now show that the answer is: "No."

We start at "EW Step #1", taking

$$Q_x = A(q,q^*,Q), \qquad (17a)$$

$$Q_t = B(q,q^*,q_x,q_x^*,q_{xx},q_{xx}^*,Q). \qquad (17b)$$

We now cross-differentiate, thereby obtaining

$$0 = A_{qqq} = A_{q^*qq} = A_{q^*q^*q} = A_{q^*q^*q^*}, \qquad (18a)$$

$$0 = [A, A_{qq}] = [A, A_{q^*q}] = [A, A_{q^*q^*}], \qquad (18b)$$

$$0 = [A, [A_q, A_{q^*}]], \qquad (18c)$$

and

$$B = q_{xx}A_q + q^*_{xx}A_{q^*} - \frac{1}{2}q_x^2 A_{qq} - q^*_x q_x A_{q^*q}$$

$$- \frac{1}{2}q_x^{*2} A_{q^*q^*} + q_x[A,A_q] + q^*_x[A,A_{q^*}] + D(q,q^*,Q), \quad (19)$$

$$D_q = [A,[A,A_q]], \quad (20a)$$

$$D_{q^*} = [A,[A,A_{q^*}]], \quad (20b)$$

$$[A,D] + i\, q^*q(qA_q - q^*A_{q^*}) = 0. \quad (21)$$

Continual differential of (21) with respect to q and q*, freely using (18) and (20), allows one to establish that

$$0 = A_{qq} = A_{q^*q^*}, \quad (22)$$

so the most general form of A is

$$A = x_0(Q) + qx_1(Q) + q^*x_2(Q) + q^*qx_3(Q). \quad (23)$$

Then from (18) and (20), the general solution for $D(q,q^*,Q)$ is

$$D = x_4(Q) + q[0,0,1] + q^*[0,0,2] + \frac{1}{2}q^2[1,0,1]$$

$$+ q^*q[1,0,2] + \frac{1}{2}q^{*2}[2,0,2]. \quad (24)$$

And (18) and (21) give the relations

$$0 = [0,3] = [1,3] = [2,3], \tag{25a}$$

$$0 = [0,1,2] = [1,1,2] = [2,1,2], \tag{25b}$$

$$0 = [0,4] = [1,1,0,1] = [2,2,0,2], \tag{25c}$$

$$[1,1,0,2] + \tfrac{1}{2}[2,1,0,1] = -i\alpha x_1, \tag{25d}$$

$$[2,1,0,2] + \tfrac{1}{2}[1,2,0,2] = i\alpha x_2, \tag{25e}$$

$$[1,0,0,1] + \tfrac{1}{2}[0,1,0,1] = 0, \tag{25f}$$

$$[2,0,0,2] + \tfrac{1}{2}[0,2,0,2] = 0, \tag{25g}$$

$$[1,4] + [0,0,0,1] = 0, \tag{25h}$$

$$[2,4] + [0,0,0,2] = 0, \tag{25i}$$

$$[3,4] + [1,0,0,2] + [2,0,0,1] + [0,2,0,1] = 0. \tag{25j}$$

This completes the EW Step #1.

In the second step, we want to see if we can close the above Lie algebra, or else show that no such Lie algebra, consistent with (25), can exist. We shall prove the latter. To do so, we shall find it convenient to order our elements in some convenient fashion. For this problem, we note that (25) is invariant under the scaling

$$x_o \to \lambda x_o,$$
$$x_1 \to \lambda x_1,$$
$$x_2 \to \lambda x_2,$$
$$x_3 \to \lambda x_3,$$
$$\alpha \to \lambda^3 \alpha,$$
$$x_4 \to \lambda^3 x_4. \qquad (26)$$

The only significance of this scaling is that we shall use it to organize our elements. Thus at first-order, we have four elements:

$$x_o, \quad x_2,$$
$$x_1, \quad x_3. \qquad (27)$$

From these elements, we may construct six brackets, three of which are determined by (25). The other three are (possibly) new second-order elements. So in second-order:

$$[0,1] = x_{01}, \quad [1,2] = x_{12},$$
$$[0,2] = x_{02}, \quad [1,3] = 0,$$
$$[0,3] = 0, \quad [2,3] = 0, \qquad (28)$$

where x_{01}, x_{02}, and x_{12} are our (possible) new elements. At third-order, x_4 enters, and also our first Jacobi identities (involving only first-order elements) are used for determining some of the brackets, along with (25b). One finds

$$[0,01] = x_{001}, \quad [2,01] = x_{102},$$
$$[0,02] = x_{002}, \quad [2,02] = x_{202},$$
$$[0,12] = 0, \quad [3,01] = 0,$$
$$[1,01] = x_{101}, \quad [3,02] = 0,$$
$$[1,02] = x_{102}, \quad [3,12] = 0,$$
$$[1,12] = 0, \quad \text{and } x_4, \qquad (29)$$

where x_{001}, x_{002}, x_{101}, x_{102}, x_{202}, and x_4 are the (possibly) new elements. At fourth-order, we have the Jacobi identities between two first-order elements and one second-order element. And all of the remaining equations in (25) come into play. So, at fourth-order:

$$[0,4] = 0, \quad [2,001] = -\tfrac{1}{3}x_{34}+x_5,$$
$$[1,4] = x_{14}, \quad [2,002] = 0,$$
$$[2,4] = x_{24}, \quad [2,101] = -\tfrac{2}{3}i\alpha x_1,$$
$$[3,4] = x_{34} \quad [2,102] = \tfrac{2}{3}i\alpha x_2,$$
$$[0,001] = -x_{14}, \quad [2,202] = 0$$
$$[0,002] = -x_{24}, \quad [01,02] = x_5,$$
$$[0,101] = 0, \quad [01,12] = 0,$$
$$[0,102] = -\tfrac{1}{3}x_{34}, \quad [02,12] = 0,$$
$$[0,202] = 0, \quad [3,001] = 0,$$
$$[1,001] = 0, \quad [3,002] = 0,$$
$$[1,002] = -\tfrac{1}{3}x_{34}-x_5, \quad [3,101] = 0,$$
$$[1,101] = 0, \quad [3,102] = 0,$$
$$[1,102] = -\tfrac{2}{3}i\alpha x_1, \quad [3,202] = 0,$$
$$[1,202] = \tfrac{2}{3}i\alpha x_2, \tag{30}$$

where the (possible) new elements are x_{14}, x_{24}, x_{34}, and x_5.

At this point, all the constraints in (25) have been applied. So from 5th-order on, it is only the Jacobi identities which can constrain the Lie algebra. In 5th-order, all brackets are determined except for

$$[0,14] = x_{014}, \quad [2,14] = x_{214},$$
$$[0,24] = x_{024}, \quad [2,24] = x_{224},$$
$$[1,14] = x_{114}, \quad [2,34] = x_{234},$$
$$[1,24] = x_{124}, \quad [3,34] = x_{334},$$
$$[1,34] = x_{134}, \tag{31}$$

which are (possible) new elements. All other brackets are uniquely determined by the Jacobi identities, which give

$$[0,34] = 0, \qquad\qquad [02,4] = x_{024},$$
$$[3,14] = x_{134}, \qquad [02,001] = \tfrac{2}{3}x_{214} + \tfrac{1}{3}x_{124},$$
$$[3,24] = x_{234}, \qquad [02,002] = x_{224},$$
$$[0,5] = \tfrac{1}{3}(x_{124} - x_{214}), \qquad [02,101] = -\tfrac{2}{3}i\alpha x_{01},$$
$$[1,5] = \tfrac{1}{3}x_{134}, \qquad [02,102] = \tfrac{1}{3}x_{234} + \tfrac{2}{3}i\alpha x_{02},$$
$$[2,5] = -\tfrac{1}{3}x_{234}, \qquad [02,202] = 0,$$
$$[3,5] = 0, \qquad\qquad [12,4] = x_{124} - x_{214},$$
$$[01,4] = x_{014}, \qquad\qquad [12,001] = 0,$$
$$[01,001] = x_{114}, \qquad\qquad [12,002] = 0,$$
$$[01,002] = \tfrac{2}{3}x_{124} + \tfrac{1}{3}x_{214}, \qquad [12,101] = 0,$$
$$[01,101] = 0, \qquad\qquad [12,102] = 0,$$
$$[01,102] = \tfrac{1}{3}x_{134} - \tfrac{2}{3}i\alpha x_{01}, \qquad [12,202] = 0.$$
$$[01,202] = \tfrac{2}{3}i\alpha x_{02}, \tag{32}$$

So far, everything has appeared to be fine. The Jacobi identities are finding no contradictions, each order is allowing new elements to appear, and the system looks like it could be closed, provided one could make the proper identifications. However, all such attempts were unsuccessful, and when we got to the next order, sixth-order, we find out why. From the four different Jacobi identities composed out of the sets of elements:

$$(1,3,14)$$

$$(3,01,001)$$

$$(1,01,102)$$

$$(1,02,101) \tag{33}$$

these Jacobi identities demand as a necessary condition that:

$$x_{101} = 0. \tag{34}$$

Now, returning to (30), we see that this implies also

$$x_1 = 0, \tag{35}$$

and then the whole system starts to break down. The <u>coup de grâce</u> is that one may similarly also show that

$$x_{202} = 0, \tag{36}$$

from which we have

$$x_2 = 0.$$

Now the algebra becomes trivial, with every single bracket vanishing. Thus no Lax pairs in the form of (17) can exist, and we must then consider (16) to be nonintegrable.

III. SUMMARY

These two examples serve to illustrate the EW method and how it can be used to obtain a Lax pair and also how it can be used to verify that no such Lax pair can exist (of a chosen assumed form). This is an extremely powerful tool, and the procedure for its utilization has been well outlined here and elsewhere [25].

The disadvantage of the EW method is that except for the simplest of systems, the calculations can become quite laborious, involving much rechecking and cross-checking, with any one sign or numerical error creating a totally wrong answer. It is more for these reasons than anything else, that the method is seldom used. However, it is exactly these types of laborious calculations which may be done by electronic computers using symbolic manipulation. And computers can do such calculations error-free, whereas hand calculations, such as I have presented here, can only be considered to be error-free after elaborate cross-checking and rechecking. So it is a hope that as electronic symbolic manipulation becomes more available, we shall see more utilization of, and results from the EW method.

In conclusion, we note that so far, the EW method has not been applied to three dimensional systems, nor has it been applied to the integral type of operators as in the intermediate-wave equation [27]. And, it may be that in the future, such applications shall be found.

REFERENCES

1. P.D. Lax, Comm. Pure Appl. Math. 21, 467-490 (1968).
2. M.J. Ablowitz, D.J. Kaup, A.C. Newell and H. Segur, Phys. Rev. Letters 31, 125-7 (1973).
3. V.E. Zakharov and A.B. Shabat, Funkts. Anal. Prilozh. 8, 43-53 (1974).
4. N.J. Zabusky and M.D. Kruskal, Phys. Rev. Letters 15, 240 (1965).
5. A. Rieman and D.J. Kaup, Phys. Fluids 24, 228-32 (1981).
6. R. Hirota, J. Math. Phys. 14, 805-9 (1973).
7. Lecture Notes in Mathematics, Vol. 515, Edited by A. Dold and B. Eckmann (Springer-Verlay, New York, 1976).
8. "Solitons", Vol. 17 of Topics in Current Physics, Editors: R.K. Bullough and P.J. Caudrey (Springer-Verlag, New York, 1980).
9. J. Clairin, Ann. Sci. École Norm. Sup. 3e Ser. Suppl. 19, 1-63 (1902).
10. J. Clairin, Ann. Fac. Sci. Univ. Toulouse 2e Ser. 5, 437-58 (1903).
11. M.J. Ablowitz, A. Ramani and H. Segur, J. Math. Phys. 21, 715-721 (April, 1980), ibid, 21, 1006-1015 (May, 1980).
12. M.J. Ablowitz, (private communication).
13. A.S. Fokas, J. Math. Phys. 21, 1318-25 (1980).
14. H.D. Wahlquist and F.B. Estabrook, J. Math. Phys. 16, 1-7 (1975).

15. F.B. Estabrook and H.D. Wahlquist, J. Math. Phys. 17, 1293-97 (1976).
16. J. Corones, J. Math. Phys. 17, 756-59 (1976); ibid, 18, 163-4 (1977).
17. J. Corones, J. Math. Phys. 19, 2431-6 (1978).
18. H.C. Morris, J. Math. Phys. 17, 1867-9 (1976); ibid, 1870-2 (1976).
19. H.C. Morris, J. Math. Phys. 18, 285-8 (1977); ibid, 530-2 (1977); ibid, 533-6 (1977).
20. H.C. Morris, J. Theor. Phys. (USA) 16, 227-31 (1977).
21. H.C. Morris, J. Math. Phys. 19, 85-7 (1978).
22. H.C. Morris, J. Phys. A. (GB) 12, 131-4 (1979); ibid, 261-7 (1979).
23. V.A. Belinsky and V.E. Zakharov, Zh. Eksp. Teor. Fiz. 75, 1953 (1978) (Sov. Phys. JETP 48, 985 (1978)); ibid, 77, 3 (1979) (ibid, 50, 1 (1979)).
24. C.M. Cosgrove, J. Math. Phys. 21, 2417-47 (1980).
25. D.J. Kaup, Physica 1D, 391-411 (1980).
26. D.J. Benney (private communication).
27. J. Satsuma, M.J. Ablowitz and Y. Kodama, Physics Letters 73 A, 283-6 (1979).

COMMENTS ON INVERSE SCATTERING FOR THE KADOMTSEV-PETVIASHVILI EQUATION

Harvey Segur

Aeronautical Research Associates of Princeton, Inc.
Princeton, NJ 08540

ABSTRACT

Manakov (1981) has given an inverse scattering formalism to solve the Kadomtsev-Petviashvili equation with positive dispersion as an initial-value problem. A consequence of this formulation is that neither plane-wave solitons nor algebraically decaying "lumps" evolve from spatially confined initial data. The present paper shows that an inverse scattering formulation similar to Manakov's can be justified for initial data that are small enough in a certain norm, and that lumps are excluded by this requirement of smallness.

1. INTRODUCTION

The equation of Kadomtsev and Petviashvili (1970; hereafter called KP)

$$(u_t + 6uu_x + u_{xxx})_x = 3u_{yy} \qquad (1.1)$$

is a natural generalization of the Korteweg-de Vries (KDV) equation to 2(space) + 1(time) dimensions. Generally, if the KDV equation can be derived in some physical context by imposing a restriction that waves travel in only one spatial dimension, then the KP equation (either as in (1.1) or with the sign of u_{xxxx} reversed) can be derived in the same context by relaxing that restriction.

That (1.1) is completely integrable was indicated by the work of Dryuma (1974), who found a Lax pair:

$$i\,\psi_y + \psi_{xx} + u\,\psi = 0 \qquad (1.2)$$

$$\psi_t + 4\psi_{xxx} + 6u\,\psi_x + 3\psi\left[u_x - i\int_{-\infty}^{x} u_y d\bar{x}\right] = 0 \qquad (1.3)$$

The reader may verify that the compatibility condition for (1.2) and (1.3) is (1.1).

As a generalization of KdV to 2 + 1 dimensions, (1.1) has two interesting aspects. i) Soliton solutions of KdV (i.e., y-independent plane waves) also solve KP, but Kadomtsev and Petviashvili (1970) showed that they are linearly unstable. (ii) The KP equation admits another class of solitons, sometimes called "lumps", that decay algebraically in both spatial dimensions (Bordag, Its, Manakov, Matveev and Zakharov, 1977; see also Ablowitz and Satsuma, 1978). One lump is given by

$$u(x,y,t) = 2\partial_x \ln\left[(x + X)^2 + Y^2\right] \qquad (1.4)$$

where

$$X(y,t) = ay - 3(b^2 - a^2)t + x_0,$$
$$Y^2(y,t) = b^2(y + 6at + y_0)^2 + b^{-2}, \quad Y > 0.$$

N-lump solutions of (1.1) also are available. Both aspects suggest that the asymptotic ($t \to \infty$) solution of the initial value problem of (1.1) may look much different from that of the KdV equation.

The first significant work on solving (1.1) as an initial value problem was done by Zakharov and Manakov (1979), with subsequent improvements by Manakov (1981). They developed an inverse scattering formalism to solve (1.1), beginning with (1.2) as a scattering problem, and ending with a linear integral equation of Gel'fand-Levitan-Marchenko-type. No indication was given of the class of initial data to which their methods apply, other than that $u(x,y; t = 0)$ must vanish rapidly as $x^2 + y^2 \to \infty$. Manakov, Santini and Takhtajan (1980) showed that no lumps ever evolve from initial data for which these methods are valid.

This last result is somewhat surprising, especially if one expects lumps to play as fundamental a role in this (2+1) dimensional problem as solitons play in (1+1) dimensional problems. The present work was motivated by the question: Were lumps excluded by some limitation of Manakov's method, or are they unstable in some sense?

To answer this, an inverse scattering method is given that differs in some respects from those of either Zakharov and Manakov (1979) or Manakov (1981). However, all three formulations depend on convergent Neumann series to solve (by iteration) the various integral equations that arise, so they all require "small enough" initial data. We show that lumps are excluded by this requirement of smallness; i. e., their absence from the initial value problem, as reported by Manakov, Santini and Takhtajan (1980), is due to a limitation of the method. No inference about nonlinear stability of the lumps can be drawn from this result.

Let us state precisely where the lumps were excluded, since it is the main point of the paper. Denote the initial data of (1.1) by u(x,y). We will repeatedly use Fourier transforms (in x only), and we define

$$\hat{u}(m,y) = \int dx\, u(x,y)\exp\{-imx\} ,$$

so that

$$u(x,y) = \frac{1}{2\pi}\int dm\, \hat{u}(m,y)\exp\{imx\} .$$

[Throughout this paper, integrals without limits are to be interpreted as ranging over $(-\infty, \infty)$]. We require that

$$U(\infty) = \frac{1}{2\pi}\int dy \int dm|\hat{u}(m,y)| < \infty . \tag{1.5}$$

This restriction seems to be fundamental. For a lump solution of (1.1) it follows from (1.4) that

$$\hat{u}(m,y) = 4\pi\, |m|\, \exp\{-|m|Y + imX\} , \tag{1.6}$$

and that

$$U(\infty) = 4\pi . \tag{1.7}$$

Lumps are quantized in this sense. Moreover, $U(\infty)$ provides a suitable measure of smallness. In particular, we will see that the appropriate eigenfunctions in the direct scattering problem are defined (by iteration) if

$$U(\infty) < 1 \ . \tag{1.8}$$

Taken together, (1.7) and (1.8) suggest that lumps are not small enough to be obtained by iteration, and that they will not evolve from initial data for which these iterative methods apply. Additional evidence for this conjecture is given below. The situation may be compared to that for the modified KdV equation,

$$v_t + 6v^2 v_x + v_{xxx} = 0 \ . \tag{1.9}$$

In that problem, every soliton satisfies

$$\int dx |v| = \pi \ ,$$

corresponding to (1.7). Ablowitz, Kaup, Newell and Segur (1974) show that the appropriate Gel'fand-Levitan equation can be solved by iteration if

$$\int dx |v| < 0.904$$

initially and that the solution that evolves from these initial data contains no solitons. For larger initial data the problem can be solved, but not by iteration.

Restrictions only on $U(\infty)$ are not sufficient to insure the well-posedness of (1.1). Another important restriction can almost be seen directly from (1.1), as noted by Ablowitz and Segur (1979). If we require that u and its derivatives vanish for all time as $x \to -\infty$, then (1.1) can be integrated to

$$u_t + 6uu_x + u_{xxx} = 3\partial_y^2 \int_{-\infty}^{x} u \, d\bar{x} \tag{1.10}$$

Even if u and its derivatives vanish initially as $x \to -\infty$, (1.10) shows that u_t does not vanish as $x \to +\infty$ unless

$$\int dx \, u(x,y,t) = A(t) y + B(t) . \tag{1.11}$$

The method given here requires a slightly stronger condition on the initial data:

$$\int dx \, u(x,y) = 0 . \tag{1.12}$$

Note that a lump solution of (1.1) satisfies (1.12).

A third restriction determines how well confined the initial data are. Eventually it is convenient to require

$$\bar{U} = \int dx \int dy \, |u(x,y)| < \infty . \tag{1.13}$$

If $u(x,y)$ represents the displacement of the surface of an incompressible fluid, then $[\int dx \int dy \, u(x,y)]$ represents the net mass of the wave, and $\bar{U}$ the total mass involved. Thus (1.13) seems quite plausible physically, although one lump violates it. A consequence of (1.13) is that necessarily $A = B = 0$ in (1.11) initially, so that (1.11) and (1.13) together imply (1.12).

Finally we should note that all questions of smoothness are ignored here.

2. DIRECT SCATTERING (AT $t = 0$) AND TIME DEPENDENCE

Because $u(x,y) \to 0$ as $y \to \pm\infty$, (1.2) admits solutions of the form

$$\psi(x,y;k) \sim \exp\{ikx - ik^2 y\}$$

in these limits. Let ψ_L and ψ_R denote two families of solutions of (1.2) defined by

$$\left. \begin{array}{c} \psi_L(x,y;k) \\ \psi_R(x,y;k) \end{array} \right\} \to \exp\{ikx - ik^2 y\} \text{ as } y \to \left\{ \begin{array}{c} -\infty \\ +\infty \end{array} \right. \tag{2.1}$$

Alternatively, we may define

$$\psi(x,y;k) = \mu(x,y;k) \exp\{ikx - ik^2y\} \quad , \tag{2.2}$$

so that (1.2) becomes

$$i\mu_y + \mu_{xx} + 2ik\mu_x + u\mu = 0 \quad . \tag{2.3}$$

Then μ_L, μ_R are defined by requiring that

$$\mu_L \to 1 \text{ as } y \to -\infty, \quad \mu_R \to 1 \text{ as } y \to +\infty \, .$$

Consider the linear integral equations:

$$\mu_L(x,y;k) = 1 + \frac{i}{2\pi} \int_{-\infty}^{y} d\eta \int d\chi \int dm \; \exp(i\phi) \; u(\chi,\eta) \; \mu_L(\chi,\eta;k), \tag{2.4a}$$

$$\mu_R(x,y;k) = 1 - \frac{i}{2\pi} \int_{y}^{\infty} d\eta \int d\chi \int dm \; \exp(i\phi) \; u(\chi,\eta) \; \mu_R(\chi,\eta;k), \tag{2.4b}$$

where

$$\phi = m(x - \chi) - m(m + 2k)(y-\eta) \, . \tag{2.4c}$$

It is easy to verify (by differentiating formally under the integral signs) that solutions of (2.4) also are formal solutions of (2.3). To solve (2.4) by iteration, let

$$\mu_L(x,y;k) = 1 + \sum_{1}^{\infty} \mu_{L,n}(x,y;k) \, ,$$

and assume (temporarily) that $\mu_{L,n}$ has a Fourier transform in x, $\hat{\mu}_{L,n}(m,y;k)$. From (2.4),

$$\hat{\mu}_{L,1}(m,y;k) = i\int_{-\infty}^{y} d\eta \ \exp\{-im(m + 2k)(y - \eta)\}\hat{u}(m,\eta) \quad (2.5a)$$

and for $n \geq 1$

$$\hat{\mu}_{L,n+1}(m,y;k) = \frac{i}{2\pi} \int_{-\infty}^{y} d\eta \ \exp\{-im(m + 2k)(y - \eta)\} \ *$$

$$* \int d\ell \ \hat{u}(m - \ell,\eta)\hat{\mu}_{L,n}(\ell,\eta;k) \ . \quad (2.5b)$$

Define

$$U(y) = \frac{1}{2\pi} \int_{-\infty}^{y} d\eta \int dm |\hat{u}(m,y)| \geq 0 \ . \quad (2.6)$$

It follows from (2.5) that

$$\frac{1}{2\pi} \int dm |\hat{\mu}_{L,1}(m,y;k)| < U(y) \ , \quad (2.7a)$$

and that

$$\frac{1}{2\pi} \int dm |\hat{\mu}_{L,n}| < \frac{|U(y)|^n}{n!} \ . \quad (2.7b)$$

But $|\mu_{L,n}| < \frac{1}{2\pi} \int dm |\hat{\mu}_{L,n}|$, so

$$|\mu_L(x,y;k)| < 1 + \sum_{1}^{\infty} \frac{[U(y)]^n}{n!} = \exp\{U(y)\} \ . \quad (2.8)$$

Result: i) If $U(\infty) < \infty$, then (2.4a) has a unique solution for all real (x,y,k). For each y, it is bounded by (2.8) uniformly in (x,k).

ii) Because $[\mu_L(x,y;k)-1]$ has a Fourier transform in x that is absolutely integrable (in m), $\mu_L(x,y;k)$ is continuous in x and $\mu_L \to 1$ as $|x| \to \infty$.

iii) Similar statements apply to $\mu_R(x,y;k)$, the solution of (2.4b). In particular, for all $(x,y;k)$,

$$|\mu_R(x,y;k)| \le 1 + \sum_1^\infty \frac{1}{2\pi}\int dm |\rho_{R,n}| < \exp\{U(\infty)\} \,. \qquad (2.9)$$

With more effort, we also may establish two more results.
i) If

$$\frac{1}{2\pi}\int dy \int dm\, m^2 |\hat{u}(m,y)| < \infty \qquad (2.10)$$

in addition to (1.5), then differentiation under the integrals is permitted in (2.4), and the solutions of (2.4) are pointwise solutions of (2.3). We will not enforce (2.10) because we are neglecting all questions of smoothness.

ii) The solutions of (2.4) are continuous functions of k.

The solutions of (2.4a,b) are not independent. Define a "scattering kernel",

$$S(k,k+m) = -\frac{i}{2\pi}\int d\eta \int d\chi \, \exp\{-im\chi + im(m+2k)\eta\}\, u(\chi,\eta)\mu_R(\chi,\eta;k). \qquad (2.11)$$

Then one may show by subtracting (2.4a) from (2.4b) that

$$\mu_R(x,y;k) = \mu_L(x,y;k) +$$

$$\int dm\ S(k,k+m)\mu_L(x,y;k+m)\exp\{imx - im(m+2k)y\} \qquad (2.12a)$$

In terms of ψ, we have

$$\psi_R(x,y;k) = \psi_L(x,y,k) + \int d\ell\ S(k,\ell)\ \psi_L(x,y;\ell) \ . \qquad (2.12b)$$

An inverse scattering theory requires a relation like (2.12) in which: (i) the scattering kernel evolves simply in time; (ii) the eigenfunctions involved are analytic functions of k in appropriate half-planes. Zakharov and Manakov (1979) showed that $S(k,\ell,t)$ does have simple time dependence. Because ψ_L and ψ_R simplify as $y \to \pm \infty$, one obtains from (1.3) that

$$S(k,\ell;t) = S(k,\ell)\ \exp\{4i(k^3 - \ell^3)t\} \qquad (2.13)$$

Unfortunately, μ_L, μ_R generally are not analytic functions of k. In fact, it is apparent from the kernels in (2.4) that the integrals there are defined only for real k if $u(x,y)$ is real, because $(y-\eta)$ is unbounded and (m) takes on both signs. Consequently, μ_L, μ_R are not appropriate eigenfunctions for the inverse problem if the initial data are real.

There is an exceptional class of initial data for which (2.12) is the basis of the inverse problem. We say that $u(x,y)$ is upper in x if it is analytic for Im x > 0, and lower in x if it is analytic in x for Im x < 0. If $u(x,y)$ satisfies (1.5) then it is upper (or lower) in x if

$$\hat{u}(m,y) = 0\ ,\quad m < 0\ (\text{or } m > 0).$$

Because $\hat{u}(m,y) = \hat{u}^*(-m,y)$ for real $u(x,y)$, it follows that nontrivial real functions satisfying (1.5) are neither upper nor lower in x. One shows easily from the preceding analysis that if $u(x,y)$ is upper in x, then $\mu_L(x,y;k)$ is upper in x and lower in k, while $\mu_R(x,y;k)$ is upper in x and upper in k. A similar statement holds if $u(x,y)$ is lower in x. For these initial data, (2.12) completes the direct scattering analysis. We shall not pursue this idea any further, however, because (1.1) requires real initial data for physical interest.

Henceforth we consider only real initial data. Following Manakov (1981), we define next another set of solutions of (2.3) that are analytic functions of k in half-planes. At t = 0, let $\mu^\uparrow$ and $\mu^\downarrow$ be defined by

$$\mu^\uparrow(x,y;k) = 1 - \frac{i}{2\pi}\int_y^\infty d\eta \int_0^\infty dm \int d\chi \, \exp\{i\phi\} u(\chi,\eta) \mu^\uparrow(\chi,\eta;k)$$

$$+ \frac{i}{2\pi}\int_{-\infty}^y d\eta \int_{-\infty}^0 dm \int d\chi \, \exp\{i\phi\} u(\chi,\eta) \mu^\uparrow(\chi,\eta;k), \qquad (2.14a)$$

$$\mu^\downarrow(x,y;k) = 1 - \frac{i}{2\pi}\int_y^\infty d\eta \int_{-\infty}^0 dm \int dx \, \exp\{i\phi\} u \mu^\downarrow +$$

$$+ \frac{i}{2\pi}\int_{-\infty}^y d\eta \int_0^\infty dm \int dx \, \exp\{i\phi\} u \mu^\downarrow , \qquad (2.14b)$$

where $\phi = m(x - \chi) - m(m + 2k)(y - \eta)$. As before, if differentiation under the integral sign is permitted, then solutions of (2.14) also solve (2.3).

We may solve (2.14) by iteration, as above. If $\mu^\uparrow = 1 + \sum_1^\infty \mu_n^\uparrow$, then it follows that

$$\frac{1}{2\pi}\int dm |\hat{\mu}_1^\uparrow(m,y;k)| \leq \frac{1}{2\pi}\left[\int_y^\infty d\eta \int_0^\infty dm|\hat{u}| + \int_{-\infty}^y d\eta \int_{-\infty}^0 dm|\hat{u}|\right]$$

Because u(x,y) is real,

$$\frac{1}{2\pi} \int dm |\hat{\mu}_1^\uparrow| \leq \frac{1}{2} U(\infty) \quad . \tag{2.15a}$$

Moreover,

$$\frac{1}{2\pi} \int dm |\hat{\mu}_n^\uparrow| \leq \frac{1}{2} \left[U(\infty) \right]^n \quad . \tag{2.15b}$$

Result:

i) If $U(\infty) < 1$, then for all real (x,y,k), (2.14a) has a unique solution that is uniformly bounded by

$$|\mu^\uparrow(x,y;k)| < 1 + \frac{1}{2} \sum_1^\infty [U(\infty)]^n = \frac{2 - U(\infty)}{2(1 - U(\infty))} \tag{2.16}$$

ii) $[\mu^\uparrow (x,y;k) -1]$ has a Fourier transform in x that is absolutely integrable. Therefore $\mu^\uparrow (x,y;k)$ is continuous in x and $\mu^\uparrow \to 1$ as $|x| \to \infty$.

iii) Similar statements hold for $\mu^\downarrow (x,y;k)$, the solution of (2.14b).

iv) $\mu^\uparrow$ may be extended to Im $(k) > 0$, and $\mu^\downarrow$ to Im $(k) < 0$.

The bounds in (2.15) are rather loose, so that although (1.8) is sufficient to solve (2.14) by iteration, it is probably smaller than necessary. On the other hand, it is evident from the Fredholm-nature of the integral equation in (2.14) that some finite bound is necessary if (2.14) is to be solved by iteration. It seems likely that lumps are excluded from the possible solutions of (1.1) at just this point, by requiring that (2.14) be solved by iteration.

Conversely, suppose $u(x,y)$ consists of exactly one lump, given by (1.4) at $t = 0$, so that (1.8) is violated. Ablowitz and Fokas (private communication) have shown that at $t = 0$,

$$\mu(x,y;k_\pm) = \frac{c_\pm}{x + Z} + \frac{d_\pm}{x + Z^*} \tag{2.17}$$

satisfies (2.3) and vanishes as $x^2 + y^2 \to \infty$, where

$$Z = X(y) + iY(y) ,$$

$$c_\pm = 1 \mp \frac{b^2(y + y_0) - i}{bY} , \quad d_\pm = c_\mp ,$$

$$2k_\pm = -a \pm ib .$$

Moreover, one also may show that $\mu(x,y;k_+)$ is a homogeneous solution of (2.14a), and that $\mu(x,y;k_-)$ is a homogeneous solution of (2.14b). Because (2.14) admits homogeneous solutions for these initial data, it follows that they cannot be solved by iteration. We anticipate that, in general, lumps may evolve from initial data satisfying (1.5) but not those satisfying (1.8)

Now return to the case when $U(\infty) < 1$. To show that $\mu^\uparrow(x,y;k)$ is analytic for Im k > 0, we differentiate (2.14a) in k, and observe that the homogeneous terms in the resulting equation are unchanged. The nonhomogeneous term is defined for Im k > 0, because if $m(y-\eta) < 0$, then

$$|2im(y - n)e^{-2ikm(y-n)}| \leq \left[\text{Im}(k)\, e\right]^{-1}$$

Thus $\mu^\uparrow$ is analytic in k if Im k > 0, and $\mu^\downarrow$ is analytic if Im k < 0.

The relation between $\mu^\uparrow(x,y;k)$ and $u(x,y)$ becomes simple as Im $(k) \to +\infty$. By integrating by parts in (2.14a), one may show that as Im $(k) \to +\infty$,

$$\mu^\uparrow(x,y;k) \sim 1 + \frac{1}{4\pi k} \int dm\, \frac{\hat{u}(m,y)e^{imx}}{m} + o(|k|^{-1}) \qquad (2.18)$$

Clearly, (2.18) is sensible only if $\hat{u}(0,y) = 0$, which is (1.12). Moreover, it shows that if we can recover $\mu^\uparrow$ from the scattering data in the inverse problem, then u may be otained from $\mu^\uparrow$, via (2.18).

The relation (2.18) also can be obtained formally from (2.3) as follows. Let $\mu = 1 + \nu$, so that (2.3) becomes

$$i\nu_y + \nu_{xx} + 2ik\nu_x + u\nu + u = 0 .$$

As Im $(k) \to \infty$, either $\nu \to 0$ or $\nu = O(1)$. The latter case yields only a trivial result. If $\nu \to 0$ along with its derivatives, then

$$2ik\nu_x + u \sim 0 ,$$

which is equivalent to (2.18).

It remains to find a relation between $\mu\uparrow$ and $\mu\downarrow$ of the form

$$\mu\uparrow(x,y;k) - \mu\downarrow(x,y;k) =$$

$$\int d\ell\, F(k,\ell)\, \mu\downarrow(x,y;\ell)\, \exp\{i(\ell-k)x - i(\ell^2 - k^2)y\} , \qquad (2.19a)$$

or equivalently,

$$\psi\uparrow(x,y;k) = \psi\downarrow(x,y;k) + \int d\ell\, F(k,\ell)\psi\downarrow(x,y;\ell) . \qquad (2.19b)$$

Here $F(k,\ell)$ must be well defined at $t = 0$, and should evolve simply in time. Subtracting (2.14a,b) does not give (2.19) directly, but it identifies

$$T(k,k+m) = -\frac{i}{2\pi}\, \text{sgn}(m) \int d\eta \int d\chi\, \exp\{-im\chi + im(m+2k)\eta\}u(\chi,\eta)\mu\uparrow(\chi,\eta;k) \qquad (2.20a)$$

and

$$T_1(k,k+m) = \frac{i}{2\pi}\int d\eta \int d\chi\, \exp\{-im\chi + im(m+2k)\eta\}u(\chi,\eta)\mu\downarrow(\chi,\eta;k) \qquad (2.20b)$$

It is not difficult to show that both T and T_1 are absolutely integrable in ℓ. Now $F(k,\ell)$ may be found in terms of $T(k,\ell)$ as follows:

i) Assume (2.19a) with $\int d\ell \ |F(k,\ell)| < \infty$;

ii) Subtract (2.14a,b) and substitute (2.19) in two places;

iii) Rewrite (2.14b) for $\mu\!\!\downarrow\!(x,y;\ell)$, multiply by $F(k,\ell)^*$ $\exp\{i(\ell-k)x - i(\ell^2 - k^2)y\}$, and integrate in ℓ;

iv) Subtract (iii) from (ii).

The result shows that a certain absolutely integrable function has a vanishing Fourier Transform, so the function itself must vanish. This vanishing function defines $F(k,\ell)$:

$\underline{k > \ell}$

$$F(k,\ell) - T(k,\ell) - \int_{-\infty}^{\ell} dp\ F(k,p)\ T_1(p,\ell) = 0; \qquad (2.21a)$$

$\underline{k < \ell}$

$$F(k,\ell) - T(k,\ell) + \int_{\ell}^{\infty} dp\ F(k,p)\ T_1(p,\ell) = 0. \qquad (2.21b)$$

One may show directly from (2.21) that $\int d\ell |F(k,\ell)| < \infty$ if $\int d\ell |T_1(p,\ell)| < 1$. But one also may show that

$$\int d\ell |T_1(p,\ell)| \leq \frac{U_\infty(2 - U_\infty)}{2(1 - U_\infty)}\ .$$

<u>Result:</u>

i) For fixed k, $F(k,\ell)$ is defined by (2.21) if $U_\infty < 2 - \sqrt{2}$. $F(k,\ell)$ is absolutely integrable in ℓ for every k.

ii) If (1.13) holds as well, then $F(k,\ell)$ is defined pointwise by (2.21).

iii) Given $F(k,\ell)$, defined by (2.21), $\psi\!\!\uparrow$ and $\psi\!\!\downarrow$ are related by (2.19).

This completes the direct scattering problem at $t = 0$.
For $t > 0$, we find first the time dependence of T and T_1, then that of $F(k,\ell;t)$. Zakarov and Manakov (1979) noted that if $u(x,y)$ is real and ψ_1, ψ_2 are any two solutions of (1.3), then

$$i\partial_y[\psi_1(\psi_2)^*] + \partial_x[(\psi_1)_x(\psi_2)^* - \psi_1(\psi_2)^*_x] = 0 .\qquad (2.22)$$

Thus, if the boundary terms vanish, then $\int dx[\psi_1(\psi_2)^*]$ is y-independent. It happens that the boundary terms do not vanish for any of (ψ_L, ψ_R, $\psi^\dagger$, $\psi^\dagger$), but we may subtract off their contributions to obtain simple results. Thus as $y \to +\infty$,
$\psi_R(x,y;\ell) \sim \exp\{i\ell x - i\ell^2 y\}$, while $\psi^\dagger(x,y;k)$ is defined by the limiting form of (2.14a). The result is that

$$\int dx\left[\psi^\dagger(x,y;k)\{\psi_R(x,y;\ell)\}^* - e^{i(k-\ell)x - i(k^2-\ell^2)y}\right]$$

$$= 2\,\theta(k - \ell)\, T(k,\ell) ,\qquad (2.23)$$

where $\theta(z) = \begin{cases} 1 & z > 0 \\ 0 & z < 0 \end{cases}$.

Similarly, by evaluating as $y \to -\infty$, we find

$$\int dx\left[\psi^\dagger(x,y;k)\{\psi(x,y;\ell)\}^* - e^{i(k-\ell)x - i(k^2-\ell^2)y}\right]$$

$$= 2\,\theta(\ell - k)\, T(k,\ell) ,\qquad (2.24)$$

Finally we compute $\int dx[\psi^\dagger(x,y;k)\{\psi_R(x,y;\ell)-\psi_L(x,y;\ell)\}^*]$, using (2.12). The result is the desired relation between T and S,

$$T(k,\ell)\,\mathrm{sgn}(k - \ell) = S^*(\ell,k) + \int_0^\infty dm\, S^*(\ell,k + m) T(k,k + m).\qquad (2.25)$$

At $t = 0$, $S^*(\ell,k)$ is defined through (2.11), $T(k,\ell)$ is defined by (2.20), and (2.25) is an identity, valid for any legitimate $u(x,y)$. If we expand both μ_R and $\mu^\dagger$ in their appropriate series (i.e., $\mu = 1 + \Sigma \mu_n$) then (2.25) becomes an infinite sequence of identities, ordered in powers of $u(x,y)$. As an independent test of (2.25), I have verified the validity of these identities at the first two powers of u.

For $t \geq 0$, let $S^*(\ell,k,t)$ evolve in accord with (2.13). It is evident that all time-dependence drops out of (2.25) if

$$T(k,\ell,t) = T(k,\ell) \exp\{4i(k^3 - \ell^3)t\} . \qquad (2.26)$$

Moreover, (2.26) is the only possibility if (2.25) has a unique solution, which can be assured by requiring (1.13) and by further restricting $U(\infty)$. By a similar argument,

$$T_1(k,\ell,t) = T_1(k,\ell) \exp\{4i(k^3 - \ell^3)t\}. \qquad (2.27)$$

Finally, given (2.26) and (2.27), all time dependence drop out of (2.21) if

$$F(k,\ell,t) = F(k,\ell) \exp\{4i(k^3 - \ell^3)t\} . \qquad (2.28)$$

Thus all of the scattering kernels (S, T, T_1, F) have the same simple time dependence. This completes the direct scattering analysis.

3. INVERSE SCATTERING FOR $t \geq 0$

In the inverse problem, the time (t) is arbitrary and fixed, and $F(k,\ell,t)$ is given in terms of fixed initial conditions, $u(x,y)$, via (2.21) and (2.28). Because $u(x,y,t)$ can be recovered from $\mu^\dagger$ via (2.18) or (2.3), the main problem is to recover $\mu^\dagger(x,y;k,t)$ from $F(k,\ell,t)$ via (2.19).

A formal procedure to solve (2.19) was given by Manakov (1981), who assumed "triangular representations" of the form

$$\psi\downarrow(x,y;k) = \exp\{ikx-ik^2y\} + \int_{-\infty}^{x} dz\, K(x,z,y)\, \exp\{ikz-ik^2y\} \tag{3.1}$$

with a similar expression for $\psi\uparrow$. If such a representation exists, then (2.19b) can be reduced to a linear integral equation of Gel'fand-Levitan-Manchenko type. Moreover, substituting (3.1) into (1.2) shows that if K exists, then

$$u(x,y) = -2\frac{d}{dx}K(x,x,y) \tag{3.2}$$

so that (2.18) is no longer needed. This procedure may be justified by answering two questions:

i) Does such a representation exist? In their original paper, Gel'fand and Levitan (1951) show explicitly that their kernel, corresponding to K in (3.1), exists using the theory of hyperbolic differential equations. No such proof is given by Manakov.

ii) Are further restrictions on $u(x,y)$ necessary to assure that the Gel'fand-Levitan equation has a unique solution?

Further analysis of the inverse problem will be deferred to a later paper in which the initial data need not be small, and lumps are not excluded a priori.

Acknowledgements: The author is grateful to M. J. Ablowitz and A. S. Fokas for very useful conversations, including the discovery of an error in an earlier version of the manuscript. This work was supported by the U. S. Army Research Office.

References

Ablowitz, M.J., D.J. Kaup, A.C. Newell and H. Segur, Stud. App. Math., Vol. 53, p. 249, 1974.

Ablowitz, M.J. and J. Satsuma, J. Math Phys., Vol. 19, p. 2180, 1978.

Ablowitz, M.J., and H. Segur, J. Fluid. Mech., Vol. 92, p. 691, 1979.

Bordag, L., A. Its, S. Manakov, V. Matveev and V. Zakharov, Phys. Lett. Vol. 63A, p. 205, 1977.

Dryuma, V.S., Sov. Phys. JETP Lett., Vol. 19, p.387, 1974.

Gel'fand, I.M. and B.M. Levitan, Izv. Akad. Nauk SSSR, Ser. Mat. 15, p. 309, 1951; translated in Amer. Math Soc. Trans., Ser. 2, Vol. 2, p. 259, 1955.

Kadomtsev, B.B. and V.I. Petviashvili, Sov. Phys. Doklady, Vol. 15, 539, 1970

Manakov,, S.V., Physica, Vol. 3D, p. 420, 1981.

Manakov, S.V., P. Santini and L. Takhtajan, Phys. Lett., Vol. 75a, p. 451, 1980.

Muskhelishvili, N.I., Singular Integral Equations, P. Noordhoff-Holland, 1953.

Zakharov, V.E. and S.V. Manakov, Sov. Sci. Rev. Phys. Rev., Vol. 1, p. 133, 1979

A DIRECT LINEARIZATION ASSOCIATED WITH THE BENJAMIN-ONO EQUATION

M.J. Ablowitz and A.S. Fokas
Department of Mathematics and Computer Science
Clarkson College of Technology
Potsdam, New York 13676

1. INTRODUCTION

The Benjamin-Ono (BO) equation [1] is a nonlinear singular-integro-differential equation which describes long internal gravity waves in stratified fluid. It has been established that the BO equation possesses: a) N soliton solutions [2], [3]; b) Bäcklund transformations, conservation laws, and a novel Lax pair [4], [5]; c) two non-local linear operators which generate its infinitely many commuting symmetries and constants of motion in involution [6].

In this note we outline a method of solution for the initial value problem of the BO equation, which we take in the form

$$u_t + 2uu_x + Hu_{xx} = 0; \quad Hv(x) = \frac{1}{\pi} \int_{-\infty}^{\infty} \frac{v(\xi)}{\xi - x} d\xi, \quad (1)$$

where H denotes the Hilbert transform and principal value integrals are assumed if needed. We make use of the Lax pair

$$i\Phi_x^+ + \lambda(\Phi^+ - \Phi^-) = -u\Phi^+ \quad (2)$$

$$i\Phi_t^\pm - 2i\lambda \Phi_x^\pm + \Phi_{xx}^\pm - 2i[u]_x^\pm \Phi^\pm = -\nu\Phi^\pm, \quad (3)^\pm$$

where $\Phi^+(x,t;\lambda)(\Phi^-)$ is the limit of a function analytic in the upper (lower) half z-plane as $z \to x$ (z is the complex extension of x); similarly $[u]^+$, $[u]^-$ are the (+) and (−) parts of $\underline{u}(x,t)$ respectively, i.e. $u = [u]^+ - [u]^-$, where $[u]^+$, $[u]^-$ are analytic in the upper and lower half z-plane respectively (λ is constant and is interpreted as a spectral parameter, ν is an arbitrary constant).

Our method differs substantially from the inverse scattering transform method as applied for example to the Korteweg-deVries equation. We think that the "inverse problem" associated with (2) is not solvable (in the usual sense) and, as a result of this, equations $(3)^\pm$ (i.e. the time-part of the Lax pair) play now a fundamental role.

The method we propose for linearizing (1) consists essentially of the following steps (we only state the results for the (+) functions, since the results for the (−) functions are similar):

a) Use equation (2) to express $[u]^+$ in terms of Φ^+ and the "reflection coefficient".

b) Use equation (3) to find how the "reflection coefficient" evolves in time and then substitute the expression obtained in a) in equation (3), to obtain a nonlinear equation for Φ^+.

c) Linearize the above equation for Φ^+ to its linear part, either by an explicit transformation (in the case of solitons only) or by a linear integral equation (in the general case).

2. THE x-EIGENVALUE PROBLEM

Equation (2) should be interpreted as a <u>differential Riemann-Hilbert problem</u> for the analytic functions $\Phi^\pm(z,t,\lambda)$. Equation (2) describes the jump condition across the real axis x; it yields unique solutions for $\Phi^\pm$ provided one imposes some boundary conditions as $z\to\infty$, say, in the upper half-plane. Here we assume that either $\Phi^+(z,t,\lambda)\to 0$, or 1, as $z\to\infty$, $IM(z) > 0$.

i) <u>Left and right eigenfunctions</u>

Consideration of the equation (2) with the above boundary conditions yields that: a) there exist continuous eigenfunctions $\Phi^\pm(x,t,\lambda)$, where λ is real and positive, and these eigenfunctions satisfy Fredholm equations of the second type; b) there exist discrete eigenfunctions $\Phi^\pm(x,t,\lambda_j)$, where λ_j are real and negative, and these eigenfunctions satisfy homogeneous Fredholm equations. More specifically, let us consider only the (+) functions and let $M,\overline{M}$ denote "left eigenfunctions, while $\overline{N},N$ denote "right" eigenfunctions. These eigenfunctions are specified by the following asymptotic behavior.

$$\begin{aligned} M &\to 1 \\ \overline{M} &\to e^{i\lambda x}; \quad x\to-\infty \end{aligned} \quad (4)$$

$$\begin{aligned} \overline{N} &\to 1 \\ N &\to e^{i\lambda x}; \quad x\to+\infty \end{aligned} \quad (5)$$

Furthermore, let Φ_j denote the descrete (+) eigenfunctions. Then

$$\begin{pmatrix} M(x,t,\lambda) \\ \overline{M}(x,t,\lambda) \end{pmatrix} = \begin{pmatrix} 1 \\ e^{i\lambda x} \end{pmatrix} + \int_{-\infty}^{\infty} G^+(x,y,\lambda)u(y,t) \begin{pmatrix} M(y,t,\lambda) \\ \overline{M}(y,t,\lambda) \end{pmatrix} dy, \quad (6)$$

$$\begin{pmatrix} \overline{N}(x,t,\lambda) \\ N(x,t,\lambda) \end{pmatrix} = \begin{pmatrix} 1 \\ e^{i\lambda x} \end{pmatrix} + \int_{-\infty}^{\infty} G^-(x,y,\lambda)u(y,t) \begin{pmatrix} \overline{N}(y,t,\lambda) \\ N(y,t,\lambda) \end{pmatrix} dy, \quad (7)$$

$$\Phi_j(x,t) = \int_{-\infty}^{\infty} G(x,y,\lambda_j) u(y,t) \Phi_j(y,t) dy. \qquad (8)$$

In the above expressions G^+, G^- are the (+) and (-) parts of the sectionally holomorphic function

$$G(x,y,\zeta) = \frac{1}{2\pi} \int_0^{\infty} \frac{e^{i(x-y)p}}{p-\zeta} dp, \qquad (9)$$

where ζ denotes the complex extension of λ, i.e.

$$G^{\pm}(x,y,\lambda) = \lim_{\varepsilon \to 0} \frac{1}{2\pi} \int_0^{\infty} \frac{e^{i(x-y)p}}{p-(\lambda \pm i\varepsilon)} dp \qquad (10)$$

Equations (6) and (7) can also be obtained from the corresponding equations associated with the intermediate long wave equation [7] in the appropriate limit.

The line of discontinuity of G is given by the positive λ axis, hence using Plemelj's formulae [8]

$$G^+ - G^- = \begin{cases} ie^{i(x-y)\lambda} & \lambda > 0 \\ 0 & \lambda < 0 \end{cases} \qquad (11)$$

In particular for the discrete eigenfunctions λ_j, $G^+ = G^- = G(x,y,\lambda_j)$.

ii) <u>The relationship between left and right eigenfunctions</u>

There exists the following relationship between left and right eigenfunctions

$$M = \bar{N} + \beta(\lambda,t) N; \lambda > 0$$

$$M = \bar{N} \qquad ; \lambda < 0 \qquad (12)$$

where

$$\beta(\lambda,t) = i\int_{-\infty}^{\infty} u(y,t) M(y,t,\lambda) e^{-i\lambda y} dy$$

PROOF

Let $\Delta \doteq M - \bar{N}$. Then using (6), (7)

$$\Delta = 1 + \int_{-\infty}^{\infty} G^+ uM\,dy - 1 - \int_{-\infty}^{\infty} G^- u\bar{N}\,dy =$$

$$= \int_{-\infty}^{\infty} G^+ uM\,dy - \int_{-\infty}^{\infty} G^- u(M-\Delta)\,dy.$$

Hence

$$\Delta - \int_{-\infty}^{\infty} G^- u\Delta\,dy = ie^{i\lambda x} \int_{-\infty}^{\infty} e^{-i\lambda y} uM\,dy, \qquad (13)$$

where we have used (11). Equation (13) implies that

$\Delta = \beta(\lambda, t) N$.

iii) Some analyticity properties

The results of Fredholm theory imply that M is a $(+)$ function in the ζ-plane, except for possible poles. These poles correspond to homogeneous solutions of equation (6). Hence

$$M = 1 + \sum_{1}^{P} \frac{c_j \phi_j}{\lambda - \lambda_j} + \tilde{M}; \quad \tilde{M} \text{ is a } (+) \text{ function in } \lambda. \qquad (14)$$

Similarly

$$\bar{N} = 1 + \sum_{1}^{P} \frac{c_j \phi_j}{\lambda - \lambda_j} + \tilde{\bar{N}}; \quad \tilde{\bar{N}} \text{ is a } (-) \text{ function in } \lambda. \qquad (15)$$

The above representations, together with knowledge of β, are inadequate for solving the "inverse problem" associated with (12). However, using equations (12), (14), (15) one obtains the following important relationship.

$$[u]^+ = \frac{1}{2\pi i} \int_0^\infty \beta(\lambda,t) N(x,t,\lambda)\,d\lambda - \sum_1^P c_j(t) \phi_j(x,t) \qquad (16)$$

iv) The time evolution

Using equation (12) in $(3)^+$ one easily establishes that

$$c_j(t) = \text{constants}, \quad \beta(\lambda,t) = \beta_0(\lambda) e^{i\lambda^2 t}, \qquad (17)$$

$$\lambda_j(t) = \text{constants}.$$

3. THE t-PART OF THE LAX PAIR

i) Solitons

Let $\beta_o = 0$. Then substituting (16) into (3) one obtains the following system of coupled nonlinear PDE's:

$$\Phi_{jt} - 2\lambda_j \Phi_{jx} - i\Phi_{jxx} + 2(\sum_{1}^{P} c_\ell \Phi_\ell)_x \Phi_j = 0. \qquad (18)$$

If one assumes that

$$\Phi_j(x,t) = \sum_{1}^{P} \frac{c_{kj}(t)}{x - x_k(t)},$$

one is led to the Calogero-Moser system [9]

$$\ddot{x}_k = 8 \sum_{\ell=1}^{P} \frac{1}{(x_k - x_\ell)^3}, \qquad k = 1, \ldots, P.$$

Hence, using (16) (and its complex conjugate) one recovers the well known P-soliton solution of the BO equation. Thus, the P-solitons correspond to the discrete spectrum of the eigenvalue equation (2).

Recently [10] we have been able to linearize equations (18): the transformations

$$\Phi_j(x,t) = V_j(x,t) - i\sum_{1}^{P} c_\ell \Phi_\ell(x,t) W_{j\ell}(x,t), \qquad (19)$$

where

$$W_{j\ell} = e^{i(\lambda_j - \lambda_\ell)x} \int_a^x V_j(\xi,t) e^{-i(\lambda_j - \lambda_\ell)\xi} d\xi \qquad (20)$$

$$+ B(t) e^{i(\lambda_j - \lambda_\ell)(x-a)},$$

$$B_t + i(\lambda_\ell^2 - \lambda_j^2) B = (\lambda_\ell + \lambda_j) V(a,t,\lambda) \qquad (21)$$

$$+ iV_x(a,t,\lambda), \text{ a arbitrary,}$$

relate equations (18) to the linear equations

$$V_{jt} - 2\lambda_j V_{jx} - iV_{jxx} = 0. \qquad (22)$$

ii) The general case
Substituting (16) into (3) one obtains

$$\hat{N}_t - 2\lambda \hat{N}_x - i\hat{N}_{xx} - 2[\frac{1}{2\pi i}\int_0^\infty \beta_o(\tau)\hat{N}d\tau - \sum_1^P c_\ell \Phi_\ell]_x \hat{N} = 0 \qquad (23)$$

$$\Phi_{jt} - 2\lambda j\Phi_{jx} - i\Phi_{jxx} - 2[\frac{1}{2\pi i}\int_0^\infty \beta_o(\tau)\hat{N}d\tau - \sum_1^P c_\ell \Phi_\ell]_x \Phi_j = 0, \qquad (24)$$

where

$$\hat{N} = Ne^{i\lambda^2 t}$$

The above equations can also be linearized to their linear parts with the aid of an integral equation [10]. This linearization can be used for solving the initial value problem of the BO equation. Here, we only prove this linearization (we actually prove a slightly stronger result than the one needed here). A complete discussion of the initial value problem is given in [10].

Theorem
Let $\psi(x,t,\lambda)$ be a solution of

$$\psi(x,t,\lambda) = V(x,t,\lambda) - i\int_L \psi(x,t,\ell)W(\lambda,\ell,x,t)d\rho(\ell), \qquad (25)$$

where $V(x,t,\lambda)$ and $W(\lambda,\ell,x,t)$ are defined by

$$\mathcal{L}(\lambda)V \doteq V_t - 2\lambda V_x - iV_{xx} = 0, \qquad (26)$$

$$W(\lambda,\ell,x,t) = e^{i(\lambda-\ell)x}\int_a^x V(\lambda,\xi,t)e^{-i(\lambda-\ell)\xi}d\xi \qquad (27)$$

$$+ B(t,\lambda,\ell)e^{i(\lambda-\ell)(x-a)},$$

$$B_t + i(\ell^2 - \lambda^2)B = (\lambda+\ell)V(a,t,\lambda) + iV_x(a,t,\lambda). \qquad (28)$$

Assume that the homogeneous equation corresponding to (25) has only the trivial solution and that one may interchange differentiations w.r.t. x,t and the integral along L. Then ψ also solves the non-linear equation

$$\Omega(\psi) = \psi_t - 2\lambda\psi_x - i\psi_{xx} + 2\left(\int_L \psi_x(\ell) d\rho(\ell)\right)\psi = 0. \quad (29)$$

PROOF

Let ψ satisfy (25). Then

$$\Omega(\psi)(\lambda) + i\int_L \Omega(\psi)(\ell)W(\lambda,\ell)d\rho(\ell) = \mathcal{L}(\lambda)V$$

$$-i\int_L \psi(\ell)\dot{\mathcal{L}}(\lambda)W(\lambda,\ell)d\rho(\ell)$$

$$-2\int_L \psi_x(\ell)[W_x(\lambda,\ell) + i(\ell-\lambda)W(\lambda,\ell) - V(\lambda)]d\rho(\ell) = 0.$$

Hence, $\Omega(\psi) = 0$ provided that

$$\mathcal{L}(\lambda)V = 0, \quad \dot{\mathcal{L}}(\lambda)W(\lambda,\ell) = 0, \quad W_x + i(\ell-\lambda)W - V = 0. \quad (30)$$

Equation (30.a) is equation (26) and equation (30.c) is equivalent to equation (27). To satisfy equation (30.b) use equation (27) and integration by parts, this readily yields equation (28).

It is clear that equations (23), (24) are a special case of (29) where $d\rho(\ell) = [\sum_1^P c_j \delta(\ell-\lambda_j) - \frac{1}{2\pi i}\beta_0(\ell)]d\ell$.

REFERENCES

1. i) T.B. Benjamin, J. Fluid Mech. 29, 559 (1967).
    ii) H. Ono, J. Phys. Soc. Japan, 39, 1082 (1975).
2. K. M. Case, Proc. Nat. Acad. Sc. U.S.A., 75, 3563 (1978); 76, 1 (1979).
3. H.H. Chen, Y.C. Lee, N.R. Pereira, Phys. Fluids 22, 187 (1979).
4. A. Nakamura, J. Phys. Soc. Jap. 47, 1701 (1979).
5. T.L. Boch, M.D. Kruskal, Phys. Lett. 74A, 173 (1979).
6. A.S. Fokas, B. Fuchssteiner, "The hierarchy of the Benjamin-Ono equation", to appear in Phys. Lett.
7. Y. Kodama, M.J. Ablowitz, J. Satsuma, Phys. Rev. Lett. 46, 687 (1981).
8. N.I. Muskhelishvili, "Singular Integral Equations", ed. by J. Radok (Noordhoff, Brouingen, 1953).
9. i) F. Calogero, J. Math. Phys. 12, 419 (1971).
    ii) J. Moser, Adv. in Math. 16, 197 (1975).
10. M.J. Ablowitz, A.S. Fokas, R.L. Anderson, "A Method of Solution of the Benjamin-Ono equation" (preprint).

Direct Linearizations of the
Korteweg-deVries Equations

by

A. S. Fokas and M. J. Ablowitz
Department of Mathematics and Computer Science
Clarkson College of Technology
Potsdam, New York 13676

1. INTRODUCTION

The use of the celebrated Gel'fand-Levitan-Marchenko (GLM) equation [1] for both obtaining potentials of the Schrödinger scattering problem as well as for solving the initial value problem of the Korteweg-deVries (KdV) equation [2] has been well established. However, in spite of its wide applicability the GLM equation has certain limitations. Namely it characterizes only those potentials-solutions which decay fast enough as $|x| \to \infty$ [3]. The following two well known examples readily illustrate the above point:

i) Consider the problem of finding solutions of the

KdV $u_t + 6uu_x + u_{xxx} = 0$ which decay like $O(\frac{1}{x^2})$ as $|x| \to \infty$. These solutions are outside the range of applicability of the GLM equation. Ablowitz and Cornille [4] analyzed such solutions by "perturbing" the GLM equation around $u_o = -2/x^2$. u_o is a solution of KdV (the degenerate solutions of this "perturbed" GLM equation are the so-called quasi-solitons). Similar "perturbed" GLM equations can be found by perturbing around any "natural state" of the KdV [5].

ii) Consider the problem of finding the self-similar solutions of the KdV (i.e. the solutions of KdV invariant under a scaling transformation). These solutions satisfy a third order ODE; this ODE is of the Painlevé type and its first integral is related, through a one to one map, to Painlevé II [6]. Using the ideas of Ablowitz and Segur [7] one may obtain, via the GLM equation (properly scaled), a one parameter family of solutions of this third order ODE. How can one obtain a three parameter family of solutions? It is quite clear that one needs a more general equation than the GLM equation.

Recently [8] we have proposed such a generalization to the GLM equation. Furthermore, we have used this generalized equation to characterize a three parameter family of solutions of the ODE mentioned in ii) above, through a system of Fredholm equations [8]. The sense in which our equation provides a generalization to the GLM equation can be best understood by recalling the "direct approach" method of Zakharov and Shabat [9]: These authors, bypassing the connection with inverse scattering and using certain linear operators, proved directly that solutions of the GLM equation (i.e. solutions of a linear integral equation of the Fredholm type) are also solutions of the KdV equation. Similarly, we have proved directly that solutions of a rather general linear integral equation (which in some cases is a singular integral equation) are also solutions of the KdV equation. The generality of our equation results from the fact that it involves an arbitrary measure-contour. Actually for a specific choice of the measure-contour it reduces to the so-called k-space equation recently introduced by Newton [10], and shown to be equivalent to the GLM equation.

It should be noted that in the linear limit our approach yields the general solution of the underlying linear equation $u_t + u_{xxx} = 0$ (this is a consequence of the so-called Ehrenpreis principle). This should be contrasted with the linear limit of the GLM equation which yield only those solutions of $u_t + u_{xxx} = 0$ which are obtained through the Fourier transform. In this sense, our equation can be thought of as the analogue of a generalized transform for solving a nonlinear PDE, in the same way that the GLM equation corresponds to the Fourier transform.

In this note we

a) present a direct proof of our new linearization and comment on its linear limit;

b) present the analogous generalization to the "perturbed" GLM equation discussed in i) above.

2. THE MAIN RESULTS

Theorem 1

Let $\phi(k,x,t)$ be a solution of

$$M\phi \doteq \phi(k;x,t) + ie^{i(kx+k^3t)} \int_L \frac{\phi(\ell;x,t)}{\ell+k} d\zeta(\ell) = e^{i(kx+k^3t)}. \quad (1)$$

Then

$$u = -\frac{\partial}{\partial x} \int_L \phi(k;x,t) d\zeta(k) \quad (2)$$

solves the KdV equation

$$u_t + 6uu_x + u_{xxx} = 0. \quad (3)$$

Before proving this result we make the following remarks.

REMARKS

1. The measure $d\zeta(\ell)$ and contour L above are quite arbitrary. The only assumption made is that equation (1) is well defined in the sense that one can interchange differentiations w.r.t. x and t and the integral along L.

2. If the nonlinearity is absent then equation (1) yields $\bar{\phi} = \exp(i(kx+k^3t))$. Hence equation (2) implies

$$\bar{u} = -\frac{\partial}{\partial x} \int_L e^{i(kx+k^3t)} d\zeta(k). \quad (4)$$

This is the general solution (Ehrenpreis principle) of the linear equation:

$$\bar{u}_t + \bar{u}_{xxx} = 0. \tag{5}$$

3. $\phi(k;x,t)$ is directly related to the Schrödinger eigenvalue problem

$$\phi_{xx} + u\phi - ik\phi_x = 0. \tag{6}$$

Let $\phi = \psi\exp(\frac{1}{2}(kx+k^3 t))$ and equation (6) reduces to the usual Schrödinger equation

$$\psi_{xx} + ((\tfrac{k}{2})^2 + u)\psi = 0. \tag{7}$$

4. For the proof, we also assume that the homogeneous equation corresponding to (1) has only the trivial solution.

PROOF

Proving that u as defined by (2) solves the KdV is equivalent to proving that ϕ satisfies

$$\Omega(\phi) \doteq \phi_t + \phi_{xxx} - 3(\int_L \phi_x(\ell) d\zeta(\ell))\phi_x = 0. \tag{8}$$

However, applying the operator $\partial_t + \partial_x^3$ to (1) one easily establishes that

$$M\Omega = 3k[i\phi_{xx} + k\phi_x - i(\int_L \phi_x(\ell) d\zeta(\ell))\phi] \doteq 3kw, \tag{9}$$

where the linear operator M is defined in (1). Similarly, applying the operator $i\partial_x^2 + k\partial_x$ to (1) one obtains

$$Mw = 0. \tag{10}$$

Hence, because of the assumption 4 above, $w = 0$ and hence equation (9) implies $\Omega = 0$.

Particular choices of the measure-contour in (1) yield [8] i) a three parameter family of solutions of the ODE mentioned in the introduction; ii) the class of solutions obtainable through the usual inverse scattering transform method.

We now present an example of another linearization of KdV corresponding to the "perturbed" GLM equation discussed in the introduction.

Theorem 2

Let $\psi(k;x,t)$ be a solution of

$$\psi(k;x,t) + ie^{i(kx+k^3t)} \int_L \frac{(1+\frac{2i}{\ell x}+\frac{2i}{kx})}{k+\ell} \psi(\ell;x,t) d\zeta(\ell)$$

$$= e^{i(kx+k^3t)}(1+\frac{2i}{kx}). \tag{11}$$

Then

$$u = -\frac{2}{x^2} - \frac{\partial}{\partial x} \int_L \psi(k,x,t)(1+\frac{2i}{kx}) d\zeta(k) \tag{12}$$

solves the KdV equation (3).

Acknowledgements:

This research was partially supported by the Air Force of Scientific Research, under Grant No. 78-3674-C and the Office of Naval Research under Grant No. N00014-76-C0867.

REFERENCES

[1] M. J. Ablowitz and H. Segur, "Solitons and the Inverse Scattering Transform", SIAM Monograph, December, 1981.
[2] C. S. Gardner, J. M. Greene, M. D. Kruskal and R. M. Miura, Phys. Rev. Lett. $\underline{19}$, 1095 (1967).
[3] P. Deift and E. Trubowitz, Comm. Pure Appl. Math., $\underline{32}$, 121 (1979).
[4] M. J. Ablowitz and H. Cornille, Phys. Lett. $\underline{72A}$, 277 (1979).
[5] M. J. Ablowitz and H. Airault, C. R. Acad. Sc. Paris, 292 (1981).
[6] A. S. Fokas and M. J. Ablowitz, "On a unified approach to transformations and elementary solutions of Painlevé equations", to appear in J. Math. Phys., December, 1981.
[7] M. J. Ablowitz and H. Segur, Phys. Rev. Lett. $\underline{38}$, 1103 (1977).
[8] A. S. Fokas and M. J. Ablowitz, Phys. Rev. Lett. $\underline{47}$, 1096 (1981).
[9] V. E. Zakharov and A. B. Shabat, Funkts. Anal. Pril. $\underline{8}$, 43 (1974), [Funct. Anal. Appl. $\underline{8}$, 226 (1974)].
[10] R. G. Newton, J. Math. Phys. $\underline{21}$, 493 (1980); Geophys. J. R. Astr. Soc. 65, 191 (1981).

ANALYTIC STRUCTURE OF THE HENON-HEILES SYSTEM*

John Weiss
Center for Studies of Nonlinear Dynamics
La Jolla Institute
La Jolla, California 92038

I. INTRODUCTION

The Henon-Heiles System originally developed to model the structure of a spiral galaxy,[1] is an oft, and perhaps over-studied example of a simple dynamical system that can display either smooth or chaotic behavior, depending on the initial conditions and other parameters of the system. In this paper we shall investigate several aspects of the "analytic structure" of the Henon-Heiles System and seek to relate this structure to the behavior of the solution when observed at real times. By "analytic structure" we mean the structure of the set of "movable" singularities for the system[2]; (a movable singularity is a singularity whose position depends on the initial conditions of the problem and is, hence, variable). This approach to the study of dynamical systems was initiated by S. Kowalevskaya in 1890.[3,10] By finding all the occurrences of single-valued movable singularities for the rigid body problem, she was able to identify a new integral of motion and integrate the equations of motion for a particular specialization of the parameters. Recently, there have been several investigations[4,5] concerning the relationship of this Painlevé property (single-valued movable singularities) to the integrability of various dynamical systems. For instance by restricting the coefficients of the Henon-Heiles System so that the movable singularities are Painlevé, it is possible to identify all the known parameter values for which the system is integrable.

Herein, we find that the movable singularities of the Henon-Heiles System are, in general, infinitely multiple-valued. The formal expansion of the solution about a singularity is developed, and it is shown that a <u>closed</u> subset of the recursion relations defined by this expansion is associated with the asymptotic behavior of the solution near a singularity. From this asymptotic form of the solution it is possible to deduce several consequences of the multiple-valuedness of the expansion about a singularity. In particular, when the expansion includes terms of the form: $(t-t_0)^\alpha$ where α is a <u>complex</u> number it is possible to predict the existence and self-similar structure of a natural boundary.[6]

In Sec. II the types of singularities that occur in the Henon-Heiles system are presented.

In Sec. III, we derive the recursion relations for the expansion about a singularity and find the asymptotic form of the singular set, in a certain parameter range, to consist of a double, equiangular spiral of singularities about <u>every</u> singularity. The set of singularities, in this parameter range, appears to be a perfect set contained in an infinitely branched Riemann surface. It has the general appearance of a self-similar, natural boundary for the solution.

In Sec. IV we examine, for a different range of parameters, the appearance of "general" and "singular" forms of the solution. (A "singular" form of solution depends on less parameters than the dimensionality of the system). The singularities of the "general" solutions are found to be finite degree branch points for a dense subset of the parameter range considered. Surface of section calculations reveal a somewhat tenuous link between the structure of this subset and the appearance of widespread chaos in this parameter range.

In Sec. V, the Henon-Heiles is reformulated by means of a Ricati-type transformation into a system of three first-order equations. The structure of this system is investigated. Numerical investigations suggest that this system has several remarkable properties which might be useful in understanding the underlying Henon-Heiles dynamics.

Finally, the results described in this paper were obtained through the collective efforts of several persons. To my co-workers, Michael Tabor, Y. F. Chang, and John M. Greene, I wish to express my thanks for a stimulating and enjoyable collaboration.

II. TYPES OF SINGULARITIES FOR THE HENON-HEILES SYSTEM

The Henon-Heiles System is defined by a Hamiltonian:

$$H = \tfrac{1}{2}(Px^2+Py^2+Ax^2+By^2)+\lambda x^2 y - \frac{y^3}{3} .$$

The equations of motion are:

$$\ddot{x} = -Ax - 2\lambda xy$$
$$\ddot{y} = -By + y^2 - \lambda x^2 .$$
(2.1)

To find the types of singularities allowed by the system, we make the "leading order" anzatz's:

$$x \sim a(t-t_0)^\alpha$$
$$y \sim b(t-t_0)^\beta$$
(2.2)

By balancing "most singular" terms here are found to be two possibilities.

Case 1	Case 2
$\alpha = -2$	$\alpha = \tfrac{1}{2}(1 \pm \sqrt{1-48\lambda})$
$\beta = -2$	$\beta = -2$
	and
	$\mathrm{Re}(\alpha) > -2$

(2.3)

To determine the form of the expansion about a singularity, it is necessary to recognize the powers of $(t-t_0)$ at which the coefficient matrix of the recursion relations is singular and, therefore, when it is possible to introduce an arbitrary constant in the solution. Following the procedure presented in Ref. 7, we set:

$$x \sim a(t-t_o)^\alpha + p(t-t_o)^{r+\alpha}$$
$$y \sim b(t-t_o)^\beta + q(t-t_o)^{r+\beta} \quad ;$$
(2.4)

and then requiring that the determinant of terms linear in p and q vanish, determines r to be:

Case 1
$$r = -1, \frac{5}{2} \pm \tfrac{1}{2}\sqrt{1-24(1+1/\lambda)}, 6$$

Case 2
$$r = -1, 0, \pm \sqrt{1-48\lambda}, 6$$

We note that $r = -1$ corresponds to the arbitrary location of the singularity. For the terms $(t-t_o)^{r+\alpha}$, $(t-t_o)^{r+\beta}$ to appear in the expansion it is necessary that $\mathrm{Re}(r) \geq 0$, since (α, β) are assumed to represent the most singular terms in the expansion. We note that Case 2 singularities have two branches: the upper branch, $\alpha = \tfrac{1}{2}(1 + \sqrt{1-48\lambda})$; and the lower branch, $\alpha = \tfrac{1}{2}(1 - \sqrt{1-48\lambda})$. Using the above information, we represent the various types of singularities in Table I. Here (Complex, Real) denotes the type of resonance that occurs; (Four, Three), the number of arbitrary constants in the expansion and (Upper, Lower), the branch of Case 2 singularities.

We remark that four-parameter expansions might be regarded as the "general" form of the solution, while three-parameter solutions are "singular."[2]

Finally, when:

$$\lambda = 0, -1/6, -1$$

the leading orders and resonances are integers in both cases, and the solutions can be shown to have the Painlevé property,[2] (single-valued movable critical points). These three-parameter values represent all of the known instances for which the Henon-Heiles system is integrable. (See Ref. 4.) Recently, we have found[4] that when $\lambda = -1/16$ and $\beta = 16A$ in Eq. (2.1), the variables (x^2, y) are single-valued for both Case 1 and Case 2 singularities. This suggests that the Henon-Heiles system is integrable for this set of parameters and that there may exist other parameter values for which the same is true. We will consider this in more detail in Sec. IV.

TABLE I. Types of resonance and parametric dependence.

λ	Case 1	Case 2 Upper $\alpha_+=\tfrac{1}{2}(1+\sqrt{1-48\lambda})$	Lower $\alpha_-=\tfrac{1}{2}(1-\sqrt{1-48\lambda})$
$1/48<\lambda<\infty$	Complex Four	Complex Four	Complex Four
$0<\lambda<1/48$	Complex Four	Real Three	Real Four
$-\tfrac{1}{2}<\lambda<0$	Real Three	Real Three	Real Four
$-\tfrac{24}{23}<\lambda<-\tfrac{1}{2}$	Real Four	Real Three	Not Defined $\mathrm{Re}(\alpha-)<-2$
$-\infty<\lambda<-\tfrac{24}{23}$	Complex Four	Real Three	Not Defined

III. SELF-SIMILAR NATURAL BOUNDARIES: $1/48 < \lambda < \infty$

In this section, we will consider the occurrence of self-similar natural boundaries when $1/48 < \lambda < \infty$. Originally, studying numerically the analytic continuation of the Henon-Heiles System when $\lambda = 1$, we found the structure shown in Fig. 1. This was somewhat fortuitous since the singular set here posseses an unusual degree of symmetry. Indeed, the entire singular set appears to arise from a isoceles triangle construction with an angle of 25°. When λ varies from one, the structure undergoes various deformations that obscure the self-similar nature of the singular set. However, it is precisely the self-similar, geometrically contracting singular set that creates the natural boundary found when $1/48 < \lambda < \infty$. To see this, we next consider in detail the form of the expansion about a singularity. In the following, it will be assumed that $A = B = 1$.

The general expansion of the solution about a singularity when $\lambda > 1/48$ (set at $t_* = 0$) takes the form of a double series. For Case 1 leading orders,

$$x(t) = t^{-2} \sum_{k=0}^{\infty} \sum_{j=0}^{\infty} a_{kj} \tau^k t^j + t^{-2} \sum_{k=1}^{\infty} \sum_{j=0}^{\infty} \bar{a}_{kj} \bar{\tau}^k t^j , \quad (3.1a)$$

$$y(t) = t^{-2} \sum_{k=0}^{\infty} \sum_{j=0}^{\infty} b_{kj} \tau^k t^j + t^{-2} \sum_{k=1}^{\infty} \sum_{j=0}^{\infty} \bar{b}_{kj} \bar{\tau}^k t^j , \quad (3.1b)$$

where

$$\tau = t^\alpha , \quad \alpha = \frac{1}{2} + \frac{1}{2}\sqrt{1 - 24\left(\frac{1}{\lambda} + 1\right)} , \quad (3.1c)$$

$$\bar{\tau} = t^{\bar{\alpha}} , \quad \bar{\alpha} = \frac{1}{2} - \frac{1}{2}\sqrt{1 - 24\left(\frac{1}{\lambda} + 1\right)} , \quad (3.1d)$$

and

$$a_{00} = \frac{\pm 3}{\lambda}\sqrt{2 + 1/\lambda} , \quad b_{00} = \frac{-3}{\lambda} .$$

For <u>case 2</u> leading orders,

248

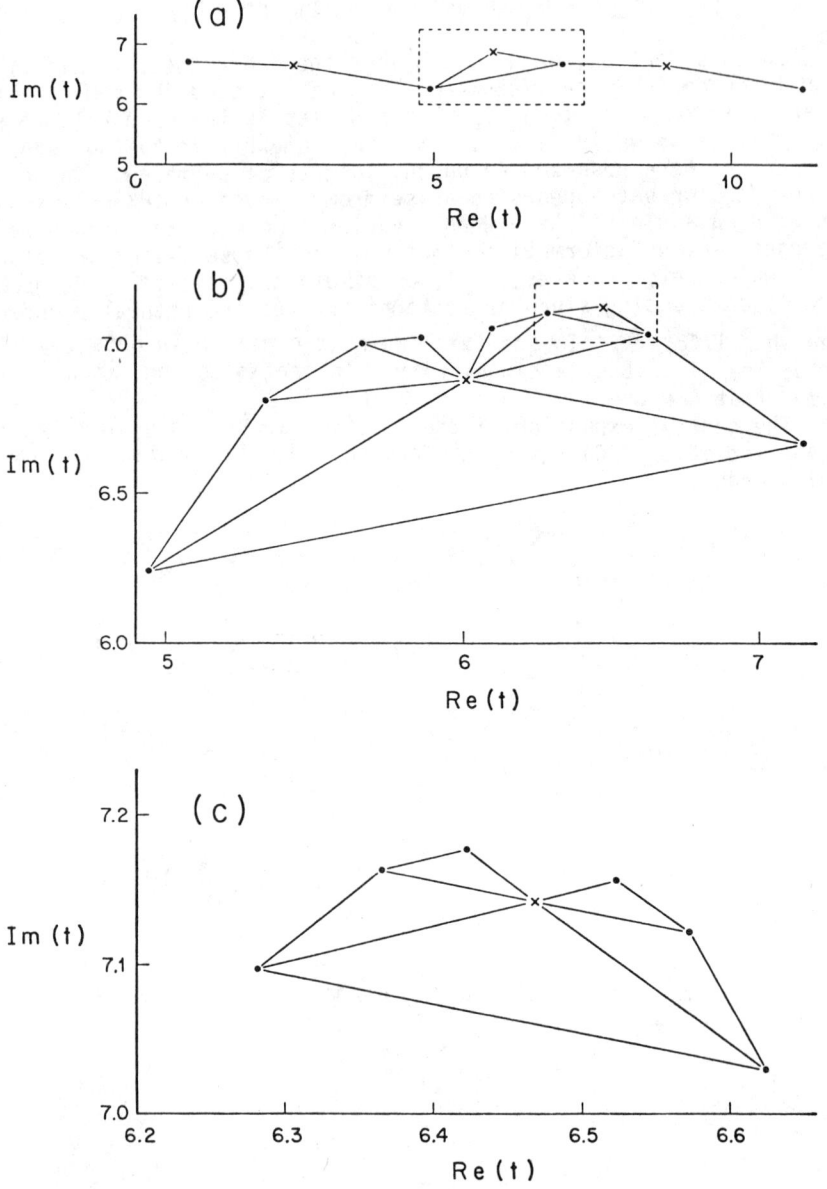

Fig. 1. Numerical calculation showing location of • case 1 and × case 2 singularities in the complex-time plane when λ = 1. View (b) shows boxed region in view (a). View (c) shows boxed region in view (b).

$$x(t) = \tau \sum_{k=0}^{\infty} \sum_{j=0}^{\infty} a_{kj} \tau^k t^j + \bar{\tau} \sum_{k=0}^{\infty} \sum_{j=0}^{\infty} \bar{a}_{kj} \bar{\tau}^k t^j \quad , \quad (3.2a)$$

$$y(t) = t^{-2} \sum_{k=0}^{\infty} \sum_{j=0}^{\infty} b_{kj} \tau^k t^j + t^{-2} \sum_{k=1}^{\infty} \sum_{j=0}^{\infty} \bar{b}_{kj} \bar{\tau}^k t^j \quad , \quad (3.2b)$$

where

$$\tau = t^{\alpha} \quad , \quad \alpha = \frac{1}{2} + \frac{1}{2} \sqrt{1 - 48\lambda} \quad (3.2c)$$

$$\bar{\tau} = t^{\bar{\alpha}} \quad , \quad \bar{\alpha} = \frac{1}{2} - \frac{1}{2} \sqrt{1 - 48\lambda} \quad (3.2d)$$

and

$$a_{00}, \bar{a}_{00} \text{ are arbitrary, } b_{00} = 6 \quad .$$

A. Recursion relations

Substitution of the series expansions for the Case 1 singularities into the equations of motion (3.2) leads (after much tedious manipulation) to the following set of recursion relations:

$$(\alpha k + j - 2)(\alpha k + j - 3) a_{kj} + a_{kj-2} + 2\lambda \sum_{\ell=0}^{k} \sum_{m=0}^{j} \bar{a}_{k-\ell, j-m} \bar{b}_{\ell m}$$

$$+ 2\lambda \sum_{n=1}^{j} \sum_{m=0}^{j-n} \left\{ a_{k+n,m} \bar{b}_{n,j-n-m} + b_{k+n,m} \bar{a}_{n,j-n-m} \right\} = 0 \quad , \quad (3.3a)$$

$$(\bar{\alpha} k + j - 2)(\bar{\alpha} k + j - 3) \bar{a}_{kj} + \bar{a}_{kj-2} + 2\lambda \sum_{\ell=0}^{k} \sum_{m=0}^{j} a_{k-\ell, j-m} \bar{b}_{\ell m}$$

$$+ 2\lambda \sum_{n=0}^{j} \sum_{m=0}^{j-n} \left\{ \bar{a}_{k+n,m} b_{n,j-n-m} + \bar{b}_{k+n,m} a_{n,j-n-m} \right\} = 0 \quad , \quad (3.3b)$$

$$(\alpha k + j - 2)(\alpha k + j - 3)b_{kj} + b_{kj-2} + \sum_{\ell=0}^{k}\sum_{m=0}^{j}$$

$$\left\{ \lambda a_{k-\ell,j-m} a_{\ell m} - b_{k-\ell,j-m} b_{\ell m} \right\}$$

$$+ \sum_{n=1}^{j}\sum_{m=0}^{j-n} \left\{ \lambda a_{k+n,m} \bar{a}_{n,j-n-m} - b_{k+n,m} \bar{b}_{n,j-n-m} \right\} = 0$$

(3.3c)

$$(\alpha k + j - 2)(\bar{\alpha} k + j - 3)\bar{b}_{kj} + \bar{b}_{kj-2} + \sum_{\ell=0}^{k}\sum_{m=0}^{j} \lambda \bar{a}_{k-\ell,j-m} \bar{a}_{\ell m}$$

$$- \bar{b}_{k-\ell,j-m} \bar{b}_{\ell m}$$

$$+ \sum_{n=0}^{j}\sum_{m=0}^{j-n} \left\{ \lambda \bar{a}_{k+n,m} a_{n,j-n-m} - \bar{b}_{k+n,m} b_{n,j-n-m} \right\} = 0 .$$

(3.3d)

It can be verified that the expansions defined by equations (3.7a-d) are consistent and well defined. The parameters of the expansion are found, in accordance with the resonance condition to be:

(i) $\quad b_{12} = \pm \sqrt{2 + 1/\lambda}\, a_{12}$, $\quad \bar{b}_{12} = \pm \sqrt{2 + 1/\lambda}\, \bar{a}_{12}$,

where a_{12} and $\bar{a}_{12}$ are arbitrary and

(ii) $\quad b_{06}$ is determined by $\quad \frac{42}{\lambda}\left(3 + \frac{1}{\lambda}\right) b_{06} = H$

(3.4)

where H is the Hamiltonian (3.1) i.e., the total energy.

By detailed consideration of the recursion relations, one finds that the non-zero coefficients form a certain pattern. This is shown in Fig. 2(a). Furthermore, one may show that the following sets of coefficients

$a_{j,2j}$ and $b_{j,2j}$ for $j = 0,1,2,...$

or

$\bar{a}_{j,2j}$ and $\bar{b}_{j,2j}$ for $j = 0,1,2$

defined a <u>closed</u> set of recursion relations. For example, setting

$$\theta_j = a_{j,2j} , \quad \psi_j = b_{j,2j} , \tag{3.5}$$

it is easy to show that

$$(\alpha j + 2j - 2)(\alpha j + 2j - 3)\theta_j + 2\lambda \sum_{m=0}^{j} \theta_{j-m}\psi_m = 0 ,$$

$$(\alpha j + 2j - 2)(\alpha j + 2j - 3)\psi_j + \sum_{m=0}^{j} \{\lambda\theta_{j-m}\theta_m - \psi_{j-m}\psi_m\} = 0 .$$
$$\tag{3.6}$$

As we shall see, these closed sets of relations may be used to study the asymptotic properties ($|t|<<1$) of the series expansions near a singularity.

• One may also derive the recursion relations for the expansions associated with the Case 2 singularities (3.2). These are:

$$[\alpha(k+1)+j][\alpha(k+1)+j-1]a_{kj} + a_{kj-2} + 2\lambda \sum_{\ell=0}^{k}\sum_{m=0}^{j} a_{k-\ell,j-m} b_{\ell m}$$
$$+ 2\lambda \sum_{n=1}^{j} \sum_{m=0}^{j-n} \{a_{k+n,m} \bar{b}_{n,j-n-m} + b_{k+n,m} \bar{a}_{n,j-n-m}\} = 0 , \tag{3.7a}$$

$$[\bar{\alpha}(k+1)+j][\bar{\alpha}(k+1)-j]\bar{a}_{kj} + \bar{a}_{k,j-2} + 2\lambda \sum_{\ell=0}^{k}\sum_{m=0}^{j} a_{k-\ell,j-m} b_{\ell m}$$
$$+ 2\lambda \sum_{n=0}^{j} \sum_{m=0}^{j-n} \{\bar{a}_{k+n,m} b_{n,j-n-m} + \bar{b}_{k+n,m} a_{n,j-n-m}\} = 0 \tag{3.7b}$$

$$(\alpha k + j - 2)(\alpha k + j - 3)b_{kj} + b_{kj-2} + \lambda \sum_{\ell=0}^{k-2} \sum_{m=0}^{j-4} a_{k-\ell-2,j-m-4} \, a_{\ell m}$$

$$- \sum_{\ell=0}^{k} \sum_{m=0}^{j} b_{k-\ell,j-m} \, b_{\ell m} + \lambda \sum_{n=0}^{j-5} \sum_{m=0}^{j-n-5} a_{k+n,m} \, \overline{a}_{n,j-n-m-5}$$

$$- \sum_{n=1}^{j} \sum_{m=0}^{j-n} b_{k+n,m} \, \overline{b}_{n,j-n-m} = 0 \, , \qquad (3.7c)$$

$$(\overline{\alpha} k + j - 2)(\overline{\alpha} k + j - 3)\overline{b}_{kj} + \overline{b}_{k,j-2} + \lambda \sum_{\ell=0}^{k-2} \sum_{m=0}^{j-4} \overline{a}_{k-\ell-2,j-m-4} \, \overline{a}_{\ell m}$$

$$- \sum_{\ell=0}^{k} \sum_{m=0}^{j} \overline{b}_{k-\ell,j-m} \overline{b}_{\ell m} + \lambda \sum_{n=0}^{j-5} \sum_{m=0}^{j-n-5} \overline{a}_{k+n,m} \, a_{n,j-n-m-5}$$

$$- \sum_{n=0}^{j} \sum_{m=0}^{j-n} \overline{b}_{k+n,m} \, b_{n,j-n-m} = 0 \, . \qquad (3.7d)$$

As before one can determine the parameters of this expansion. These are:

(i) a_{00} and $\overline{a}_{00}$ are arbitrary and

(ii) b_{06} is determined by $-84 b_{06} = H$. $\qquad (3.8)$

The pattern of non-vanishing coefficients is shown in Fig. 2(b). The closed sets of recursion relations are now associated with the coefficients

$$a_{2j,4j} \text{ and } b_{2j,4j} \quad \text{for} \quad j = 0, 1, 2, \ldots \, ,$$

or

$$\overline{a}_{2j,4j} \text{ and } \overline{b}_{2j,4j} \quad \text{for} \quad j = 0, 1, 2, \ldots \, .$$

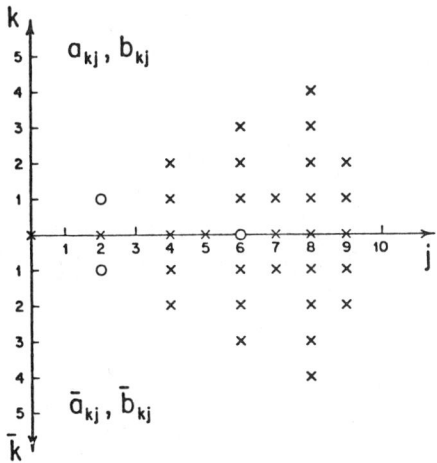

Fig. 2(a). Pattern of non-zero coefficients for the expansion about a case 1 singularity.

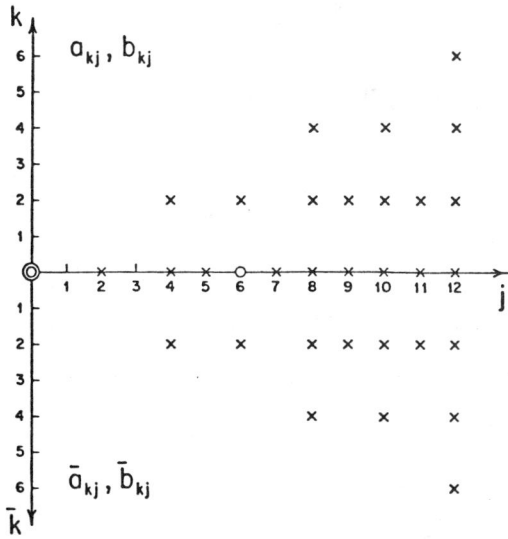

Fig. 2(b). Pattern of non-zero coefficients for the expansion about a case 2 singularity. We note that open circles denote the location of resonances.

For example, setting

$$\theta_j = a_{2j,4j}, \quad \psi_j = b_{2j,4j} \qquad (3.9)$$

one finds that

$$(\alpha(2j+1) + 4j)(\alpha(2j+1) + 4j - 1)\theta_j + 2\lambda \sum_{m=0}^{j} \theta_{j-m}\psi_m = 0,$$

$$(\alpha(2j) + 4j - 2)(\alpha(2j) + 4j - 3)\psi_j + \sum_{m=0}^{j} \{\lambda\theta_{j-m}\theta_m - \psi_{j-m}\psi_m\} = 0.$$

$$(3.10)$$

In order to investigate the asymptotic properties of the series expansions we introduce certain generating functions. (Such an analysis was carried out in an earlier paper on the Lorenz system[9]). For Case 1 these are

$$\Theta(X) = \sum_{j=0}^{\infty} a_{j,2j} X^j, \qquad (3.11a)$$

$$\Psi(X) = \sum_{j=0}^{\infty} b_{j,2j} X^j, \qquad (3.11b)$$

where

$$X = t^{\alpha+2}, \quad \alpha = \frac{1}{2} + \frac{1}{2}\sqrt{1-24(1/\lambda+1)}, \qquad (3.12)$$

and for case 2

$$\Theta(X) = \sum_{j=0}^{\infty} a_{2j,4j} X^j, \qquad (3.13a)$$

$$\Psi(X) = \sum_{j=0}^{\infty} b_{2j,4j} X^j, \qquad (3.13b)$$

where

$$X = t^{2(\alpha+2)}, \quad \alpha = \frac{1}{2} + \frac{1}{2}\sqrt{1-48\lambda}. \qquad (3.14)$$

Using the (closed) recursion relations for the coefficient sets $(a_{j,2j}, b_{j,2j})$ and $(a_{2j,4j}, b_{2j,4j})$, the generating functions may be shown to satisfy the equation for Case 1

$$(\alpha + 2)^2 X(X\Theta')' - 5(\alpha + 2)X\Theta' + 6\Theta + 2\lambda\Theta\Psi = 0 \quad , \tag{3.15a}$$

$$(\alpha + 2)^2 X(X\Psi')' - 5(\alpha + 2)X\Psi' + 6\Psi + \lambda\Theta^2 - \Psi^2 = 0 \quad , \tag{3.15b}$$

and for <u>case 2</u>

$$4(\alpha + 2)^2 X(X\Theta')' + 4(\alpha + 2)\left(\alpha - \frac{1}{2}\right)X\Theta' + \alpha(\alpha - 1)\Theta + 2\lambda\Theta\Psi = 0 \tag{3.16a}$$

$$4(\alpha + 2)^2 X(X\Psi')' - 10(\alpha + 2)X\Psi' + 6\Psi + \lambda X\Theta^2 - \Psi^2 = 0 \quad , \tag{3.16b}$$

where prime denote differentiation with respect to X. These differential equations may now be used to analytically continue the functions $\Theta(X)$ and $\Psi(X)$.

The equations (3.15) and (3.16) may also be obtained in a more direct way. For Case 1, substitution of

$$x(t) = \frac{1}{t^2} \Theta(X) \quad \text{and} \quad y(t) = \frac{1}{t^2} \Psi(X), \tag{3.17}$$

where the variable X is defined in (3.12), into the equations of motion (3.2), yields exactly equations (3.15); the contribution from the linear terms (which vanish in the limit $|t|\to 0$ being ignored. Similarly, for Case 2, the substitution into (3.2) of

$$x(t) = t^\alpha \Theta(X) \quad \text{and} \quad y(t) = \frac{1}{t^2} \Psi(X). \tag{3.18}$$

where now the variable X is defined by (3.14), yields equations (3.16). Again the linear terms in (3.2) are ignored. (It is amusing to note that the term λx^2 contributes to the equation for $\Psi(X)$ although it does not contribute to the resonance calculation.

The types of singularities that $\Theta(X)$ and $\Psi(X)$ can display are easily determined by applying a standard leading order analysis. Considering first the Case 1 equations (3.15), we set

$$\Theta(X) \simeq A(X - X_0)^\gamma \quad , \quad \Psi(X) \simeq B(X - X_0)^\delta \quad ,$$

where X_0 is the singularity position, and find that there are two possible cases

case a $\quad \gamma = -2 \quad , \quad A = \pm \frac{3}{\lambda} \sqrt{2 + 1/\lambda} \ X_0^2 (2 + \alpha)^2$,

$\quad \quad \quad \delta = -2 \quad , \quad B = -\frac{3}{\lambda} X_0^2 (2 + \alpha)^2$, (3.19)

case b $\quad \gamma = \frac{1}{2} \pm \frac{1}{2} \sqrt{1 - 48\lambda} \quad , \quad A = $ arbitrary ,

$\quad \quad \quad \delta = -2 \quad , \quad B = 6 X_0^2 (2 + \alpha)^2$ (3.20)

The identical two cases are also found for the Case 2 equations (3.16), except for slight changes in the leading order coefficients A and B. Thus, $\Theta(X)$ and $\Psi(X)$ display exactly the same sort of singularities in the X-plane as do $x(t)$ and $y(t)$ in the t-plane. Therefore, close to a given singularity, the singularity structure of $x(t)$ and $y(t)$ may be determined by studying the singularities of $\Theta(X)$ and $\Psi(X)$. The key is to correctly map the singularities from the X-plane to the t-plane.

To see how this is done, we first consider the Case 2 equations (3.16) which correspond to the asymptotics about an irregular pole (which is placed at the origin of the X-plane and will, of course, map to the "origin" of the t-plane set at $t = t_*(=0)$. Since we are unable to solve equations (3.16) analytically, we have to resort to a numerical solution; this always indicates that the singularity nearest $X = 0$ is of the regular variety. This singularity position, X_0, can be multiplied by the unit phase factor, i.e., we can say that the singularity is at

$$X = X_0 \equiv X_0 e^{2\pi i n} \quad , \quad n = 0, 1, 2, \ldots$$

We recall that for Case 2, the variable X is $X = t^{2(2+\alpha)}$ and $\alpha = + \frac{i}{2} \sqrt{48 - 1}$ ($\lambda > 1/48$). Therefore, the corresponding singularity position in the t-plane is given by

$$t_0 = X_0^{1/2(2+\alpha)} \exp\left[\frac{\pi \min}{2(2+\alpha)}\right] \qquad (3.21)$$

$$= X_0^{1/2(2+\alpha)} \exp\left\{n\pi\left[\frac{5i - \sqrt{48\lambda-1}}{2(12\lambda+6)}\right]\right\} .$$

Thus, each pole in the X-plane yields an equiangular spiral of poles in the t-plane; one pole for each value of n = 0,1,2... These poles have an angular displacement about the central (irregular) pole given by

$$\Delta\theta = \frac{5\pi}{2(12\lambda+6)} , \qquad (3.22a)$$

and with radial decrement

$$\Delta|t| = \exp\left[\frac{-n\pi\sqrt{48\lambda-1}}{2(12\lambda+6)}\right] , \qquad n = 0,1,2... \quad (3.22b)$$

In the canonical resonance case $\lambda = 1$, these quantities are

$$\Delta\theta = \frac{5\pi}{36} = 25^\circ , \qquad (3.23a)$$

$$\Delta|t| = \exp\left[-\frac{n\pi\sqrt{47}}{36}\right] . \qquad (3.23b)$$

This mapping is shown in Fig. 3 and can be compared with some numerical results in Fig. 3(c). (The reason why we observe double spirals will be explained shortly.)

We notice in these figures that the singularities all seem to lie on the corners of exactly isosceles triangles. That this is indeed almost exactly so comes about as the result of an amusing coincidence. Returning to Fig. 3, in order to demonstrate that the triangle OAB is

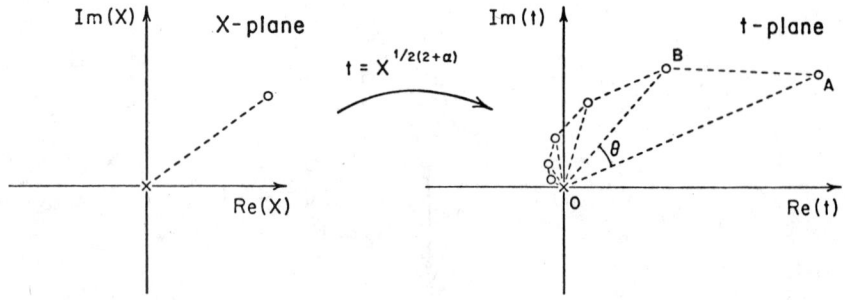

Fig. 3(a). Mapping of singularity in x plane into spiral in t plane for case-2 singularity.

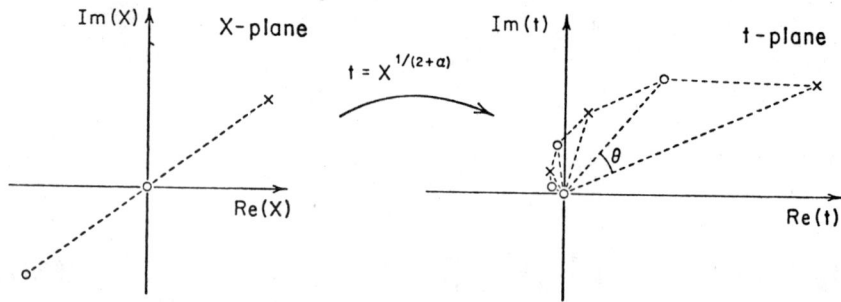

Fig. 3(b). Same for case-1 singularity. We note that this figure is representative for $\lambda \simeq 1$.

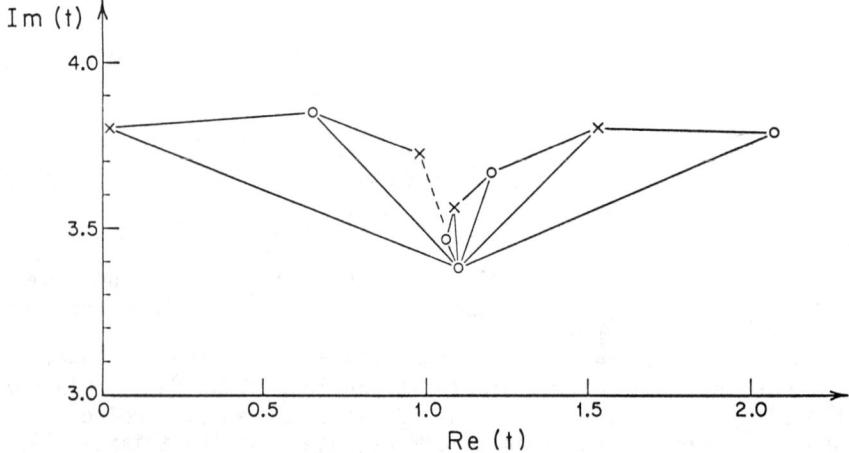

Fig. 3(c). Numerical calcualtion of alternating, double spirals about a case 1 singularity.

isosceles, we require that $OB \cos \Theta = \frac{1}{2} OA$. From (3.21) we deduce that this can only be so if $\cos\left(\frac{5\pi}{36}\right) = \frac{1}{2} \exp\left[\frac{\pi\sqrt{47}}{36}\right]$. The actual numerical values are $\cos\left(\frac{5\pi}{36}\right) = 0.90631...$ and $\frac{1}{2}\exp\left[\frac{\pi\sqrt{47}}{36}\right] = 0.90948...$!.

Thus, for all practical purpose the triangles may indeed be taken to be isosceles.

The same arguments may be applied to singularities of (X) and (X) associated with the Case 1 equations (3.15). However, here, recall that the variable is now $X = t^{2+\alpha}$ and $\alpha = \frac{1}{2} + \frac{i}{2} \sqrt{24(1+1/\lambda)-1}$. Thus, about the central (regular) pole the singularity positions map onto the t=plane as

$$t_o = X^{1/(2+\alpha)} \exp\left\{2n\pi\left[\frac{5i+\sqrt{24(1+1/\lambda)-1}}{(12+6/\lambda)}\right]\right\} \quad (3.24)$$

In the canonical resonance case $\lambda = 1$ we therefore obtain a spiral with angle

$$\Delta\Theta = \frac{5\pi}{18}, \quad (3.25a)$$

and radial decrement

$$\Delta|t| = \exp\left\{-n\pi\sqrt{47}\over 18\right\}, \quad n = 0,1,2... \quad (3.25b)$$

It would appear, then, that around a regular pole the spiral angle is twice that about an irregular pole. However, when we investigate the situation numerically, in the X-plane, we find that near X = 0 regular and irregular singularities always appear in diametrically opposite pairs. Thus when one of these <u>pairs</u> is mapped onto the t-plane we again obtain the highly symmetric 25° spiral; but now with alternating regular and irregular poles. This is shown in Fig. 3(b) and can be compared with some numerical results shown in Fig. 3(c).

We note that the whole of the above analysis can be repeated using the variables $\overline{X} = t^{2+\overline{\alpha}}$, with $\overline{\alpha} = \frac{1}{2} - \frac{i}{2}\sqrt{24(1+1/\lambda)-1}$, for the asymptotics about the regular poles and $\overline{X} = t^{2(2+\overline{\alpha})}$, with $\overline{\alpha} = \frac{1}{2} - \frac{i}{2}\sqrt{48\lambda-1}$

for the asymptotics about the irregular poles. Everything is the same as before except that the spirals are now in the opposite direction. Thus around any given pole there is a double spiral of singularities. This is exactly what we observe.[4]

An asymptotic analysis may also be carried out on the singularities of $\Theta(X)$ and $\Psi(X)$ in the X-plane. An illustrative computation is shown in Fig. 4 for solutions of equations (3.15) with $\lambda = 1$. The two nearest singularities are spaced equally and opposite across the singularity at the origin, but, as discussed above, are of Case 1 and Case 2 type respectively. Together these two singularities form a spiral in the t-plane as illustrated in Figs. 3(b) and 3(c). Each of the singularities in the X-plane has its own spiral of singularities around it. The singularities along the spirals around the Case 1 singularities are alternately Case 1 and Case 2, while only Case 1 singularities form the inner spiral around Case 2 singularities. Since the nearest singularities in the X-plane, shown here, form the spirals in the t-plane, and each of these spirals around it, the full hierarchy of spirals is contained in the asymptotic equations as illustrated in this figure.

In Table II there are listed several values of the angle and radial decrement. Figures 5 through 6 show the corresponding singular sets that were found numerically. Although these calculations show singularities that are not in the asymptotic region, $|t| \ll 1$, the agreement with the values found by the asymptotic analysis is quite good. Indeed, the above analysis appears to provide not only a "local" description of the singular set, but indicates the type of deformation the singular set undergoes when λ changes from $\lambda=1$.

Finally, it is interesting to conjecture what influence a self similar, natural boundary will have on the real time behavior. To investigate this point, we are currently calculating the power spectra of the Henon-Heiles System.

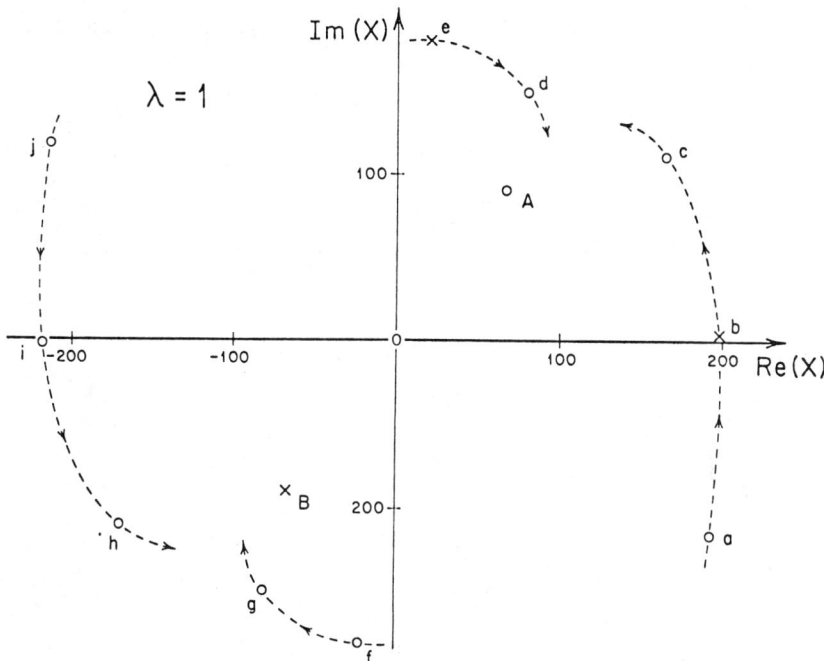

Fig. 4. Numerical solution ($\lambda = 1$) of Eq. (3.15) (case 1) showing spirals about singularities in x plane. We note that the solution is regular in an open region about the origin and (A,B) are equidistant ($\lambda =1$) pair of singularities closest to the origin.

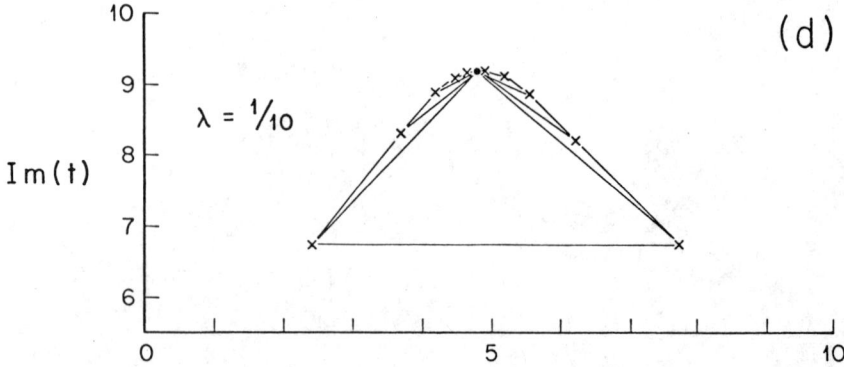

Fig. 5(a). Numerical calculation of the spirals about a case 1 singularity when λ = 1/10.

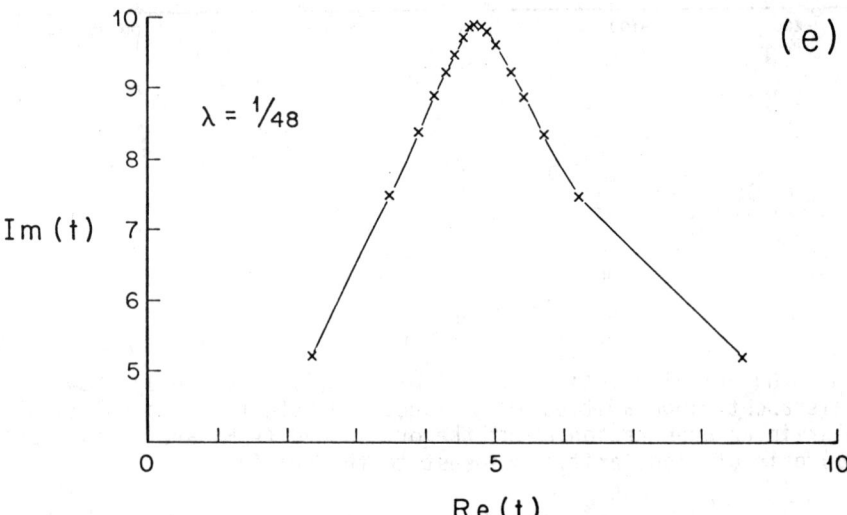

Fig. 5(b). Same, when λ = 1/48.

TABLE II. Angular and radial decrements.

	Case 1		Case 2	
	$\Delta\Theta \quad \dfrac{5\pi}{12+6/\lambda}$		$\Delta\Theta \quad \dfrac{5\pi}{2(12\lambda+6)}$	
	$\Delta\|t\| = \dfrac{-\pi\sqrt{24(1+1/\lambda)-1}}{12+6/\lambda}$		$\Delta\|t\| = \dfrac{-\pi\sqrt{48\lambda-1}}{2(12\lambda+6)}$	
λ	$\Delta\Theta$	$LN\Delta\|t\|$	$\Delta\Theta$	$LN\Delta\|t\|$
1	$50°$	-1.1965	$25°$	-0.5983
1/10	$\dfrac{5\pi}{72} = 12.5°$	-0.7076	$\dfrac{25\pi}{72} = 62.5°$	-0.4253
1/48	$\dfrac{\pi}{60} = 3°$	-0.3590	$\dfrac{2\pi}{5} = 72°$	-0.0000
4	$\dfrac{10\pi}{27} \quad 66.6°\ldots$	-1.2532	$\dfrac{5\pi}{108} \quad 8.4°$	-0.4020
$+\infty$	$75°$	-1.2556	0	0.000

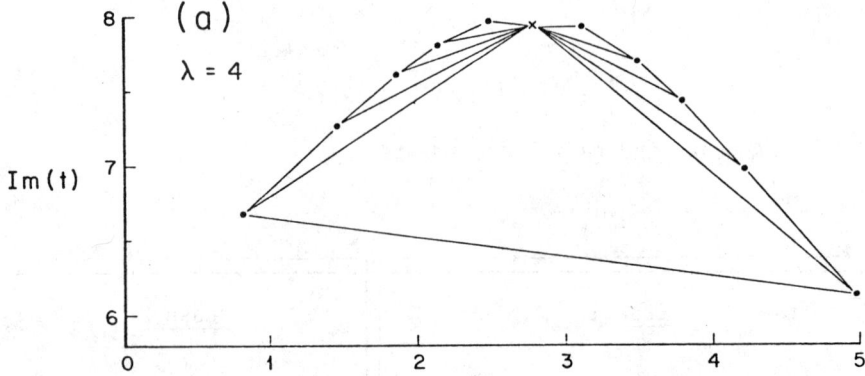

Fig. 6(a). Spiral about case 2 singularity. $\lambda = 4$.

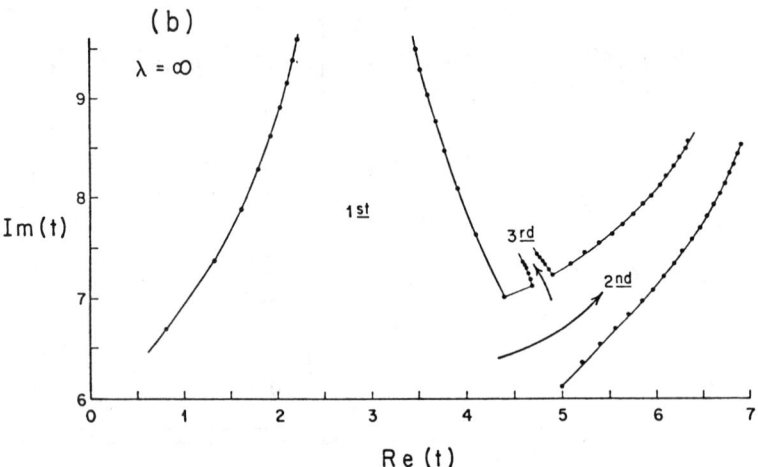

Fig. 6(b). $\lambda = \infty$.

IV. GENERAL AND SINGULAR SOLUTIONS: $-\frac{1}{2}<\lambda<0$

When $-\frac{1}{2}<\lambda<0$ the Case 1 singularity represents a singular form of the solution that depends on just three parameters, while the lower branch, Case 2 singularities represent the general four-parameter form of the solution. Both types of solution have real resonances. (See Table 1). Numerical investigations of the singular structure in this range of λ has found only the general form (Case 2). Whether the singular solution (Case 1 singularity) may appear as part of the solution for bounded motion is not known. For the present, we will concentrate on the structure of the general form of solution. Since the resonances are real there are values of λ for which the resonances are rational numbers.

These are:

$$\lambda = \frac{1}{48}\left(1 - \frac{m^2}{n^2}\right) \qquad (4.1)$$

where

$$1 < \frac{m}{n} < 5 \quad . \qquad (4.2)$$

For these values of , we find:

$$\alpha = \frac{1}{2}\left(1-\frac{m}{n}\right)$$

$$\beta = -2$$

and

$$r = 0, -1, \frac{m}{n}, 6.$$

The expansion about a singularity may take the form: (to=0)

$$x = t^{(m-n)/2n} \sum_{j=0}^{\infty} a_j\, t^{j/n}$$

$$y = t^{-2} \sum_{j=0}^{\infty} b_j\, t^{j/n} \qquad (4.3)$$

if the compatibility conditions on the coefficients (a_j, b_j) are satisfied at the resonances $j=0,m,6n$.

Otherwise, Lnt terms must be introduced into the expansion. Since the resonance at $j=0$ is trivial (a_0 arbitrary), there are, in effect, two compatability conditions to be satisfied. We find[11] that if $\frac{m}{n}$ does not assume the following values, the above expansion is valid.

First Resonance

$$\frac{m}{n} \neq \frac{5K}{K+1}, \frac{5K+2}{K+1}, \frac{5K+4}{K+1} \qquad (4.4)$$

Second Resonance

$$\frac{m}{n} \neq \frac{5K-1}{K+1}, \frac{5K+1}{K+1}, \frac{5K+3}{K+1}, \frac{5K+4}{K+1} \qquad (4.5)$$

where $0 \leq K \leq \infty$.

We note that when $\frac{m}{n}$ is equal to one of the above values, the expansion may take the above form. However, <u>each</u> case must be checked separately. The first few excluded values of $\frac{m}{n}$ are:

$$\frac{m}{n} = 1 < 2 < 5/2 < 3 < 10/3 < 7/2 < 11/3 < \ldots$$

We find: (yes: compatibility condition satisfied)

$\frac{m}{n}$	λ	1st Res.		2nd Res.	
1	0	yes		yes:	System Integrable.
2	-1/16	yes	B = 16A	yes:	System Integrable.
5/2	-7/64	no		yes	
3	-1/6	yes		yes:	System Integrable.
10/3	-3/16	yes		no	
7/2	-15/64	no		no	

The Henon-Heiles is known to be integrable and Painleve when $\lambda = 0, -1/6, -1$. When $\lambda = -1/16$ and $\beta = 16A$, the "general" form of the solution will take the form:

$$x = t^{-1/2} \sum_{j=0}^{\infty} a_j t^j$$

$$y = t^{-2} \sum_{j=0}^{\infty} b_j t^j \qquad (4.6)$$

Thus, (x^2,y) are single-valued about the "general" singularity. In addition, the three-parameter form of the solution can be shown to be Painlevé. Indeed, we have recently learned that L. Hall[8] has found, by another method, a second integral of motion at these parameter values, thus demonstrating the system's integrability. Surface of sections for neighboring values of λ also show smooth behavior ($\lambda = -1/14$, Fig. 7). Indeed all surface of section calculated show behavior until $\lambda \le -15/64$. (See Figs. 7-8).

It is interesting to note that $\lambda = -15/64$ ($\frac{m}{n} = 7/2$) is the first value of λ for which Lnt terms must be introduced at <u>both</u> resonances. This observation is less compeling than it might be since for differing sets of linear frequencies (A,B) smooth behavior was observed throughout the range $-\frac{1}{2} < \lambda < 0$

However, no widespread chaos was found for $-\frac{15}{64} < \lambda < 0$. This suggests that there may be a connection between the finite branch point structure and the regular behavior observed in this range of parameter values. To further develop this link, it appears necessary to a) determine the role of the singular solution, b) find an algorithm for converting statements about analytic structure (single-valuedness) into statements concerning the existence of integrals of motion. i.e. Louiville's theorem.

Finally, we note that when $-\frac{24}{23} < \lambda < -\frac{1}{2}$ and $\lambda = -\frac{24}{23 + \frac{m^2}{n^2}}$ the general solution (case 1) again may have a finite-branch point structure. Again, there are two compatibility conditions. Detailed calculations reveals that one compatibility condition may be satisfied for a subset of the above values of λ, while the other compatibility condition is, in general, not satisfied. Thus, finite branch point solutions do not exist in this range of λ. ($\lambda \ne -1$). However, for $\lambda < -27/29$ the first compatability condition will be satisfied for the above values of λ. Surface of sections (Fig. 8) show a transition near this λ value from wideapread chaos.

V. RICATI TRANSFORM OF THE HENON-HEILES SYSTEM

The Henon-Heiles system [Eqs. (2.1), (2.2)] can be transformed into the following system of equations by means of the Ricati transformation:

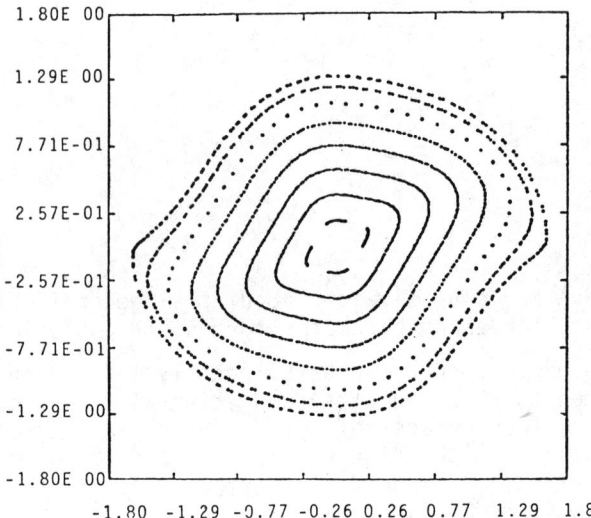

Fig.7(a). Surface of section calculation. λ = -1/16. B = 16A.

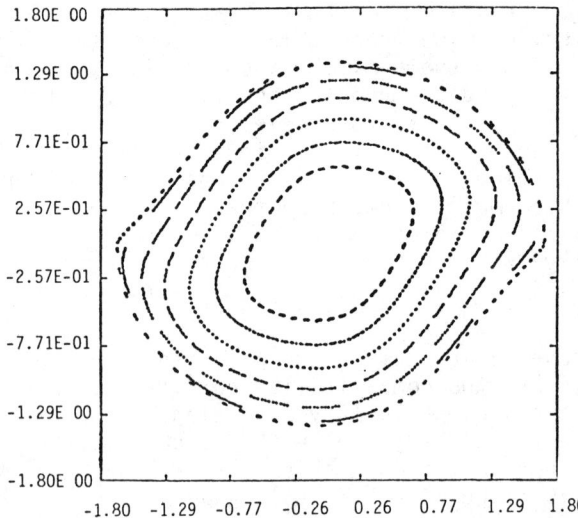

Figure 7(b). Surface of section. λ = 1/14. B = 14A.

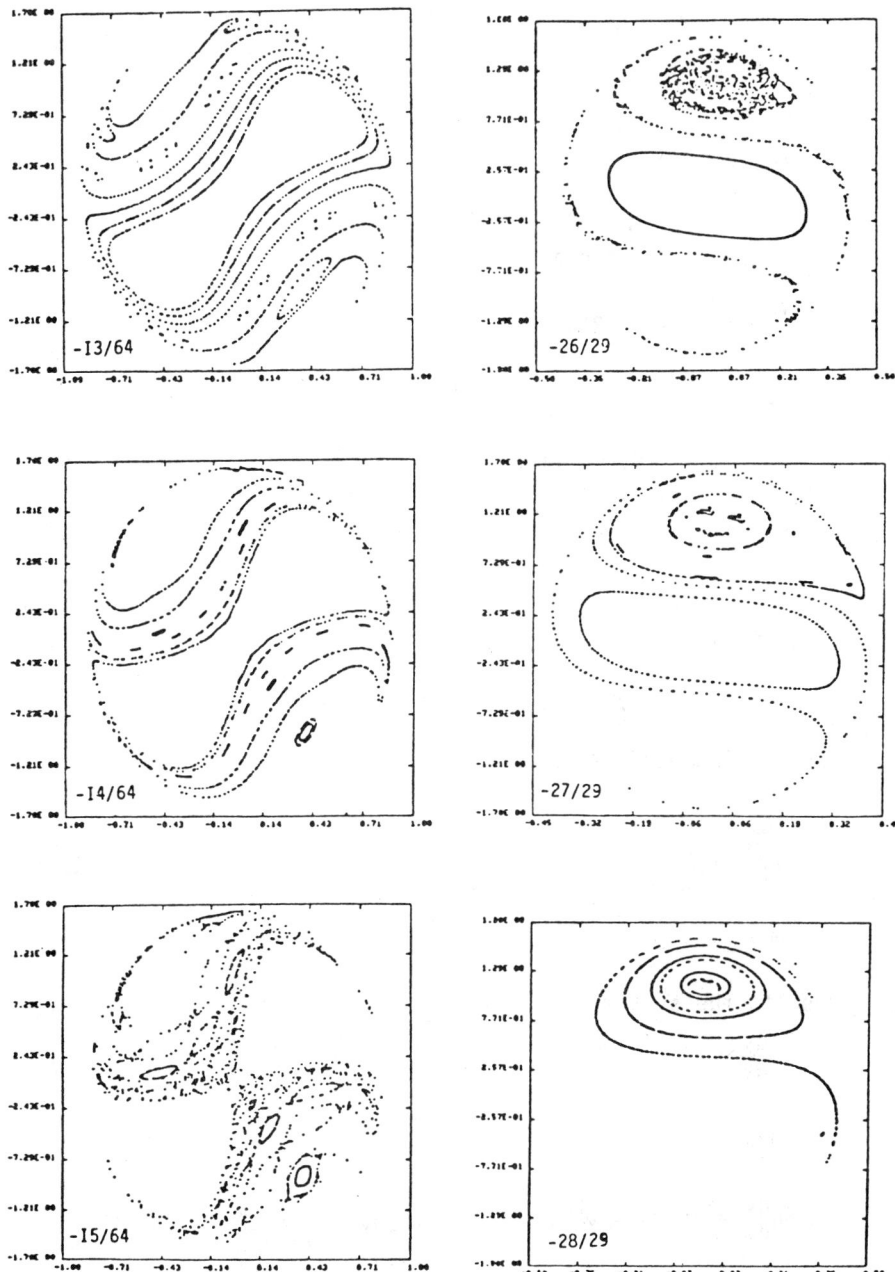

Fig. 8

$$\psi = \frac{\dot{x}}{x}. \qquad (5.1)$$

We find:

$$\dot{\psi} + \psi^2 = -A - 2\lambda y,$$
$$\ddot{y} + By - y^2 = -\lambda x^2, \qquad (5.2)$$
$$\dot{x} = \psi x.$$

This system may be written as:

$$\dot{\psi} + \psi^2 = -A - 2\lambda y$$
$$(\frac{d}{dt} - 2\psi)(\ddot{y} + By - y^2) = 0. \qquad (5.3)$$

Letting: $u = \dot{y}$, $v = \ddot{y}$, we find that:

$$\dot{\psi} = -\psi^2 - A - 2\lambda y,$$
$$\dot{y} = u,$$
$$\dot{u} = v, \qquad (5.4)$$
$$\dot{v} = 2\psi(By - y^2 + v) - Bu + 2uy.$$

Now, changing independent variable,
$$t \to \psi,$$

$$\frac{d}{dt} = h\frac{d}{d\psi}, \qquad (5.5)$$

$$h = -A - \psi^2 - 2\lambda y,$$

there results the system

$$h\frac{dy}{d\psi} = u,$$
$$h\frac{du}{d\psi} = v,$$
$$h\frac{dv}{d\psi} = 2\psi(By - y^2 + v) - Bu + 2uy, \qquad (5.6)$$

where

$$h = -A - 2\lambda y - \psi^2.$$

This system may be further reduced by using the Hamiltonian to eliminate x^2 in Eq. (5.2). That is, using

$$H = \frac{1}{2}(\dot{x}^2 + \dot{y}^2 + Ax^2 + By^2) + \lambda x^2 y - \frac{y^3}{3},$$

where

$$\dot{x} = \psi x, \quad \dot{y} = u, \quad h = -A - \psi^2 - 2\lambda y.$$

It is found that

$$x^2 = -\frac{1}{h}\left(2H - u^2 - By^2 - \frac{2y^3}{3}\right). \tag{5.7}$$

Substitution into Eq. (5.2) and changing variables (5.3) results in the system

$$h\frac{dy}{d\psi} = u,$$

$$h\frac{du}{d\psi} = -By - y^2 + \frac{\lambda}{h}\left(2H - u^2 - By^2 - \frac{2y^3}{3}\right), \tag{5.8}$$

where

$$h = -A - 2\lambda y - \psi^2.$$

Since Eqs. (5.6) and (5.8) are essentially equivalent, we shall concentrate on Eqs. (5.6). The effect of the transformation

$$t \to \psi$$

is to throw all of the complex-t zeros/singularities of the Henon-Heiles system into the point at infinity in the ψ plane.

By an analysis of the singularities of Eqs. (5.6) it is found that:

(1) There are no singularities in the finite ψ plane, except where h is zero (i.e. singularities of the transformation $t \to \psi$).

(2) No singularities can occur for real ψ when:

$$H < H_{dis} = \frac{A^2}{8\lambda^2}\left(B + \frac{A}{3\lambda}\right).$$

That is, no real singularities of system (5.6) can occur for bounded motion of the Henon-Heiles unless the energy is equal to the energy of disassociation.

(3) When $H = H_{dis}$, real ψ singularities must occur at $\psi = 0$ (if

they occur at all) and take the following form:

$$y = -\frac{A}{2\lambda} + y_1 \psi^{2/3} + \ldots,$$

$$u = u_1 \psi^{1/3} + \ldots,$$

$$v = v_0 + v_1 \psi^{2/3} + \ldots$$

(4) In general, complex-ψ singularities take the form

$$y = \sum_{j=0}^{\infty} Y_j (\psi - \psi_0)^{j/2},$$

$$u = \sum_{j=0}^{\infty} u_j (\psi - \psi_0)^{j/2},$$

$$v = \sum_{j=0}^{\infty} v_j (\psi - \psi_0)^{j/2},$$

where

$$A + 2\lambda y_0 + \psi_0^2 = 0$$

and

$$u_0^2 + By_0^2 - \frac{2y_0^3}{3} = 2H.$$

We remark that statements 1) and 4) are simply calculations, while statements 2) and 3) follow from the observation that, when $H = H_{dis}$, $Y = -\frac{A}{2\lambda}$ is a contour of the Henon-Heiles potential energy, V, when $V(x,y) = H_{dis}$. At a singularity, ψ_0;

$$A + 2\lambda y_0 + \psi_0^2 = 0$$

follows from the vanishing of h. From the above observation:

  a) $y > -\frac{A}{2\lambda}$ when $H < H_{dis}$,

And:

b) when $H = H_{dis}$ and $y = -\frac{A}{2\lambda}$,

then $\dot{x} = \dot{y} = 0$

or $\psi_0 = u_0 = 0$.

And statements 2) and 3) are obtained. Furthermore, when $y = -\frac{A}{2\lambda}$ and $\dot{x} = \dot{y} = 0$, the trajectory of the Henon-Heiles system reverses (doubles back) and "locally:"

$$x(t) = x_0 - \frac{d}{12}x_0 \ddot{y}_0 t^4 + \text{even powers of } t,$$

$$y(t) = -\frac{A}{2\lambda} + \frac{\ddot{y}_0}{24} t^2 + \text{even powers of } t.$$

Thus, real ψ singularities that occur for system (5.6) when $H = H_{dis}$ merely reflect this reversing of trajectories and introduce no further complications.

A more serious problem arises when one considers that the portion of the real-time axis between any two zeros of x is mapped by the transformation

$$t \to \psi = \frac{\dot{x}}{x}$$

into the entire real ψ axis. Thus, there does not appear to be any means for representing the solution of the Henon-Heiles by a solution to system (5.6) beyond one cycle of the variable, x.

However, by numerical solution of (5.6) for complex ψ, the following behavior was observed.

Near the origin of the ψ plane two order ½ branches were observed in the upper-half plane (with complex conjugates in the lower-half plane). If the path of integration was not closed and did not pass between the pair of singularities the solution rapidly became asymptotic with increasing distance from the origin to a constant value. If the path of integration passed between the pair of singularities the solution appeared to increase algebraically with increasing distance from the origin. Furthermore, by integrating along a closed curve that enclosed the pair of singularities the solution was observed to continue into the next cycle of the variable, $x(t)$, of the original Henon-Heiles system. In the process, since the path crosses the branch lines of the singularities, the original pair of singularities was transformed into a different pair appropriate to the next cycle. (See Fig. 9.)

Finally, an interesting difference in the structure of the singularities for Eqs. (5.6) was observed when the Henon-Heiles system did or did not have a natural boundary. When $-\frac{1}{2} < \lambda < 0$ and the Henon-Hieles solution does not appear to have a natural boundary,

274

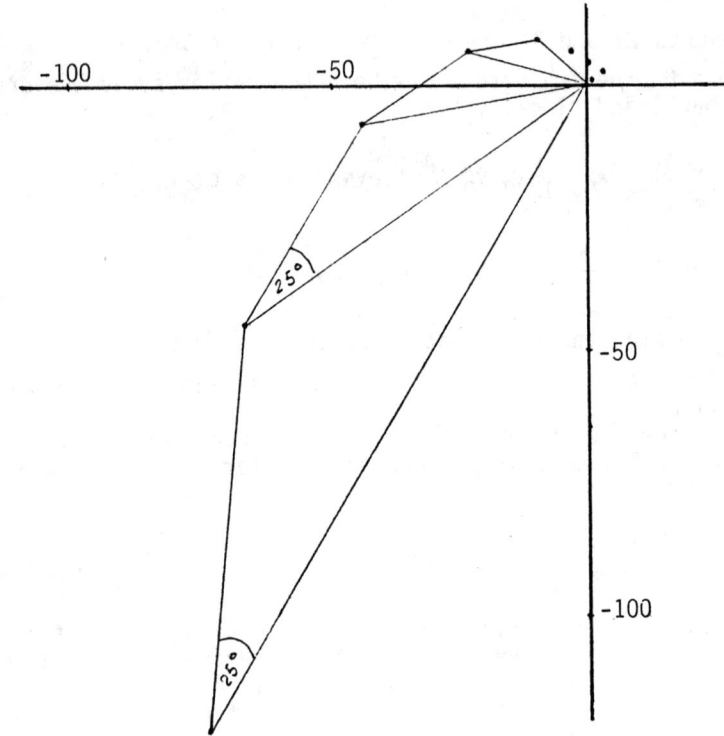

Fig. 10(b). λ=1. Singularities spiraling from left branch point. Note resemblance to complex-t singularities. (λ=1.)

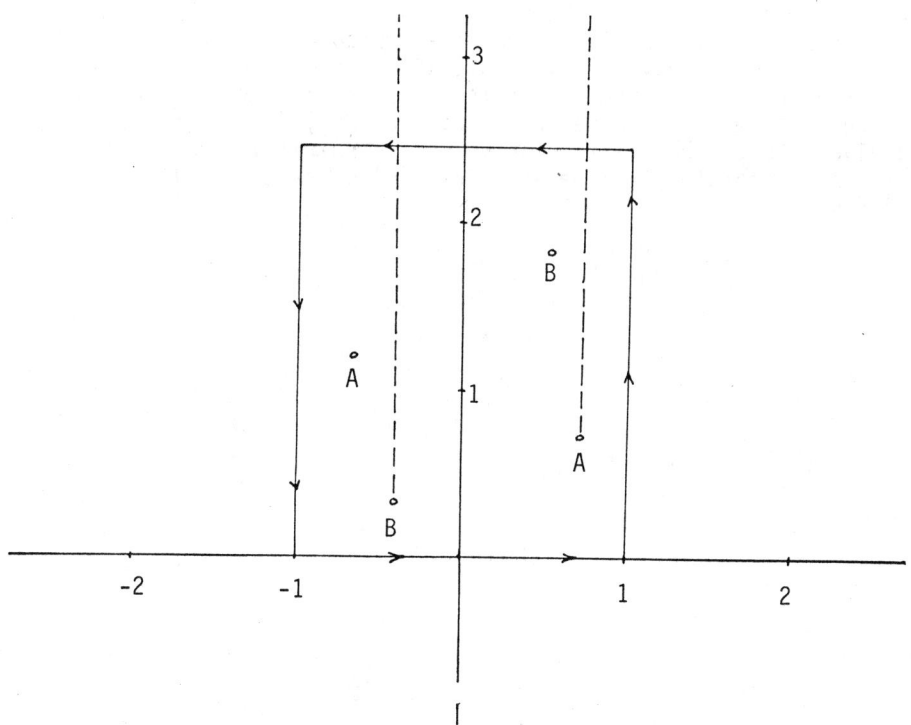

Fig. 9. Continuation of solution to next cycle by complex-ψ integration. Dotted lines show branch lines encountered during integration. A denotes basic pair of branch points found at beginning of integration. B — branch points at end of integration.

only a single pair of singularities was found on any sheet of the Riemann surface (in the upper-half plane). On the other hand, when λ = 1 and the Henon-Heiles has the natural boundary whose structure is illustrated in Fig. 1, the singularities of Eqs. (5.6) found by integrating between the original pair of branch points and then to the left or right is shown in Fig. 10. From the right branch point there is observed to emanate a straight line of singularities that extends into the upper, right-half plane. The distance between these singularities can be observed to increase with the distance form the origin. From the left branch point there is observed to develop an equiangular spiral (with measured angle of 25^0) which appears to spiral out to the point at infinity. This spiral is similar to that found in the complex t plane about each singularity for the Henon-Heiles system when λ = 1.

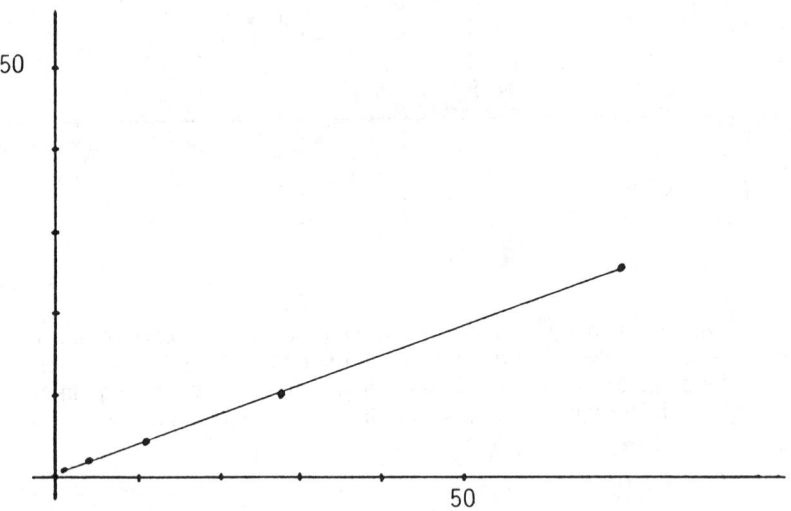

Fig. 10(a). λ=1. Singularities in upper, right-half plane that begin with right branch point.

REFERENCES

1. M. Henon and C. Heiles, Astron. J. 69, 73-79 (1964).
2. E. L. Ince, Ordinary Differential Equations (Dover, New York, 1956).
   H. Davis, Introduction to Nonlinear Differential and Integral Equations (Dover, New York, 1960).
   E. Hille, Ordinary Differential Equations in the Complex Plane (Wiley, New York, 1976).
3. S. Kowalevskaya, Acta Math. Acad. Sci. Hung 14, 81 (1890).
4. Y. F. Chang, M. Tabor, and J. Weiss, "Analytic structure of the Henon-Heiles Hamiltonian in integrable and non-integrable regimes," to be published, J. Math. Phys.
5. T. Bountis, H. Segur, and F. Vivaldi, "Integrable Hamiltonian Systems and the Painlevé Property (1981), submitted to Phys. Rev. A.
6. Y. F. Chang, J. M. Greene, M. Tabor, and J. Weiss, "The analytic structure of dynamical systems and self-similar natural boundaries," La Jolla Institute Report No. LJI-R-81-152, 1981.
7. M. J. Ablowitz, A. Ramani, and H. Segur, J. Math. Phys. 21, 715 (1980).
8. L. S. Hall, "On the existence of a last invariant of conservative motion," Lawrence Livermore Lab, 1981, preprint UCID-18980.
9. M. Tabor and J. Weiss, Phys. Rev. A 24, 2157 (1981).
10. M. Tabor, "On the Analytic Structure of Dynamical Systems: Painlevé Revisited," to be published in Exactly Solvable Classical and Quantum Models and Parallel Arithmetical Problems. Eds. David and Gregory Chudnovsky. Marcel Dekker, 1982.
11. J. Weiss, "Analytic Structure of the Henon-Heiles System", La Jolla Institute Report No. LJI-R-82-176, 1982.

*This work supported by Department of Energy, Contract No. DOE-10923 and by the Office of Naval Research, Contract No. N-00014-79-C-0537.

Logarithmic Singularities and Chaotic Behavior in Hamiltonian Systems

Tassos Bountis
Mathematics and Computer Science, Clarkson College of Technology
Potsdam, N.Y. 13676

Harvey Segur
Aeronautical Research Association of Princeton, P.O. Box 2229
Princeton, N.J. 08540

ABSTRACT

Several connections have been recently discovered between the real time behavior of the solution of dynamical systems and their movable singularities in the complex time plane. One of them has led to a direct method for identifying integrable Hamiltonian and dissipative systems by requiring that they possess the Painlevé property, i.e. that their general solutions have no movable singularities other than poles. In this paper we further explore these connections by partially lifting the Painlevé property and admitting logarithmic singularities. We find in a number of interesting Hamiltonian examples, that admitting only $\ln(t-t_0)$ terms still implies globally "regular" motion and very little chaos, whereas more complicated singularities of the type $\ln\ln(t-t_0)$ are associated with the presence of large scale regions of chaotic behavior.

1. INTRODUCTION

Every student of classical mechanics knows that in order to determine the positions and velocities of a system of mass particles for all time, one has to *integrate* a set of ordinary differential equations, which are in general coupled and nonlinear. It is perhaps less widely known that only a precious few of these systems *can be* integrated by the presently known methods of analysis.[1-6]

Since Newton, there have been a number of celebrated examples where the equations of motion can be separated and explicitly solved, e.g. the Kepler Two-Body Problem, Jacobi's solution of geodesic motion on an ellipsoid[7] and Euler's and Kovalevskaya's special cases of a rotating rigid body.[8] In recent years, several integrable Hamiltonian systems have been discovered by inspired application of the concepts of Inverse Scattering Transforms (IST).[9-13]

One of the main difficulties here is how to *recognize* an integrable dynamical system. So far, in most cases, this was done indirectly and required some degree of inspiration. Recently, however, a *direct* method has been used to identify a variety of integrable dynamical systems, be they Hamiltonian[14] or not.[15] This method, originally due to Kovalevskaya,[8] looks for all parameter values (masses, spring constants, etc.) for which the solutions have the *Painlevé property*: their *only* movable singularities[16] (in the complex time plane) are poles, and they have as many arbitrary constants as there are initial conditions.

In this paper we explore further the connection between the (complex time) singularities of the solution of Hamiltonian systems and their general behavior, "regular" vs. "chaotic", in *real* time. In particular, we ask what happens when the Painlevé property is partially lifted, e.g. when not enough arbitrary constants are present and only classes of (particular) solutions have only movable poles. Analyzing several interesting Hamiltonians, we find that if, besides poles, there are logarithmic singularities of the form $[\ln(t-t_0)]^p$ (with integer p) then most orbits still appear "regular" and very little chaos is evident. However, "worse" singularities of the form $\ln\ln(t-t_0)$ appear to be connected with dramatically different behavior in real time, giving rise to large scale chaotic regions in phase space.

In Section 2 we review briefly our recent results[14] on integrable Hamiltonian systems and illustrate the singularity analysis on a dissipative system of three coupled Lotka-Volterra equations. Section 3 contains our analysis of the logarithmic singularities of a Quartic Lattice, a Free End Toda Lattice, and a special case of the Hénon Heiles. [For a very thorough and interesting analysis of the singularities of the Hénon-Heiles system see Reference 17 and related papers in this volume.] Finally, in Section 4 we summarize our main results and offer some concluding remarks.

2. THE PAINLEVÉ PROPERTY AS AN INTEGRABILITY CRITERION

In this section we do two things: First we list several examples of integrable Hamiltonian systems, which were identified directly by requiring that their solutions possess the Painlevé property. Then we apply our method to a dissipative system of 3 coupled Lotka-Volterra equations, obtain an integrable case, and show that in one example, where the Painlevé property is partially satisfied, it is possible to find two integrals of the motion.

One remark is in order here concerning the word "integrable". When referring to an N degree of freedom Hamiltonian system, it has a precise meaning, which we also adopt in this paper: It means that there exist N analytic, single valued, global integrals, which do not explicitly depend on t and are in involution[1-7]. In the case of dissipative systems the situation is not so clear. Here, we call integrals analytic, global constants of the motion which may explicitly depend on time, and a dissipative system of N first order o.d.e's is referred to as integrable if it possesses N such integrals of the motion.

A. Integrable Hamiltonian Systems With The Painlevé Property

The examples listed below have been recently identified as having the Painlevé property and in most cases integrability was independently verified. We omit the details since they have already been presented elsewhere.[14] The interested reader will find an outline of the main steps of the analysis in our treatment of a dissipative system in Section 2.B.

1. The Hénon Heiles Hamiltonian[18,19]

$$H = \frac{1}{2}(\dot{x}^2 + \dot{y}^2 + Ax^2 + By^2) - x^2 y - \frac{\varepsilon}{3} y^3 \quad (2.1)$$

is Painlevé in the following cases:

(a) $A = B$, $\varepsilon = 1$: The equations are trivially separable in $(x+y)$, $(x-y)$ coordinates, as was observed long ago.[20]

(b) Any A, B, $\varepsilon = 6$: This case is less widely known. On being informed of our result, John Greene was able to derive the second integral

$$x^4 + 4x^2 y^2 + 4\dot{x}(\dot{x}y - \dot{y}x) - 4Ax^2 y + (4A-B)(\dot{x}^2 + Ax^2) = \text{const.}$$

(c) $B = 16A$, $\varepsilon = 16$: This case was recently shown to possess the Painlevé property[17] and the second integral has been obtained by an independent method[30].

2. The Periodic Toda Lattice[21]

$$H = \frac{p_1^2}{2m_1} + \frac{p_2^2}{2m_2} + \frac{p_3^2}{2} + e^{\delta(q_1-q_2)} + e^{\varepsilon(q_2-q_3)} + e^{q_3-q_1} \quad (2.2)$$

is Painlevé only if $m_1 = m_2 = \varepsilon = \delta = 1$. This case has been extensively studied and shown rigorously to be integrable by many authors.[22,10,23]

3. The Fixed End Toda Lattice

$$H = \frac{p_1^2}{2m_1} + \frac{p_2^2}{2m_2} + e^{-\delta q_1} + e^{\varepsilon(q_2-q_3)} + e^{q_2} \quad (2.3)$$

has the Painlevé property in the following three cases:

(a) $m_1/m_2 = 1$, $\delta = \varepsilon = 1$

(b) $m_1/m_2 = 1$, $\delta = 1$, $\varepsilon = 1/2$

(c) $m_1/m_2 = 1/3$, $\delta = 1$, $\varepsilon = 1/2$

It is interesting that Bogoyavlenski[13] using group theoretical methods found, in the two degree of freedom case, exactly the three cases we listed above and no more.

4. The Free End Toda Lattice

$$H = \frac{p_1^2}{2m_1} + \frac{p_2^2}{2m_2} + \frac{p_3^2}{2} + e^{\varepsilon(q_1-q_2)} + e^{q_2-q_3} \qquad (2.4)$$

was found to be Painlevé in three cases (in each case they are all equivalent within scaling):

(a) $m_1 = \frac{\varepsilon(2\varepsilon-1)}{2-\varepsilon}$, $m_2 = 2\varepsilon - 1$, $1/2 < \varepsilon < 2$

(b) $m_1 = \frac{\varepsilon(\varepsilon-1)}{2-\varepsilon}$, $m_2 = \varepsilon - 1$, $1 < \varepsilon < 2$

(c) $m_1 = \frac{3\varepsilon(2\varepsilon-1)}{2-3\varepsilon}$, $m_2 = 2\varepsilon - 1$, $1/2 < \varepsilon < 2/3$

The integrability of 4(a) has been proved by Moser[7] and Bogoyavlenski[13]. Cases 4(b),(c) are the only ones for which no independent verification has been found. Note that since no bounded motion is possible here the surface of section method does not immediately apply.

5. The Coupled Quartic Oscillators

$$H = \frac{1}{2}(\dot{x}^2 + \dot{y}^2 + Ax^2 + By^2) + \frac{x^4}{4} + \frac{\sigma y^4}{4} + \frac{\rho}{2} x^2 y^2 \qquad (2.5)$$

has the Painlevé property only if:

(a) $A = B$, $\sigma = 1$, $\rho = 1$: In this case since the potential is a function of $(x^2 + y^2)$ the system has an angular momentum integral

$$\dot{x}y - \dot{y}x = \text{const.}$$

(b) $A = B$, $\sigma = 1$, $\rho = 3$: Here the equations of motion uncouple in $(x+y)$, $(x-y)$ variables and a second integral is easily found

$$\ddot{xy} + A\dot{xy} + xy(x^2+y^2) = \text{const.}$$

as was also observed by Yoshida.[24]

B. Integrability in a Dissipative System

In simple models of turbulence,[25] or population biology,[26] one is often confronted with systems of coupled nonlinear o.d.e's, which in some cases are known to exhibit wild and chaotic behavior. One of the simplest systems studied in this context is a set of three coupled Lotka Volterra equations of the form [25,26]

$$\dot{y}_1 = (1 - y_1 - By_2 - Cy_3) y_1 \;, \tag{2.6a}$$

$$\dot{y}_2 = (1 - y_2 - By_3 - Cy_1) y_2 \;, \tag{2.6b}$$

$$\dot{y}_3 = (1 - y_2 - By_3 - Cy_1) y_2 \;, \tag{2.6c}$$

where B,C are constants. The solutions of these equations behave very differently for different B,C values. Our purpose here is to analyze their singularities and find values of B,C for which we can integrate (2.6).

Looking for Painlevé solutions $y_k \sim a_k(t-t_o)^p$ (p is a negative integer) as $t \to t_o$ we find that the most general "dominant" behavior is

$$y_k \sim a_k \tau^{-1} + \ldots \;, \quad k = 1,2,3 \;;\; \tau \equiv t - t_o \;, \tag{2.7}$$

and upon substitution of (2.7) in (2.6) two types of singularities are distinguished

(i) all $a_k \ne 0$ in (2.7), or,

(ii) one is zero, say $a_1 = 0$.

Starting with singularity type (i) we immediately observe that something special is happening for B = C = 1. In fact, in that case, the equations for the a_k's reduce to a single one:

$$a_1 + a_2 + a_3 = 1 \;,$$

and a 3 parameter t_o, a_1, a_2 - family of Painlevé solutions is found, which suggests that the system may be integrable for B = C = 1. This is indeed true since, in that case, (2.6) yields

$$\dot{y}_1/y_1 = \dot{y}_2/y_2 = \dot{y}_3/y_3$$

whence $y_1 = ay_2$ and $y_1 = by_3$ as is verified by inspection (a,b,t_0 are the arbitrary constants).

Turning to the type (ii) singularity (with $a_1 = 0$) we have, as $\tau \to 0$:

$$y_1 = d\tau^p + \alpha\tau^{p+r}, \quad y_2 = \frac{1-B}{1-CB}\tau^{-1} + \beta\tau^{-1+r}, \quad y_3 = \frac{1-C}{1-CB}\tau^{-1} + \gamma\tau^{-1+r}$$
(2.8)

where

$$p = (B^2 + C^2 - B - C/(1-CB))$$
(2.8a)

must be an integer for Painlevé and d is as yet undetermined. Substituting (2.8) in (2.6) we get linear equations for α, β, γ and require that the determinant vanish so that arbitrary constants may enter. This leads to the "resonance" equation[27]

$$(r+1)r[r-1+(2-C-B)/(1-BC)] = 0 ,$$
(2.9)

which is to be solved for *integer* r. One possible choice is

$$B + C = 2$$
(2.10)

which also gives integer $p = 2$ in (2.8a). There are 3 arbitrary constants here entering at $r = 0$ [i.e. d is arbitrary in (2.8)] and $r = 1$, cf. (2.9). Thus, near the singularity (ii), the solutions have Laurent series expansions and as $\tau \to 0$:

$$y_1 = d\tau^2 \ldots , \quad y_2 = \frac{\tau^{-1}}{1-B} + f + \ldots, \quad y_3 = \frac{\tau^{-1}}{B-1} + \frac{1-Cf}{B} + \ldots$$

The above solution, however, do *not* have only poles near the type (i) singularity (unless $B = C = 1$). Hence the system is not expected to be integrable for all B,C satisfying (2.10). Still, the case (2.10), identified by our Painlevé analysis, has the remarkable property of possessing two integrals[26]

$$y_1 + y_2 + y_3 = [(a-1)e^{-t}+1]^{-1} , \quad y_1 y_2 y_3 = b[(a-1)e^{-t}+1]^{-3}$$

where a,b are free constants determined by initial conditions at $t = t_0$.

3. LOGARITHMIC SINGULARITIES IN HAMILTONIAN SYSTEMS

We now turn to non-integrable systems and attempt to find connections between the chaotic behavior of Hamiltonian systems and the (movable) singularities of their solutions in the complex t-plane. As a first step in that direction we have investigated the following three Hamiltonians:

I. The Quartic Lattice:

$$H = \frac{1}{2}(\dot{x}^2+\dot{y}^2) + \frac{1}{4}[x^4+y^4 + \eta(x-y)^4] , \quad (3.1)$$

II. The Free End Toda Lattice With Linear Terms:

$$H = \frac{1}{2}(p_1^2 + p_2^2 + p_3^2) + e^{q_1-q_2} + e^{q_2-q_3} - q_1 + q_3 , \quad (3.2)$$

III. The Hénon Heiles $\varepsilon = 2$ Case:

$$H = \frac{1}{2}(\dot{x}^2 + \dot{y}^2 + Ax^2 + By^2) - x^2 y - \frac{2}{3}y^3 . \quad (3.3)$$

Our main result is that in Examples I and II, which show very little evidence of chaotic behavior (see Figs. 1-3) the solutions involve only $\tau^p(\ln\tau)^q$ terms, (with p,q integers) in their asymptotic expansions. On the other hand, in Example III, where large chaotic regions exist (see Figs. 4-6) more complicated singularities of the type $\ln(\ln\tau)$ are present.

We outline below the analysis of Example I in some detail and simply state the results for Examples II and III since they can be obtained by a very similar procedure. As described in Section 2, we start by writing the equations of motion for (3.1):

$$\ddot{x} = -x^3 - \eta(x-y)^3 \quad (3.4a)$$

$$\ddot{y} = \eta(x-y)^3 - y^3 \quad (3.4b)$$

where η is an as yet undetermined parameter. Looking for Painlevé solutions we require at leading order

$$x \sim a\tau^p , \quad y \sim b\tau^q \quad \tau \to 0 \quad (3.5)$$

with p,q negative integers. Inserting (3.5) in (3.4) we find that two choices are possible:

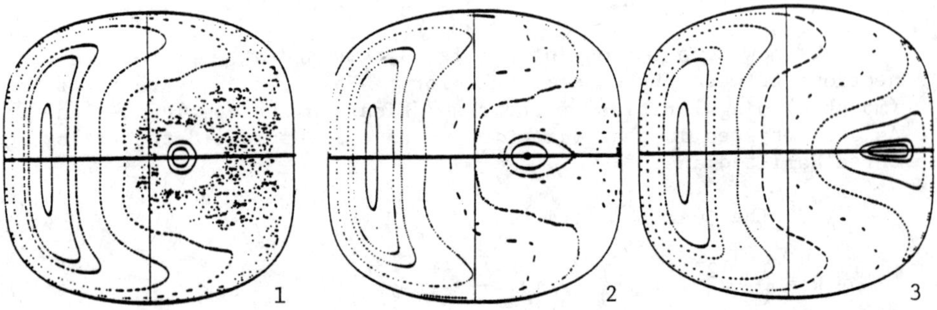

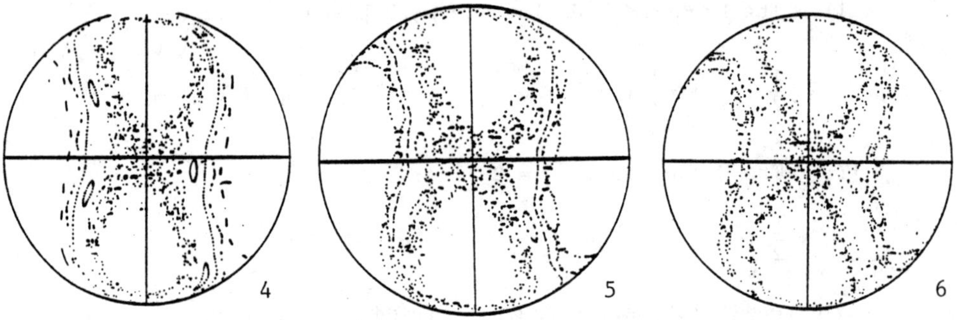

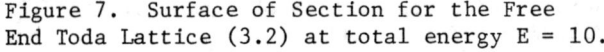

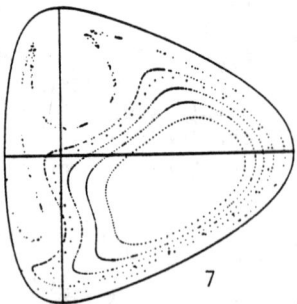

Figures 1-3. Surfaces of Section for the Quartic Lattice (3.4) at η = 0.15, 0.2 and 0.25 resp. and total energy E = 1000.

Figures 4-6. Surfaces of Section for the Hénon Heiles (2.1) with ε = 2.5, 2.0 and 1.5 resp. and total energy 70% of the escape energy in each case.

Figure 7. Surface of Section for the Free End Toda Lattice (3.2) at total energy E = 10.

(i) $p = -1$, $q = -1$

(ii) $p = q < -1$

It is straightforward to show that case (ii) leads to $a = b = 0$ for all p,q. Case (i), on the other hand, gives three possibilities:

(a) $a = -b = -i[2/(1+8\eta)]^{1/2}$

(b) $a = b = i\sqrt{2}$ \hfill (3.6)

(c) $a = [(2\eta-1+\sqrt{1-4\eta})/(1-\eta)]^{1/2}$, $a = 2/[b(1-\eta)]$

As usual we now write for the higher order terms

$$x = a\tau^{-1} + \alpha\tau^{-1+r} \quad , \quad y = b\tau^{-1} + \beta\tau^{-1+r}$$

and obtain, as in Section 2.B, a "resonance" equation:

$$s^2 + 3s[a^2+b^2+2\eta(a-b)^2] + 9[a^2b^2+\eta(a-b)^2(a^2+b^2)] = 0 \quad , \quad (3.7)$$

with

$$s \equiv (r-1)(r-2) \quad , \quad (3.7a)$$

giving us the r values at which arbitrary constants may enter [see also (2.9)].

Substituting in (3.7) the a,b values listed in (3.6) we find that case (b) yields integers r for all η, while the only <u>common</u> η values for which cases (a) and (c) give integers r are

$$\eta = 0 \quad \text{and} \quad \eta = 1/4 \quad .$$

The first choice, $\eta = 0$, leads to 4 arbitrary constants and is integrable but this is hardly surprising since the equations of motion (3.4) uncouple in that case. The second choice is much more interesting: For $\eta = 1/4$, (3.6a,c) lead to the <u>same</u> "resonance" equation

$$(r+1)r(r-3)(r-4) = 0 \quad , \quad (3.8)$$

and, at first sight, there are 4 arbitrary constants possible entering at $r = -1$ (corresponds to t_0), 0,3,4. This could mean that the solutions of (3.4) for $\eta = 1/4$ have the Painlevé property and that (3.1) is an integrable Hamiltonian system.

Upon closer investigation, however, we observe that the arbitrary constant at leading order, corresponding to r = 0, cannot be captured by admitting only algebraic singularities, c.f.(3.6). It turns out that in order to find the missing constant we need to introduce logarithmic singularities. The final result is that the complete asymptotic expansions of x,y take the form

$$x = i\left(\frac{2}{3}\right)^{1/2}\frac{1}{\tau}\left[1 + \sum_{k=1}^{\infty} a_k(\ln\tau)^{-k}\right] + \frac{\sqrt{3}}{2\tau}\sum_{k=0}^{\infty} b_k(\ln\tau)^{-k-\frac{1}{2}}$$
$$+ f\tau^2 + g\tau^3 + \ldots$$

$$y = -i\left(\frac{2}{3}\right)^{1/2}\frac{1}{\tau}\left[1 + \sum_{k=1}^{\infty} a_k(\ln\tau)^{-k}\right] + \frac{\sqrt{3}}{2\tau}\sum_{k=0}^{\infty} b_k(\ln\tau)^{-k-\frac{1}{2}} \quad (3.9)$$
$$+ f\tau^2 - g\tau^3 + \ldots$$

where $a_1 = 9/16$, $b_0 = 1$ and t_0, b_1, f, g are the 4 arbitrary constants.

Interestingly enough, as $\eta \to 1/4$, large regions of chaotic behavior disappear in the surfaces of section of (3.1) and at $\eta = 1/4$ the system appears to be integrable (see figs. 1-3). Fig. 3, however, is deceptive. Choosing more and more closely spaced sets of initial conditions - in some sense increasing the "resolution" of the surfaces of section - in between the invariant curves of fig. 3 chains of islands are found, which are separated by, admittedly small scale, but nevertheless present chaotic regions!

We have seen, in the above example, how the singularity analysis may be used to aid numerical experiment in establishing the non-integrability of a Hamiltonian system. Moreover, our analysis of the solutions of (3.1) indicates that if the only singularities (in complex time) are of the form $\tau^p(\ln\tau)^q$, chaotic behavior in real time may be confined to small scale regions of phase space.

Whether these observations are valid in general is still an open question. We did, however, find a <u>second example</u> where they also hold true: The Free End, 3 particle <u>Toda lattice</u> described by Hamiltonian (3.2). Here again, preliminary investigations indicated that the system is integrable. Recently, however, prompted by our singularity analysis, more careful numerical studies revealed the presence of small scale chaotic behavior, of the type observed in Example I (see fig. 7).

In analyzing the system (3.2) we begin as we did in our search for the integrable Toda lattices: Transforming to Flaschka's variables

$$a_k = \frac{1}{2}e^{\frac{1}{2}(q_k - q_{k+1})} \quad , \quad b_k = \frac{1}{2}p_k \quad , \quad k = 1, 2 \quad ,$$

and taking zero total momentum we obtain the equations of motion

$$\dot{a}_1 = a_1(b_2-b_1) \quad \text{(a)} \qquad \dot{b}_1 = 2a_1^2 - \frac{1}{2} \quad \text{(c)}$$
$$\dot{a}_2 = -a_2(2b_2+b_1) \quad \text{(b)} \qquad \dot{b}_2 = 2(a_2^2 - a_1^2) \quad \text{(d)} \qquad (3.10)$$

As before, the asymptotic expansions start with simple poles, i.e. as $\tau \to 0$,

$$a_1 \sim c_1 \tau^{-1} + \ldots \;,\; a_2 \sim c_2 \tau^{-1} + \ldots \;,\; b_1 \sim c_3 \tau^{-1} + \ldots,$$
$$b_2 \sim c_4 \tau^{-1} \qquad (3.11)$$

and three different cases are distinguished:

(i) $c_1 = 0, \; c_2 = i/\sqrt{2}, \; c_3 = 0, \; c_4 = \frac{1}{2}$

(ii) $c_1 = \frac{i}{2}, \; c_2 = 0, \; c_3 = \frac{1}{2}, \; c_4 = -\frac{1}{2}$

(iii) $c_1 = c_2 = i/\sqrt{2}, \; c_3 = 1, \; c_4 = 0$

Had it not been for the constant term $-\frac{1}{2}$ in (3.10c) [which is due to the linear terms in (3.2)] it would have been possible to develop the asympotic series (3.11) with only integer powers of τ coming in at every order. In fact case (ii) above would have yielded the four arbitrary constants needed for the Painlevé property and hence for integrability.

In order to develop series expansions (3.11) which satisfy (3.10), it turns out that we must include logarithmic terms of the form $\tau^p \ln \tau$, as well as τ^q terms (p,q integers). For example, in case (ii) above we find

$$a_1 = \frac{i}{2\tau} + \left(\frac{i\tau}{12} + \ldots\right) \ln \tau + h\tau + \ldots$$

$$a_2 = g\tau^{\frac{1}{2}} \left(1 + \frac{3}{2} f\tau + \ldots\right)$$

$$b_1 = \frac{1}{2\tau} + f + \left(-\frac{\tau}{6} + \ldots\right) \ln \tau + \left(2ih - \frac{1}{3}\right)\tau + \ldots$$

$$b_2 = \frac{-1}{2\tau} + f + \left(\frac{\tau}{6} + \ldots\right) \ln \tau + \left(-2ih - \frac{1}{6}\right)\tau + \ldots$$

where f, g, h and t_o are the 4 arbitrary constants and dots represent higher integer powers of τ. Similar series expansions with $\tau^p \ln \tau$ terms are also obtained for cases (i) and (iii) above, (but only with 3 arbitrary constants). Finally, note that the $\tau^{\frac{1}{2}}$ in the expansion of a_2 can be transformed away by a simple change of variables: $a \equiv a_2^2$, and the problem in a_1, a, b_1, b_2 involves only $\ln \tau$ and integer powers of τ, as claimed.

Now, consider the Hénon-Heiles Hamiltonian (3.3) of Example III: The Painlevé analysis yields the "resonance" equation

$$(r+1) \; r \; (r-5)(r-6) = 0 \qquad (3.12)$$

but here again, as with Example I, the $r = 0$ arbitrary constant cannot be captured by admitting only poles and (at most) 3 parameter family of Painlevé solutions exist.[17] Moreover, the leading order behavior of the solutions near $\tau \to 0$ involves $(\ln \tau)^{-\frac{1}{2}}$ which is again reminiscent of Example I [if one changes variables there to $s \equiv x+y$, $d \equiv x-y$, cf. (3.9)].

So, as with (3.1), one might expect the surfaces of section for (3.3) to look quite "regular", without any large size chaotic regions present. But - surprise! - nothing of the sort happens, as figs. 4-6 clearly show. Unlike Example I, there doesn't seem to be anything special about the $\varepsilon = 2$ Hénon-Heiles case.

The difference between Examples I, II and Example III, from the point of view of our analysis, is that the solutions of (3.3) have more complicated singularities than $\tau^p (\ln \tau)^q$: One actually needs to include terms of the form $\ln(\ln \tau)$ in order to capture the fourth arbitrary constant entering at $\tau = 0$, cf. (3.12). The final result is[28]

$$x = \tau^{-2} \left[3 + \frac{5}{4 \ln \tau} + \frac{3 + 20 f \ln(\ln \tau)}{8 (\ln \tau)^2} + \ldots \right] + \frac{A}{4} + \frac{A^2 \tau^2}{80} + g \tau^4 + \ldots$$

$$y = \left(-\frac{15}{2} \right)^{\frac{1}{2}} \tau^{-2} (\ln \tau)^{-\frac{1}{2}} \left[1 + \frac{120 f \ln(\ln \tau) - 7}{120 \ln \tau} + \ldots \right] + h \tau^3 + \ldots$$

(3.13)

where f, g, h, (and t_o) are the 4 free constants determined by initial conditions.

4. DISCUSSION

We have demonstrated on a number of examples that certain interesting connections appear to exist between the real time behavior of the solutions of dynamical systems and their movable singularities in the complex t plane. First we saw that if the Painlevé property is satisfied [i.e. if the asymptotic expansions of the general solutions involve only (movable) poles], the system is integrable and the motion is "regular" for all initial conditions. Second, we gave an example where partial fulfillment of the Painlevé property leads to partial integrability of the equations of motion. Finally, we presented evidence that the presence of $\ln(t-t_0)$ terms is connected with small scale chaos while $\ln\ln(t-t_0)$ singularities seem to imply large scale chaotic behavior in Hamiltonian systems.

Our results, together with those of other researchers,[17,29] suggest that the singularity analysis of the solutions of dynamical systems can shed some light on very important and long standing questions such as: when is a system integrable, when is it chaotic, to what "degree" is it chaotic and so on. The results described in this paper are far from being complete and even further from being part of a rigorous theory. They are, however, encouraging in the sense that they offer the promise of an analytical treatment in a subject which is as analytically intractable as it is fascinating.

5. ACKNOWLEDGEMENTS

We wish to thank Michael Tabor and his colleagues for organizing a very stimulating and informative Workshop. We also thank Franco Vivaldi for kindly providing us with Figures 1-7 and for many interesting and useful conversations. One of us (Tassos Bountis) gratefully acknowledges many helpful discussions with Robert Helleman, Michael Berry, Mark Ablowitz and Thanassis Fokas. This study was supported in part by the D.O.E. grant DE-AC03-77ER01538 and the Army Research Office.

6. REFERENCES

1. J. Moser, "Near Integrable and Integrable Systems", in "Topics in Nonlinear Dynamics", ed. S. Jorna, A.I.P. Conf. Proc., Vol. 46, A.I.P., New York (1978).
2. M. V. Berry, "Regular and Irregular Motion", in same volume as ref. 1.
3. A. J. Lichtenberg and M. A. Lieberman, "Regular and Stochastic Motion", to appear (E.E. Dept., U. Cal., Berkeley, Calif.) 1981.
4. J. Ford, in Fundamental Problems in Statistical Mechanics, Vol. 3, ed. E.G.D. Cohen, North Holland (1975), and references listed therein.
5. J. Moser, "Lectures on Hamiltonian Systems", Mem. Am. Math. Soc. $\underline{81}$, 1-60 (1968).
6. R.H.G. Helleman "Self-Generated Chaotic Behavior in Deterministic Systems", in Fundamental Problems in Statistical Mechanics, Vol. 5, ed. E.G.D. Cohen, North Holland (1980).
7. V. I. Arnold and A. Avez, "Ergodic Problems of Classical Mechanics", Benjamin, New York (1968).

8. V. V. Golubev, "Lectures on the Integration of the Equations of Motion of a Rigid Body about a Fixed Point" State Publishing House, Moscow (1953); S. Kovalevskaya, Acta Mathematica 14, 81 (1890).
9. J. Moser, Adv. in Math., 16, 197 (1975).
10. H. Flaschka, Phys. Rev. B, 9, 1924 (1974); and Progr. of Theor. Phys. 51, 703 (1974).
11. M. Adler, "Some Finite Dim. Integrable Systems and Their Scattering Behavior", Math. Res. Center Tech. Summary Report, Univ. of Wisconsin (1977).
12. M. A. Olshanetsky and A. M. Perelomov, Inv. Math. 37, 93 (1976).
13. O. I. Bogoyavlenski, Commun. in Math. Phys., 51, 201 (1976); see also B. Kostant, "The Solution to a Generalized Toda Lattice and Representation Theory", preprint (Math. Dept., M.I.T., 1981).
14. T. Bountis, H. Segur and F. Vivaldi, "Integrable Hamiltonian Systems and the Painlevé Property", Phys. Rev. A, to appear (1982).
15. H. Segur, Lectures at International School of Physics, "Enrico Fermi", Varenna, Italy (July 1980).
16. E. L. Ince, "Ordinary Differential Equations", Dover, New York (1944).
17. Y. F. Chang, M. Tabor and J. Weiss, "Analytic Structure of the Hénon-Heiles Hamiltonian in Integrable and Non-Integrable Regimes", preprint (La Jolla Institute, P.O. Box 1434, La Jolla, Calif.); also with G. Corliss Phys. Lett. A to appear (1982).
18. M. Hénon and C. Heiles, Astron. J., 69, No. 1, 73 (1964).
19. G. Contopoulos, Astrophys. J., 138, 1297 (1963); also Astron. J., 68, 1 (1963).
20. Y. Aizawa and N. Saitô, J. Phys. Soc. of Japan, 32, 1636 (1972).
21. M. Toda, J. Phys. Soc. of Japan 23, 501 (1967); also Progr. Theor. Phys. Suppl., 45, 174 (1970).
22. M. Hénon, Phys. Rev. B, 9, 1921 (1974).
23. S. V. Manakov, Soviet Phys. JETP, Vol. 40, 269 (1975).
24. H. Yoshida, "Integrability of Generalized Toda Lattice Systems and Singularities in the Complex T Plane", preprint (Dept. of Astronomy, Univ. of Tokyo, 113 Tokyo, Japan).
25. F. Busse and Heikes in Reports, Science, Vol. 208, 173, (April 1980) and references listed therein.
26. R. May and W. Leonard, SIAM J. Appl. Math., 29, No. 2, 243 (September 1975).
27. M. J. Ablowitz, A. Ramani and H. Segur, Lettere al Nuovo Cimento, Vol. 23, 9, 333 (1978); also J. Math. Phys. 21 (4), 715 and 21 (5), 1006 (1980).
28. A. Ramani, private communication.
29. M. Tabor and J. Weiss, "Analytic Structure of the Lorentz System", Phys. Rev. A, to appear.
30. L. S. Hall, "On the Existence of a Last Invariant of Conservative Motion", preprint (Lawrence Livermore Laboratory, P.O. Box 808-L630, Livermore, CA 94550).

INTEGRALS OF THE TEST WAVE HAMILTONIAN:
A SPECIAL CASE*

J. D. Meiss
Institute for Fusion Studies
The University of Texas at Austin
Austin, Texas 78712

The theory of weakly nonlinear waves in a homogeneous, conservative medium can often be developed in terms of a Hamiltonian which retains only cubic nonlinearities. This approach has been stressed by Ken Watson in studies of the interactions among internal waves[1,2] and between surface and internal waves[3] in the ocean.

The process of relaxation of a single wave mode in a bath of ambient modes can be described by a model of this type: The test wave Hamiltonian. This model was proposed by Watson[3] for the study of the interaction between a single internal wave and a spectrum of surface waves--and thus as a mechanism for the transfer of energy from the ocean surface to its interior.

In this model the wave actions are represented by $(J_T; J_i, J'_i$ $i = 1,2,\ldots M)$ where J_T represents the test wave and the M pairs (J_i, J'_i) are ambient waves which form M interacting triads with J_T. The subscripts refer to wavenumbers and interactions are allowed only when

$$\underline{k}_T + \underline{k}_i \pm \underline{k}'_i . \tag{1}$$

We refer to the two possible triads in Eq. (1) as sum and difference interactions.

The test wave Hamiltonian may be written

$$\mathcal{H} = \omega_T J_T + \sum_{i=1}^{M} (\omega_i J_i + \omega'_i J'_i)$$

$$- \sum_{n=1}^{M} \varepsilon_n^{\pm} \sqrt{J_T J_n J'_n} \cos(\theta_n \pm \theta'_n - \theta_T) . \tag{2}$$

Here the θ's represent wave phases, and the ε's are coupling coefficeints which are generally functions of the three wavenumbers. To linear order ($\varepsilon \to 0$) the phases evolve at the linear frequency:

---

*Presented at the "Mathematical Methods in Hydrodynamics" Workshop in La Jolla, December 7-9, 1981.

$\dot{\theta} = \omega + O(\varepsilon)$, where ω of course depends upon $\underline{k}$.

Numerical integration of the equations of motion for the Hamiltonian (2) indicates that it is completely integrable.[4] The evidence is threefold. First, two orbits initially close together separate only linearly in time (Lyapunov exponents are zero). In a nonintegrable system the separation is typically exponential. Second, the Poincaré surface of section for the two triad (M=2) system (which is, in this case, two-dimensional) apparently consists of smooth level curves indicating the existence of an additional integral. Finally, a quantitative test of the smoothness of the curves (the residue method of J. Greene) shows that to double precision accuracy they are indeed smooth. A discussion of these issues is given in greater detail in Ref. 4 and in my dissertation[5] (see, however, Ref. 8).

The numerical evidence presented above seems insensitive to the parameters of the Hamiltonian. I am therefore lead to the conjecture that Eq. (2) is integrable for arbitrary ω's and ε's. I should note, however, that Ref. 4 deals only with the difference interactions in Eq. (2). As we will see below, the Hamiltonian with only sum interactions also appears integrable; however, when there is a mixture of sum and difference interactions the behavior is unknown.

Integrability in the sense of Liouville[6] means the existence of N=2M+1 integrals--that is, one integral for each degree of freedom. I use the term integrals for a set of functions on phase space which are functionally independent and in involution (all Poisson brackets zero). Constants of motion are merely time independent ($\{F, \mathcal{H}\} = 0$) functions. Liouville's theorem on integrability shows that once the N integrals are known the equations of motion can be integrated by quadrature. Essentially, the integrals can be used as canonical momenta, and the Hamiltonian expressed in terms of these momenta is independent of the conjugate variables.

For the test wave Hamiltonian, we can immediately find M+2 of the integrals. The first is the Hamiltonian itself and the others are

$$I_n = J_n - sJ_n', \qquad n = 1, 2, \ldots M,$$

$$I_T = J_T + \sum_{n=1}^{M} J_n . \qquad (3)$$

Here, s = +1(-1) for sum (difference) triads. These integrals are related to symmetries of $\mathcal{H}$. Each of the I_n is obtained by noting that the transformation

$$\theta_n \to \theta_n + \psi ,$$

$$\theta_n' \to \theta_n' - s\psi ,$$

leaves $\mathcal{H}$ invariant. The remaining integral results from the symmetry

$$\theta_T \to \theta_T + \psi ,$$

$$\theta_n \to \theta_n + \psi , \text{ for } n = 1,2,\ldots M$$

To make further progress it is convenient to introduce the action-amplitude variables (see Watson's discussion in Ref. 1)

$$a_T = \sqrt{J_T} \exp^{-i\theta_T} \quad a_n = \sqrt{J_n} \exp^{-i\theta_n}, \quad b_n = \sqrt{J_n'} \exp^{-i\theta_n'} \quad (4)$$

These variables are not quite canonical, obeying the Poisson bracket relations

$$\{a_n^*, a_m\} = i\delta_{m,n} \quad \{b_n^*, b_m\} = i\delta_{n,m} \quad \{a_T^*, a_T\} = i \quad (5)$$

with all other brackets zero. If the Hamiltonian, Eq. (2), is written in terms of Eq. (4)

$$\mathcal{H} = \omega_T a_T^* a_T + \sum_{n=1}^M \left(\omega_n a_n^* a_n + \omega_n' b_n^* b_n\right)$$

$$- \frac{1}{2} \sum_{n=1}^M \left\{ \begin{array}{l} \varepsilon_n^+ a_T a_n^* b_n^* \\ \varepsilon_n^- a_T a_n^* b_n \end{array} + \text{complex conjugate} \right\} \quad (6)$$

then the equations of motion become

$$\dot{a}_n = \{a_n, \mathcal{H}\} = \{a_n, a_m^*\} \frac{\partial \mathcal{H}}{\partial a_m^*} = -i \frac{\partial \mathcal{H}}{\partial a_n^*}$$

or, more explicitly

$$\dot{a}_T = -i\omega_T a_T + \frac{i}{2} \sum_{n=1}^M \left\{ \begin{array}{l} \varepsilon_n^{+*} a_n b_n \\ \varepsilon_n^{-*} a_n b_n^* \end{array} \right\}$$

$$\dot{a}_n = -i\omega_n a_n + \frac{i}{2} \left\{ \begin{array}{c} \varepsilon_n^+ a_T b_n^* \\ \varepsilon_n^- a_T b_n \end{array} \right\}$$

$$\dot{b}_n = -i\omega_n' b_n + \frac{i}{2} \left\{ \begin{array}{c} \varepsilon_n^+ a_T a_n^* \\ \varepsilon_n^{-*} a_T^* a_n \end{array} \right\} \quad (7)$$

Here we have allowed complex valued coupling coefficients, ε_n. In Eqs. (6) and (7) the upper expressions are used if the nth triad is a sum triad and the lower if it is a difference triad.

The integrals for Eq. (6) can be obtained for a special set of parameter values by generalizing a result of Hald.[7] Hald found constants of motion in addition to those in Eq. (3) for a system which is equivalent to our M=2 case if all the frequencies are zero and if the coupling coefficients are equal. For Hald's system these constants are quadratic in the amplitudes (a_n, b_n).

When the frequencies are nonzero, it turns out that the magnitudes of Hald's constants are time independent.[5] Special values of the parameters are required for this result:

$$\varepsilon_n^2 = \varepsilon^2,$$

$$\Delta_n^\pm = \omega_n \pm \omega_n' - \omega_0 = \Delta. \quad (8)$$

Here ε is any complex constant and Δ is a real constant representing the resonance mismatch.

For the Hamiltonian, Eq. (6), with Eq. (8), the new constants are

$$|I_{ij}|^2 = \left\{ \begin{array}{c} |\varepsilon_i^+ a_i a_j^* - \varepsilon_j^+ b_i^* b_j|^2 \\ |\varepsilon_i^- a_i a_j^* + \varepsilon_j^- b_i b_j^*|^2 \end{array} \right\}, \quad i,j = 1,2,\ldots M \quad (9)$$

Here we require the triads labeled i and j to be either both sum or both difference interactions. From now on we consider only the cases where the triads are either all sum or all difference triads. The mixed case requires additional constants for integrability. Since $|I_{ij}|^2 = |I_{ji}|^2$ and $|I_{ii}|^2 = I_i^2$ there are ½M(M - 1) new constants in Eq. (9).

These constants are not integrals, however, since they are not in involution. It is easy to see that the Poisson brackets $\{|I_{ij}|^2, I_0\}$ and $\{|I_{ij}|^2, I_n\}$ are zero. The bracket of two of the new constants is

$$\{|I_{ij}|^2, |I_{k\ell}|^2\} = i\varepsilon_i \varepsilon_j \varepsilon_k \varepsilon_\ell \Big\{ \varepsilon_i \delta_{i,\ell} \left(I_{ij} I_{jk} I_{ki} - I_{ik} I_{kj} I_{ji} \right)$$

$$+ \varepsilon_j \delta_{j,k} \left(I_{ji} I_{i\ell} I_{\ell j} - I_{j\ell} I_{\ell i} I_{ij} \right)$$

$$+ \varepsilon_k \delta_{k,i} \left(I_{kj} I_{j\ell} I_{\ell k} - I_{k\ell} I_{\ell j} I_{jk} \right)$$

$$+ \varepsilon_\ell \delta_{\ell,j} \left(I_{\ell i} I_{ik} I_{k\ell} - I_{\ell k} I_{ki} I_{i\ell} \right) \Big\}$$

(10)

It follows from Eq. (10) that if none of the four indices, i, j, k, ℓ are equal, the bracket is zero. Furthermore, it is easy to see that

$$\{\varepsilon_i |I_{ij}|^2 + \varepsilon_k |I_{kj}|^2, |I_{ik}|^2\} = 0 \ . \tag{11}$$

The integrals can be constructed from linear combinations like that in Eq. (11) and we obtain

$$C_n = \sum_{j=1}^{n} \varepsilon_j |I_{jn+1}|^2 \ , \quad n = 1, 2, \ldots, M-1 \ . \tag{12}$$

These integrals are involutive

$$\{C_n, C_m\} = 0 \ .$$

Furthermore, it is easy to see that these integrals are all independent. Each of the I_n depends on a new variable J_n' making them independent from I_T and I_m, and similarly each of the C_n is a function of a new phase θ_{n+1}.

A complete set of 2M+1 integrals for the test wave Hamiltonian, Eq. (6), given the parameters of Eq. (8), and either all-sum or all-difference interactions is given by

$$\mathcal{H}, I_T, I_1, I_2, \ldots, I_M, C_1, C_2, \ldots, C_{M-1}$$

As the simplest example, when M = 2, the new integral is

$$\frac{1}{\epsilon_1} C_1 = |I_{12}|^2 = J_1 J_2 + J_1' J_2' - 2s\epsilon_1^s \epsilon_2^s \sqrt{J_1 J_2 J_1' J_2'}$$

$$\times \cos\left[\theta_1 - \theta_2 + s(\theta_1' - \theta_2')\right]$$

While these integrals prove integrability for the equal coupling coefficient, equal resonance mismatch model, numerical evidence indicates integrability more generally. As an example, I present a surface of section when $\Delta_1 \neq \Delta_2$, $\epsilon_1^2 \neq \epsilon_2^2$ with difference interactions (Fig. 1). There is no visible evidence of stochasticity. It remains a challenge to discover the integrals for this case, if indeed they exist.[8]

Finally we note that additional interactions will typically destroy the integrability of the test wave system. For example, when a triad involving J_1, J_1' and J_2 is added, the orbits become obviously stochastic.[9]

ACKNOWLEDGEMENTS

I am very grateful to N. Pomphrey for helpful discussions about this work and to J. Ford for encouragement and suggestions. This work was partially supported by ONR contract N00014-78-C-0050 and United States Department of Energy Contract DE-FG05-80ET-53088.

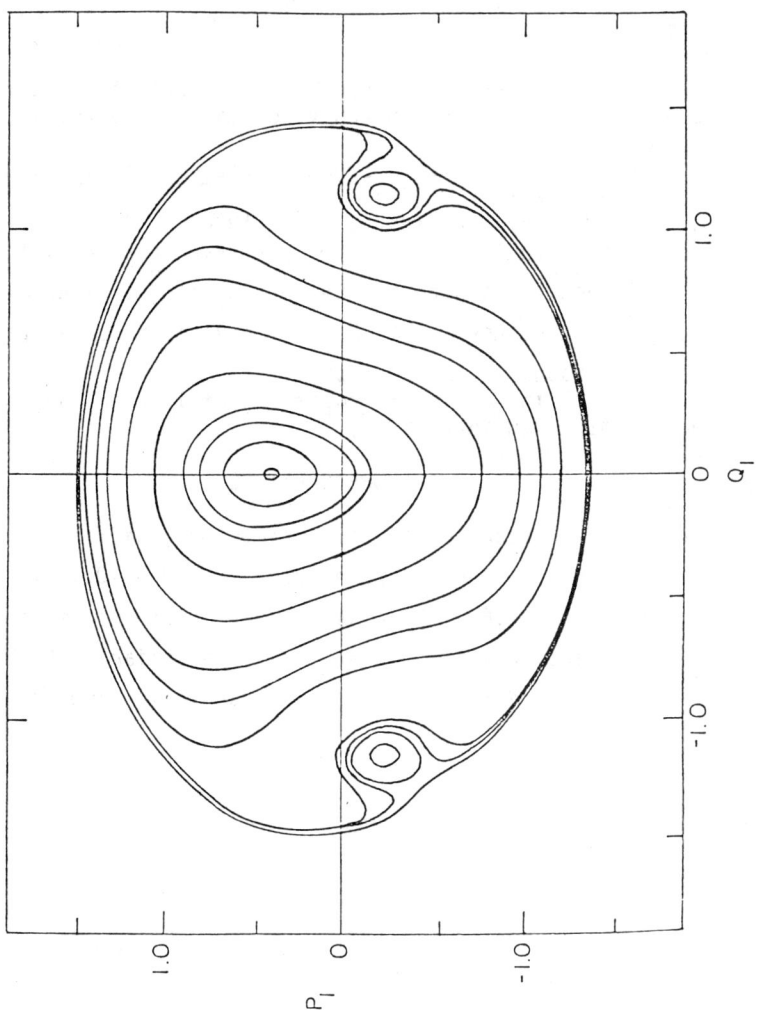

Fig. 1

REFERENCES

1. J. D. Meiss and K. M. Watson, in <u>Topics in Nonlinear Dynamics</u>, S. Jorna (ed.), (AIP, New York, 1978).
2. J. D. Meiss, N. Pomphrey and K. M. Watson, Proc. Nat'l Acad. Sci. USA, <u>76</u>, 2109, (1979).
3. K. M. Watson, B. J. West and B. I. Cohen, J. Fluid Mech. <u>77</u>, 185 (1976).
4. J. D. Meiss, Phys. Rev. A<u>19</u>, 1780 (1979).
5. J. D. Meiss, "Statistical Dynamics of Weakly Nonlinear Internal Waves", PhD. Dissertation (Univ. of Cal. at Berkeley, 1980), unpublished.
6. V. I. Arnold and A. Avez, <u>Ergodic Problems of Classical Mechanics</u>, (Benjamin, New York, 1968).
7. O. H. Hald, Phys. Fluids <u>19</u>, 914 (1976).
8. Y. C. Lee, H. H. Chen and C. Menyuk have apparently found numerical evidence for stochasticity when $\varepsilon_1^2 \neq \varepsilon_2^2$. When $\Delta_1 \neq \Delta_2$, $\varepsilon_1^2 = \varepsilon_2^2$ Y. C. Lee and H. H. Chen have found a Lax pair for the system (personal communication).
9. J. D. Meiss, in <u>Nonlinear Properties of Internal Waves</u>, B. West (ed.), (AIP, New York, 1981).

FIGURE CAPTIONS

1. Surface of section for the two triad (M=2) difference Hamiltonian. The four known integrals have the values $H = -0.1$, $I_T = 2.01$, $I_1 = 1.1$, $I_2 = 1.6$. The coupling parameters are $\varepsilon_1 = -0.37$, $\varepsilon_2 = -1.0$ and $\Delta_1 = 0.2$, $\Delta_2 = 0.13$. The variables P, Q are defined by $\sqrt{2a} = P + iQ$. The surface shown is given by $Q_2 = 0$.

ON THE STRUCTURE OF THE HYDRODYNAMICAL EQUATIONS
FOR TWO-DIMENSIONAL FLOWS OF AN INCOMPRESSIBLE FLUID:
THE ROLE OF INTEGRAL INVARIANCE

Philip Duncan Thompson
National Center for Atmospheric Research[1]
Boulder, Colorado 80307

ABSTRACT

This paper is concerned mainly with the equivalence of the principles of local vorticity conservation and invariance of total enstrophy for inviscid two-dimensional flows--in the sense that either is derivable from the other. A short and fairly complete summary of the development is given in Section 5.

1. INTRODUCTION

Earlier observations that two-dimensional flows are very peculiar are almost too numerous, ancient and well-known to cite. Perhaps the most remarkable and, indeed, distinguishing feature is that the vorticity of each individual element in the two-dimensional flow of an incompressible, inviscid fluid is locally conserved following its motion. In the case of steady flows, this fact has the immediate consequence that vorticity is constant along each streamline: if, in particular, each streamline originated in a region where the vorticity were zero, the motion would be irrotational everywhere. In two-dimensional flow, this implies that the streamfunction and velocity potential satisfy the Cauchy-Riemann conditions, whence the flow is described by analytic functions of a complex variable. Much of the theory of airfoils is founded on these simple properties of two-dimensional, almost inviscid flows, and there is an enormous body of theory in other fields that ultimately stems from the principle of vorticity conservation.

The consequences of vorticity conservation in non-steady two-dimensional flows, however, have not been fully appreciated until relatively recently. It has long been known, of course, that the principle of vorticity conservation alone implies that the total kinetic energy of a bounded two-dimensional inviscid flow is invariant from one time to next.[2] Noting that the integral of squared vorticity in such flows (now called "enstrophy,": after Leith, 1968) is also invariant, Fjörtoft (1953) pointed out that kinetic energy must

---

[1] The National Center for Atmospheric Research is sponsored by the National Science Foundation

[2] We reserve the expression "invariant" to refer to integrals taken over the entire flow. The term "conserved" will apply to quantities that are conserved locally, in the Lagrangian sense.

be transferred both upscale and downscale by nonlinear interactions between motions of different scales. As a consequence, Batchelor (1969) later speculated that "turbulent" two-dimensional flows (i.e., initially stirred in some random fashion) might, under certain conditions, exhibit two more or less distinct "inertial sub-ranges": one, maintained by the transfer of kinetic energy to large scales and, the other, in which the enstrophy of the increasingly tangled flow cascades to small scales. Indeed, arguing from this simple view of turbulent two-dimensional flow, Batchelor predicted that the spectrum of kinetic energy per unit scalar wave number k in the small scale sub-range would fall off as k^{-3}. Since then, the constraints of energy and enstrophy invariance, from which the behavior described above derives, have been "built-in" to most successful closure schemes for predicting the statistical behavior of turbulent two-dimensional flows--e.g., the DIA of Kraichnan (1967) and the Markovian model of Leith (1968)--and that's about all the physics they contain. The predicted k^{-3} spectrum in the small scale sub-range has been fairly well confirmed experimentally by Tsinober (1972), by the numerical simulations of Lilly (1969) and many others, and from direct observation of quasi-two-dimensional atmospheric flows by Kao (1970) and Wiin-Nielson (1967).

There is something a little puzzling about the success of relatively simple closure schemes based on the invariance of total kinetic energy and enstrophy. On reflection, one sees that the integral of any differentiable function of vorticity, taken over an entire inviscid bounded flow, is invariant. Why, then, is it sufficient to single out the principle of enstrophy invariance (in addition to energy invariance) as the main dynamical constraint on the statistical behavior of the flow? In fact, it would appear that we suffer from an embarrassment of riches. With an infinite number of constraints of invariance, how can the flow change?

The simplest resolution of this paradox and the one I believe to be correct is that the invariance of total kinetic energy and enstrophy of an inviscid bounded flow implies the local conservation of vorticity, following the motions of individual fluid elements. If this strong consequence is indeed true, the invariance of energy and enstrophy further imply the invariance of all integrals of differentiable functions of vorticity. In that event, since the invariance of more general integrals of motion are not in this sense independent of the invariance of energy and enstrophy, there is no paradox.

In an earlier effort, Thompson (1973) attempted to derive this result by an indirect argument, but without conscious realization that there was an underlying assumption--namely, that the instantaneous local changes in vorticity in an inviscid flow are literally "local," in the sense that they do not depend on functionals or otherwise depend on the state of flow at points far removed from the point where the changes take place. In the meanwhile, it has occurred to me that the same result can be shown by a more direct route, and one that reveals more clearly the connection between the principles of invariance and the detailed mathematical structure of the hydrodynamical equations for two-dimensional incompressible flow.

For reasons that will become apparent in Section 2, we shall first consider the local change of vorticity as a general local operator, and then see how it must be restricted in order to satisfy certain a priori conditions on the form of any universal physical law: these are primarily conditions of universality, uniqueness and invariance of form under rotation and translation of an arbitrary coordinate system. This reduces the original operator to a series of symmetrical and antisymmetrical differential operators whose coefficients depend at most on physical properties of the fluid (e.g., its viscosity) or, in the case of inviscid flow, on universal dimensionless constants. With the introduction of the condition of dimensional homogeneity, the form of the operator is further restricted to that of a very small subset of the class of local operators. The form of the vorticity equation for two-dimensional inviscid flow is then completely determined (to within an arbitrary time scale), by requiring invariance of total kinetic energy and enstrophy. It therefore appears that these invariance principles alone imply a great deal about the detailed dynamics of the flow.

The first integral of the vorticity equation in this case is seen to be the Eulerian equation of motion, with a scalar function of integration determined by the condition for incompressibility: that function, of course, is identified with "energy" but is not essential to the description of the flow.

2. THE LOCALNESS OF VORTICITY CHANGE

One's view of hydrodynamics is, of course, strongly colored by already having a fairly complete and sophisticated knowledge of a body of theory that has been built up over some three centuries, and originating in the physical laws of ballistics and celestial mechanics. Generalized to apply to fluids (through Newton's second law), the theory often refers to "forces," a notion that might not occur to anyone who had never encountered a moveable or freely moving solid object. In what follows, it may be useful for the reader to pretend that he knows nothing of hydrodynamical theory and to take an extremely naive and primitive view of the behavior of two-dimensional flows--such as that of a mathematically educated fish in a sea of Abbott's Flatland.

We assume, however, that we do have some means of measuring instantaneous fluid velocity $\underset{\sim}{V}$ and have experimentally established that two-dimensional flow is nondivergent. I.e.,

$$\nabla \cdot \underset{\sim}{V} = 0 \qquad (1)$$

from which we would certainly deduce that

$$\underset{\sim}{V} = \underset{\sim}{K} \times \nabla\psi \qquad (2)$$

where $\underset{\sim}{K}$ is a unit vector directed normal to the plane of motion, and ψ is some scalar (containing an arbitrary additive function of time only) that describes the flow at a particular time.

We may also have observed more or less distinct and persistent entities in the flow, structures that we may have chosen to call "eddies" or "vortices,"[3] and that these are associated with rather sharp extremes of some measure of the rate of spin or "vorticity" of the fluid, defined as

$$\zeta = \underset{\sim}{K} \cdot \nabla \times \underset{\sim}{V} = \nabla^2 \psi \qquad (3)$$

We may even have been led to investigate experimentally what determines the motion of these peculiar entities and, more particularly, how the changes of vorticity at a fixed point depend on the structure of the flow.

It is not <u>a priori</u> evident that the instantaneous change of vorticity at a fixed point depends only on the structure of the flow in the immediate neighborhood of that point and not at points elsewhere in the flow . One can, however, imagine a simple experiment that would test this hypothesis. Suppose we set up a symmetrical laminar flow in a straight channel with constant speed U across the broad central portion of the flow and shear zones near the walls. Along the center line we place an array of uniformly spaced vorticity meters: initially, of course, they register zero vorticity. We next create an intense circular vortex somewhere along the centerline upstream from the vorticity meters, but small enough across that the associated circulation is weak. Subsequently, we record the readings of the vorticity meters at a succession of times.

If we plotted the vorticity on a "world diagram" as a function of time t and distance x along the centerline, it would look something like Fig. 1. Looking at one such diagram, we would undoubtedly be struck by the fact that the "vorticity vs. time" curves look remarkably similar to the "vorticity vs. distance" curves. One might even have noticed that, if x is scaled in units of L and t is scaled in units of L/U, those curves would be virtually identical. This

---

[3] See, for example, Da Vinci's description and sketches of flows in canals.

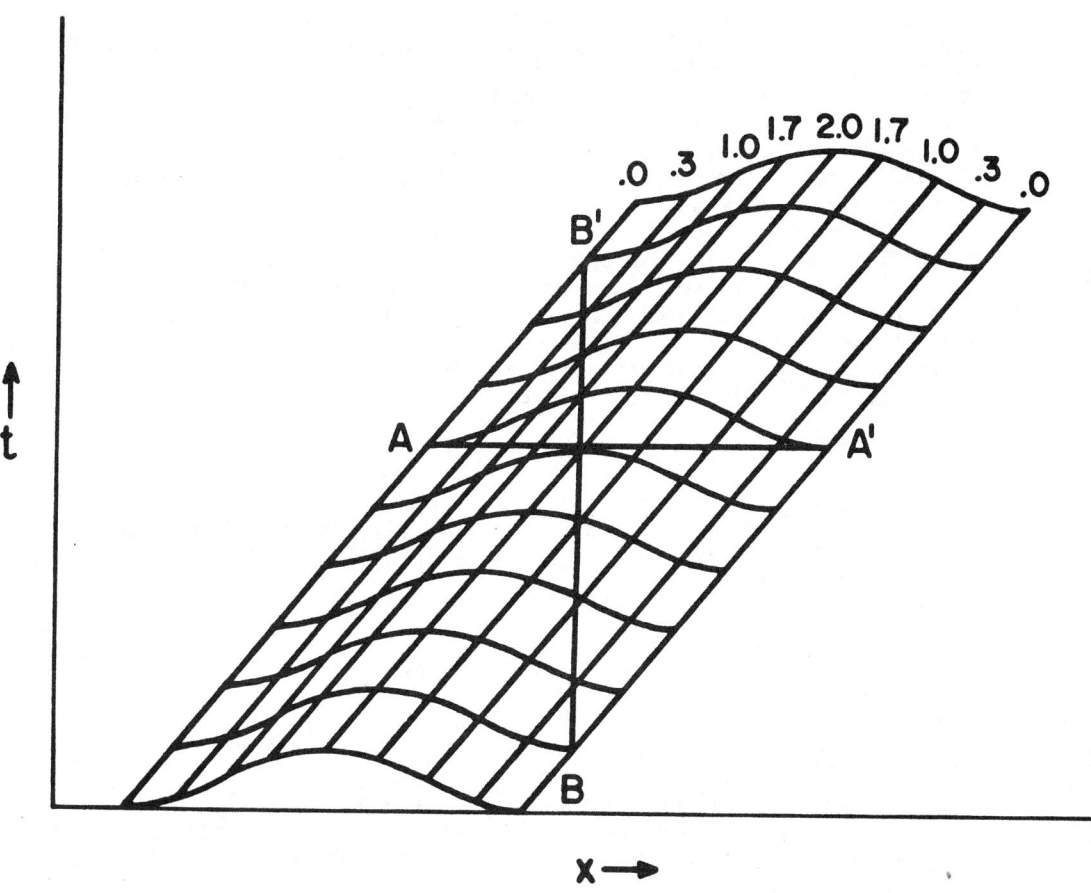

FIG. 1

alone would strongly suggest that the changes of vorticity at a fixed point depend only on the structure of the flow in the immediate neighborhood of that point.

To reassure ourselves about this latter surmise, we might then repeat the experiment several times under conditions that are otherwise identical, except that the central zone of constant speed U is made narrower and narrower. But the readings of the vorticity meters would remain unchanged from one experiment to the next. This would appear to confirm that the changes of vorticity at a fixed point do not depend on the flow at points far removed from it.

Such an imaginary experiment could not, of course, <u>prove</u> the hypothesis: that is not in the nature of experiments. They may test an hypothesis and, indeed, prove its negation, but they cannot strictly prove its affirmation. Accordingly, the statement that local changes of vorticity depend only on the structure of the flow in the immediate neighborhood does not have the force of <u>a priori</u> knowledge, but is to be taken as qualitative physical law, a generalized conception of what one perceives in a limited number of situations.

Knowing what we do actually know, one would be tempted to conclude from this imaginary experiment that vorticity is conserved. At this point, however, we stop short of drawing this strong quantitative inference, but deliberately choose to draw only the much weaker qualitative inference of "localness." The reason for this choice is that the consequences of <u>a priori</u> requirements on the form of the hydrodynamical equations become clearer.

3. CONSEQUENCES OF LOCALNESS

Our next concern is to exploit the condition of "localness"--i.e., that the time derivative of vorticity at a fixed point depends only on properties of the ψ-field that are definable in the neighborhood of that point. I.e.,

$$\frac{\partial}{\partial t} \nabla^2 \psi = N(\psi, \frac{\partial \psi}{\partial x}, \frac{\partial \psi}{\partial y}, \frac{\partial^2 \psi}{\partial x^2}, \frac{\partial^2 \psi}{\partial x \partial y}, \frac{\partial^2 \psi}{\partial y^2}, \text{etc.}) \tag{4}$$

where $\partial \nabla^2 \psi / \partial t$ and the operator N apply at the point (x,y), and (x,y) are cartesian coordinates in the plane of motion, with arbitrary orientation and origin. We may write the operator N more compactly as

$$N = N(\psi, D_{mn}) \tag{5}$$

where D_{mn} is defined by

$$D_{mn} = \frac{\partial^m \psi}{\partial x^n \partial y^{m-n}}$$

with the convention that n takes on all integer values from zero to m, and m takes on all integer values from one to infinity.

We suppose that N is a differentiable function of its arguments ψ and D_{mn} (though not that the latter are continuous and differentiable functions of x and y), and that it may be regarded as a Maclaurin expansion around the state of rest, when $N = 0$, $D_{mn} = 0$, and $\psi = \psi_0$ (a constant). In general, then,

$$N = \sum_{m=1}^{\infty} \sum_{n=0}^{m} D_{mn} \left(\frac{\partial N}{\partial D_{mn}}\right)_0 + \frac{1}{2} \sum_{m=1}^{\infty} \sum_{n=0}^{m} \sum_{p=1}^{\infty} \sum_{q=0}^{p} D_{mn} D_{pq} \left(\frac{\partial^2 N}{\partial D_{mn} \partial D_{pq}}\right)_0$$

$$+ \frac{1}{6} \sum_{m=1}^{\infty} \sum_{n=0}^{m} \sum_{p=1}^{\infty} \sum_{q=0}^{p} \sum_{k=1}^{\infty} \sum_{\ell=0}^{k} D_{mn} D_{pq} D_{k\ell} \left(\frac{\partial^3 N}{\partial D_{mn} \partial D_{pq} \partial D_{k\ell}}\right)_0$$

+ terms of quartic and higher degree in D_{mn}

+ terms containing factors of $(\psi-\psi_0)$ (6)

in which the differential coefficients of N with subscript zero denote their values near the state of rest. These are at most functions of ψ_0: however, since ψ_0 contains an arbitrary additive function of time, those derivatives must be constants that are independent of the flow--either dimensional physical constants of the medium (e.g., kinematic molecular viscosity) or universal dimensionless constants. Otherwise, Eq. (4) could not qualify as a unique and universal physical law. For the same fundamental reason, N can have no terms containing factors of $(\psi-\psi_0)$. Accordingly, N is of the form:

$$N = \sum_{m=1}^{\infty} \sum_{n=0}^{m} L_{mn} D_{mn}$$

$$+ \sum_{m=1}^{\infty} \sum_{n=0}^{m} \sum_{p=1}^{\infty} \sum_{q=0}^{p} Q_{mnpq} D_{mn} D_{pq}$$

$$+ \sum_{m=1}^{\infty} \sum_{n=0}^{m} \sum_{p=1}^{\infty} \sum_{q=0}^{p} \sum_{k=1}^{\infty} \sum_{\ell=0}^{k} C_{mnpqk\ell} D_{mn} D_{pq} D_{k\ell}$$

$$+ \text{ terms of quartic and higher degree in } D_{mn} \quad (7)$$

where the L's, Q's, and C's are constants.

It is immediately evident that N, as given by Eq. (7), is invariant under <u>translation</u> of the origin of the coordinate system, simply because N depends only on derivatives of ψ. It may also be invariant under <u>rotation</u> of the coordinate axes, but only if certain conditions on the L's, Q's, and C's are satisfied.

For the purpose of analyzing these conditions and for other reasons that will appear later, it is convenient to group the terms of (7) according to their <u>degree</u> and <u>order</u>. The "degree" of a term is the number of factors D_{mn} it contains: the "order is the <u>total</u> number of differentiations of ψ. For example, the degree of $(\partial\psi/\partial x)$ $(\partial^2\psi/\partial x\partial y)$ is 2, and its order is 3. The terms of (7) in each category (without their coefficients) are listed in Table 1, for degree up to 3 and order up to 6. This table will prove sufficient for the present purposes.

Now, the rotational transformation is:

$$x' = x \cos\theta + y \sin\theta$$
$$y' = -x \sin\theta + y \cos\theta \quad (8)$$

where x and y are the original cartesian coordinates: x' and y' are the coordinates after the axes have been rotated through an angle θ. From Eq. (8), we have

$$\frac{\partial}{\partial x} = \frac{\partial x'}{\partial x}\frac{\partial}{\partial x'} + \frac{\partial y'}{\partial x}\frac{\partial}{\partial y'} = \cos\theta \frac{\partial}{\partial x'} - \sin\theta \frac{\partial}{\partial y'}$$

$$\frac{\partial}{\partial y} = \frac{\partial x'}{\partial y}\frac{\partial}{\partial x'} + \frac{\partial y'}{\partial y}\frac{\partial}{\partial y'} = \sin\theta \frac{\partial}{\partial x'} + \cos\theta \frac{\partial}{\partial y'}$$

Introducing these relationships into the set of terms in each category, we see that each set transforms into a set of the same form, but possibly with different coefficients. The problem, then, is to determine the coefficients in such a way that they remain the same after rotation of the coordinate axes. This condition must hold for each category separately, in order that (7) remain <u>identical in form</u> under the rotational transformation. It is, in fact, a very strong condition and one which immediately eliminates many terms as admissible candidates for inclusion in N.

As an example, let us consider a category of terms of degree 1 and of odd order N. I.e.,

$$S_{1,N} = A_0 \frac{\partial^N \psi}{\partial x^N} + A_1 \frac{\partial^N \psi}{\partial x^{N-1} \partial y} + A_2 \frac{\partial^N \psi}{\partial x^{N-2} \partial y^2} + \ldots + A_N \frac{\partial^N \psi}{\partial y^N} \quad (9)$$

where the subscripts on S indicate the degree and order of the sum $S_{1,N}$, respectively. Now, if S_{1N} is to be invariant under rotation through <u>any</u> angle θ, it must at least be invariant under a rotation of $\theta = \pi/2$, in which case

$$\frac{\partial}{\partial x} = -\frac{\partial}{\partial y'}, \qquad \frac{\partial}{\partial y} = \frac{\partial}{\partial x'}$$

Introducing these relationships into (9) and noting that N is odd, we find that

$$S_{1,N} = -A_0 \frac{\partial^N \psi}{\partial y'^N} + A_1 \frac{\partial^N \psi}{\partial y'^{N-1} \partial x'} - A_2 \frac{\partial^N \psi}{\partial y'^{N-2} \partial x'^2} + \ldots$$

$$- A_{N-1} \frac{\partial^N \psi}{\partial y' \partial x'^{N-1}} + A_N \frac{\partial^N \psi}{\partial x'^N}$$

Thus, in order to maintain identity of form with (9), we must take:

$$A_0 = A_N \qquad\qquad A_N = -A_0$$
$$A_1 = -A_{N-1} \qquad\qquad A_{N-1} = A_1$$
$$A_2 = A_{N-2} \qquad\qquad A_{N-2} = -A_2$$
$$\text{etc.} \qquad\qquad \text{etc.}$$

In short, all of the A's are zero, and $S_{1,N}$ cannot be an admissible group of terms of (7), <u>if N is odd</u>. This immediately eliminates 3 of the 18 categories listed in Table 1.

As another example, let us consider the sum

$$S_{2,4} = A_0 \frac{\partial \psi}{\partial x}\frac{\partial^3 \psi}{\partial x^3} + A_1 \frac{\partial \psi}{\partial x}\frac{\partial^3 \psi}{\partial x^2 \partial y} + A_2 \frac{\partial \psi}{\partial x}\frac{\partial^3 \psi}{\partial x \partial y^2} + A_3 \frac{\partial \psi}{\partial x}\frac{\partial^3 \psi}{\partial y^3}$$

$$+ B_0 \frac{\partial \psi}{\partial y}\frac{\partial^3 \psi}{\partial x^3} + B_1 \frac{\partial \psi}{\partial y}\frac{\partial^3 \psi}{\partial x^2 \partial y} + B_2 \frac{\partial \psi}{\partial y}\frac{\partial^3 \psi}{\partial x \partial y^2} + B_3 \frac{\partial \psi}{\partial y}\frac{\partial^3 \psi}{\partial y^3}$$

As in the previous case, we first rotate the coordinate system through $\theta = \pi/2$ and equate the coefficients of like terms. Imagining that we replace x by -y and y by x, then by inspection we see that

$$B_0 = -A_3 \qquad B_1 = A_2 \qquad B_2 = -A_1 \qquad B_3 = A_0$$

whence $S_{2,4}$ reduces to

$$S_{2,4} = A_0 \left(\frac{\partial \psi}{\partial x}\frac{\partial^3 \psi}{\partial x^3} + \frac{\partial \psi}{\partial y}\frac{\partial^3 \psi}{\partial y^3} \right) + A_2 \left(\frac{\partial \psi}{\partial x}\frac{\partial^3 \psi}{\partial x \partial y^2} + \frac{\partial \psi}{\partial y}\frac{\partial^3 \psi}{\partial x^2 \partial y} \right)$$

$$+ A_1 \left(\frac{\partial \psi}{\partial x}\frac{\partial^3 \psi}{\partial x^2 \partial y} - \frac{\partial \psi}{\partial y}\frac{\partial^3 \psi}{\partial x \partial y^2} \right) + A_3 \left(\frac{\partial \psi}{\partial x}\frac{\partial^3 \psi}{\partial y^3} - \frac{\partial \psi}{\partial y}\frac{\partial^3 \psi}{\partial x^3} \right)$$

To find the invariant subset, we next rotate the coordinate system through $\theta = \pi/4$, letting

$$\frac{\partial}{\partial x} = \frac{1}{\sqrt{2}}(\frac{\partial}{\partial x'} - \frac{\partial}{\partial y'}) \qquad \frac{\partial}{\partial y} = \frac{1}{\sqrt{2}}(\frac{\partial}{\partial x'} + \frac{\partial}{\partial y'})$$

Transforming and equating the coefficients of $(\partial\psi/\partial x)(\partial^3\psi/\partial x^3)$ and $(\partial\psi/\partial x')(\partial^3\psi/\partial x'^3)$, we find that $A_2 = A_0$. Similarly, equating the coefficients of $(\partial\psi/\partial y)(\partial^3\psi/\partial x^3)$ and $(\partial\psi/\partial y')(\partial^3\psi/\partial x'^3)$, we get $A_3 = A_1$. Thus, the only rotationally invariant form of $S_{2,4}$ is:

$$S_{2,4} = A_0 \, \nabla\psi \cdot \nabla(\nabla^2\psi) + A_1 \, \kappa \cdot \nabla\psi \times \nabla(\nabla^2\psi)$$

There are, of course, more general and straightforward procedures for finding the invariant subsets, but the algebra is simpler if one judiciously chooses certain particular values of θ. In any event, one can easily find the invariant groups of terms in each of the categories of degree and order listed in Table 1. These are shown in Table 2, and are all clearly invariant under rotation. N is then just a linear combination of those invariant scalar operators, with coefficients that are either dimensional physical constants or universal dimensionless constants.

4. THE CONDITION OF DIMENSIONAL HOMOGENEITY

The form of N that is admissible on very general grounds is further limited by requiring that Eq. (4) be dimensionally homogeneous—i.e., that each term of N have the same dimensions as $(\partial/\partial t)\nabla^2\psi$. The streamfunction ψ has dimensions $(\text{length})^2/\text{time}$, so that each term of N must have the dimensions $(\text{time})^{-2}$.

Let us first examine the case when the flow does not depend on any physical constants. It is, in fact, observed that the gross features and behavior of the flow are very nearly independent of the physical properties of the medium at very low Mach numbers and very high Reynolds numbers. In this case, the coefficients in the various terms of N are all universal <u>dimensionless</u> constants. We next remark that N contains no time-derivatives. Thus, the only admissible terms of N are those that are quadratic in ψ. Further, in order that they have the dimensions $(\text{time})^{-2}$, they must also be of fourth order. In the case of greatest interest, therefore, N must be of the form:

Table 1

	Degree →		
	1	2	3
1	$\dfrac{\partial \psi}{\partial x}, \dfrac{\partial \psi}{\partial y}$	Empty	Empty
2	$\dfrac{\partial^2 \psi}{\partial x^2}, \dfrac{\partial^2 \psi}{\partial x \partial y}, \dfrac{\partial^2 \psi}{\partial y^2}$	$\left(\dfrac{\partial \psi}{\partial x}\right)^2, \dfrac{\partial \psi}{\partial x}\dfrac{\partial \psi}{\partial y}, \left(\dfrac{\partial \psi}{\partial y}\right)^2$	Empty
3	$\dfrac{\partial^3 \psi}{\partial x^3}, \dfrac{\partial^3 \psi}{\partial x^2 \partial y}, \dfrac{\partial^3 \psi}{\partial x \partial y^2},$ $\dfrac{\partial^3 \psi}{\partial y^3}$	$\dfrac{\partial^2 \psi}{\partial x^2}\dfrac{\partial \psi}{\partial y}, \dfrac{\partial^2 \psi}{\partial x \partial y}\dfrac{\partial \psi}{\partial x}, \dfrac{\partial^2 \psi}{\partial y^2}\dfrac{\partial \psi}{\partial x},$ $\dfrac{\partial^2 \psi}{\partial x^2}\dfrac{\partial \psi}{\partial y}, \dfrac{\partial^2 \psi}{\partial x \partial y}\dfrac{\partial \psi}{\partial y}, \dfrac{\partial^2 \psi}{\partial y^2}\dfrac{\partial \psi}{\partial y}$	$\left(\dfrac{\partial \psi}{\partial x}\right)^3, \left(\dfrac{\partial^2 \psi}{\partial x}\right)^2\dfrac{\partial \psi}{\partial y},$ $\dfrac{\partial \psi}{\partial x}\left(\dfrac{\partial \psi}{\partial y}\right)^2, \left(\dfrac{\partial \psi}{\partial y}\right)^3$
4	$\dfrac{\partial^4 \psi}{\partial x^4}, \dfrac{\partial^4 \psi}{\partial x^3 \partial y}, \dfrac{\partial^4 \psi}{\partial x^2 \partial y^2},$ $\dfrac{\partial^4 \psi}{\partial x \partial y^3}, \dfrac{\partial^4 \psi}{\partial y^4}$	$D_{3n}D_{1q}, D_{2n}D_{2q}$	$D_{1n}D_{1q}D_{2\ell}$
5	$\dfrac{\partial^5 \psi}{\partial x^5}, \dfrac{\partial^5 \psi}{\partial x^4 \partial y}, \dfrac{\partial^5 \psi}{\partial x^3 \partial y^2},$ $\dfrac{\partial^5 \psi}{\partial x^2 \partial y^3}, \dfrac{\partial^5 \psi}{\partial x \partial y^4}, \dfrac{\partial^5 \psi}{\partial y^5}$	$D_{1n}D_{4q}, D_{2n}D_{3q}$	$D_{1n}D_{1q}D_{3\ell}, D_{1n}D_{2q}D_{2\ell}$
6	$\dfrac{\partial^6 \psi}{\partial x^6}, \dfrac{\partial^6 \psi}{\partial x^5 \partial y}, \dfrac{\partial^6 \psi}{\partial x^4 y^2},$ $\dfrac{\partial^6 \psi}{\partial x^3 \partial y^3}, \dfrac{\partial^6 \psi}{\partial x^2 \partial y^4}, \dfrac{\partial^6 \psi}{\partial x \partial y^5},$ $\dfrac{\partial^6 \psi}{\partial y^6}$	$D_{1n}D_{5q}, D_{2n}D_{4q}$ $D_{3n}D_{3q}$	$D_{1m}D_{1q}D_{4\ell}, D_{1n}D_{2q}D_{3\ell}$ $D_{2n}D_{2q}D_{2\ell}$

25

Table 2

Degree →

	1	2	3	
1	Empty	Empty	Empty	
2	$\nabla^2\psi$	$\nabla\psi\cdot\nabla\psi$	Empty	
3	Empty	Empty	Empty	
4	$\nabla^4\psi$	$(\nabla^2\psi)^2$ $\nabla\psi\cdot\nabla(\nabla^2\psi)$ $\underline{K}\cdot\nabla\psi\times\nabla(\nabla^2\psi)$ $\nabla^2(\nabla\psi\cdot\nabla\psi)$	$\nabla^2\psi(\nabla\psi\cdot\nabla\psi)$ $\nabla\psi\cdot\nabla(\nabla\psi\cdot\nabla\psi)$ $\underline{K}\cdot\nabla\psi\times\nabla(\nabla\psi\cdot\nabla\psi)$	
5	Empty	Empty	Empty	
6	$\nabla^6\psi$	$\nabla^2\psi\nabla^4\psi$ $\nabla\psi\cdot\nabla(\nabla^4\psi)$ $\underline{K}\cdot\nabla\psi\times\nabla(\nabla^4\psi)$ $\nabla^4(\nabla\psi\cdot\nabla\psi)$ $\nabla(\nabla^2\psi)\cdot\nabla(\nabla^2\psi)$	$\nabla^4\psi(\nabla\psi\cdot\nabla\psi)$ $\nabla\psi\cdot\nabla[\nabla\psi\cdot\nabla(\nabla^2\psi)]$ $\underline{K}\cdot\nabla\psi\times\nabla[\nabla\psi\cdot\nabla(\nabla^2\psi)]$ $\nabla\psi\cdot\nabla[\underline{K}\cdot\nabla\psi\times\nabla(\nabla^2\psi)]$	$\underline{K}\cdot\nabla\psi\times\nabla[\underline{K}\cdot\nabla\psi\times\nabla(\nabla^2\psi)]$ $\nabla\psi\cdot\nabla[\nabla^2(\nabla\psi\cdot\nabla\psi)]$ $\underline{K}\cdot\nabla\psi\times\nabla[\nabla^2(\nabla\psi\cdot\nabla\psi)]$ $\nabla^2\psi[\nabla\psi\cdot\nabla(\nabla^2\psi)]$ $\nabla^2\psi[\underline{K}\cdot\nabla\psi\times\nabla(\nabla^2\psi)]$ $(\nabla^2\psi)^3$ $\nabla^2\psi[\nabla^2(\nabla\psi\cdot\nabla\psi)]$ $\nabla(\nabla^2\psi)\cdot\nabla(\nabla\psi\cdot\nabla\psi)$ $\underline{K}\cdot\nabla(\nabla^2\psi)\times\nabla(\nabla\psi\cdot\nabla\psi)$	

$$N = Q_{22} (\nabla^2\psi)^2 + Q_{112} \nabla\psi\cdot\nabla(\nabla^2\psi) + Q^*_{112} \underset{\sim}{K}\cdot\nabla\psi\times\nabla(\nabla^2\psi)$$

$$+ Q_{211} \nabla^2(\nabla\psi\cdot\nabla\psi) = \frac{\partial}{\partial t} \nabla^2\psi \qquad (10)$$

A point to be stressed is that this result, as yet, does not contain anything about the detailed dynamics of the flow, beyond the very general and qualitative requirements of localness, uniqueness, universality, invariance under translation and rotation, and dimensional homogeneity.

Now, let us impose the constraint that the integral of squared vorticity, taken over any bounded domain, be invariant for all possible ψ-fields. Singling out the third term of (10), we see that if $Q_{22} = Q_{112} = Q_{211} = 0$,

$$\frac{\partial}{\partial t} (\nabla^2\psi)^2 = Q^*_{112} \underset{\sim}{K}\cdot\nabla\psi\times\nabla(\nabla^2\psi)^2$$

$$= -Q^*_{112} \underset{\sim}{K}\cdot\nabla\times(\nabla^2\psi)^2\nabla\psi$$

Integrating both sides over the area A bounded by the curve B, we then have

$$\frac{\partial}{\partial t} \int_A (\nabla^2\psi)^2 dA = -Q^*_{112} \oint_B (\nabla^2\psi)^2 \frac{\partial\psi}{\partial s} dB$$

where $\partial\psi/\partial s$ is the derivative along the boundary. But $\partial\psi/\partial s$ vanishes everywhere on the boundary, and

$$\frac{\partial}{\partial t} \int_A (\nabla^2\psi)^2 dA = 0$$

for all possible ψ-fields--i.e., identically. Thus, the constraint of enstrophy invariance is satisfied if $Q^*_{112} \neq 0$, $Q_{22} = 0$, $Q_{112} = 0$ and $Q_{211} = 0$. Unless, however, $Q_{22} = Q_{112} = Q_{211} = 0$, enstrophy is <u>not</u> invariant for <u>all</u> ψ-fields, as can be shown by simple counter-examples.

In short, the requirement of enstrophy invariance alone implies that

$$\frac{\partial}{\partial t} \nabla^2 \psi = Q^*_{112} \underset{\sim}{K} \cdot \nabla\psi \times \nabla(\nabla^2\psi) \tag{11}$$

We also note that the equation above leads to invariance of total kinetic energy. Beyond that, however, Eq. (11) says a great deal about the dynamics of the detailed flow. The universal dimensionless constant Q^*_{112}, to be sure, is not determined (nor can it be determined by constraints of invariance alone), but it can be absorbed in the time variable. That is, the evolution of non-steady flows governed by Eq. (11) will be like that of ordinary two-dimensional inviscid flows ($Q^*_{112} = -1$), but on an arbitrary time-scale. But if one is concerned with steady-state flows or nonsteady flows in statistical equilibrium, the time-scale is not important. Perhaps this accounts for the remarkable success of statistical closure schemes that are based essentially (indeed, almost solely) on the invariance of total kinetic energy and enstrophy, without regard to details of the flow.

Before leaving the case when the flow does not depend on physical constants, we should point out that Eq. (11) implies the invariance of the integrals of all differentiable functions of vorticity: hence there is no paradox. If the integral of squared vorticity is invariant, so are all the other integrals.

Let us now examine another case. It is an observed, but qualitative fact that real flows with geometrically similar boundaries are also geometrically similar, provided that the fluid medium is the same and the <u>product</u> of d (some measure of the linear dimension of the flow) and U (a measure of characteristic speed of entry or initial speed) is constant, even if d and U individually vary. This is also found to be true if the fluid medium and (presumably) its properties are different, but the product Ud is maintained at a new constant value. Despite their qualitative character, these observations strongly suggest that the flow depends on some physical constant of the medium, with the dimensions of Ud. We postulate the existence of such a constant, specific to each medium, and call it ν.

It is now necessary to regard the coefficients of the terms in N as possibly containing powers of ν. Going through the same dimensional argument as before and noting that ν has the same dimensions as ψ, we see that the form of N is again severely limited. According to Table 2, it is:

$$N = L_1 \nu \nabla^4 \psi + Q_{22}(\nabla^2\psi)^2 + Q_{112}\nabla\psi \cdot \nabla(\nabla^2\psi)$$

$$+ Q^*_{112}\underset{\sim}{K} \cdot \nabla\psi \times \nabla(\nabla^2\psi) + Q_{211}\nabla^2(\nabla\psi \cdot \nabla\psi)$$

$$+ \frac{1}{\nu} \text{(terms of third degree and fourth order)}$$

$$+ \frac{1}{\nu^2} \text{(terms of fourth degree and fourth order), etc.}$$

where L_1 is some universal dimensionless constant. Again, this result stems only from qualitative considerations of the nature of physical law.

Now, if we are to approach the limit of enstrophy invariance as ν becomes very small, the terms of third and higher degree cannot appear. We would thus conclude that the equation governing real two-dimensional flows is of the form:

$$\frac{\partial}{\partial t} \nabla^2 \psi = Q^*_{112} \underline{K} \cdot \nabla\psi \times \nabla(\nabla^2\psi) + L_1 \nu \nabla^4 \psi$$

For $Q^*_{112} = -1$ and $L_1\nu$ = kinematic molecular viscosity, this equation will be recognized as the vorticity equation corresponding to the Navier-Stokes equations for two-dimensional flow. Before now, however, we have not so much as mentioned the Navier-Stokes equations.

5. SUMMARY AND CONCLUSIONS

The main result of this paper is that the form of the Eulerian vorticity equation for inviscid two-dimensional flow is derivable from the principle of invariance of total enstrophy, to within a universal, but undetermined time-scale. This is shown by expressing the instantaneous local time derivative of vorticity as a <u>local</u> operator on the current ψ-field, a qualitative restriction that is supported by observation. We then analyze the structure of the general local operator, to find the conditions under which it satisfies the <u>a priori</u> requirements of uniqueness, universality, invariance under translation and rotation of the coordinate system, and dimensional homogeneity. With the assumption that the flow does not depend on any physical constants of the medium, the only admissible local operator is a linear combination of four nonlinear operators, whose coefficients are all universal dimensionless constants. Finally, by requiring invariance of total enstrophy (the only quantitative constraint on the dynamics of the system), the form of the local operator is fixed. As might be expected, it corresponds to displacement of the local vorticity field with the local velocity.

It appears that the principles of invariance of total kinetic energy and enstrophy imply much more about the detailed dynamics of the flow than one might have expected or imagined. It is, for example, not now so surprising that relatively simple stochastic closure schemes—whose physical content is based almost solely on invariance of energy and enstrophy—have been successful in describing and predicting the statistical structure and evolution of turbulent two-dimensional flows. An attempt at extending the use of invariance principles and an inkling of how this might be done is contained in a recent paper by Thompson (1981).

A byproduct of this derivation is the form of the vorticity equation for two-dimensional flows that depend on a physical constant ν whose dimensions are $(length)^2/(time)$. To within the values of two universal, but undetermined constants, the form is fixed by requiring the invariance of enstrophy in the limit of very small ν, and agrees with the form derivable from the Navier-Stokes equations. In short, the qualitative constraints inherent in the nature of physical law are very strong: it takes only one (apparently weak) dynamical constraint to fix almost all other dynamical properties of the system.

REFERENCES

1. Batchelor, G.K., 1969, Phys. Fluids, 12 (suppl. II), II-233
2. Fjortoft, R., 1953, On the changes in the spectral distribution of kinetic energy for two-dimensional nondivergent flow. Tellus, 5, 225.
3. Kao, S.K. and Wendell, L.L., 1970, The Kinetic energy of the large-scale atmospheric motion in wavenumber-frequency space, I. Northern hemisphere, J. Atmos. Sci., 27, 359.
4. Kraichnan, R., 1967, Inertial ranges in two-dimensional turbulence, Phys. Fluids, 10, 1417.
5. Leith, C.E., 1968, Diffusion approximation for two-dimensional turbulence, Phys. Fluids, 11, 671.
6. Lilly, D.K., 1969, Numerical simulations of two-dimensional turbulence, Phys. Fluids, 12 (suppl. II), II-240.
7. Thompson, P.D., 1981, Nonequilibrium probability distributions for randomly forced two-dimensional flows, Phys. Fluids, (in press).
8. Tsinober, . ., 1972, Proc. Inst. of Phys. Acad. Sci. Latvian SSR, Riga.
9. Wiin-Nielson, A.C., 1967, On the annual variation and spectral distribution of atmospheric energy. Tellus, 19, 540.

A GALERKIN METHOD TO STRONGLY NONLINEAR
KdV EQUATIONS AND SCHRÖDINGER EQUATIONS

Fengsu Chen*
Division of Applied Mathematics
Brown University
Providence, Rhode Island 02912
U.S.A.
November 1981

I. INTRODUCTION

Recently, growing attention has been devoted to the study of solutions of highly nonlinear KdV equations and Schrödinger equations, where the full nonlinearity is retained. In contrast to the classical case of the standard equations, exact analytic solutions do not generally exist and the equations have to be solved numerically.

The variational method has been used to obtain approximate solutions of the nonlinear Klein Gordon equation and the KdV equation by Hsieh[1] and the approximate soliton solutions of nonlinear Schrödinger equations by Anderson, et. al.[2] In the present note, we shall employ the Galerkin method to obtain approximate solutions of strongly nonlinear KdV equations and Schrödinger equations.

II. GENERAL SCHEME OF THE GALERKIN METHOD

The method to be developed is a generalization of the Galerkin method[3] from the perspective of the variational method. The essence of the Galerkin method may be described as follows. Take the differential equation:

$$L[x(t)] = 0 .\qquad(1)$$

A trial solution is taken in the form

$$x = \sum_{i=1}^{N} c_i x_i(t) .\qquad(2)$$

where $\{x_i(t)\}$ is a set of given functions. Then choose a set of weighting functions $\{W_i(t)\}$. The parameters $\{c_i\}$ are to be determined by the following set of algebraic equations:

---

* On leave from Wuhan Institute of Hydraulic and Electric Engineering, Wuhan, People's Republic of China

$$\int (L[\sum_{i=1}^{N} C_i x_i])W_j dt = 0 \quad , \; j = 1, \ldots, N \; . \tag{3}$$

The weighting functions $\{W_i(t)\}$ were originally chosen by Galerkin to be identical to $\{x_i(t)\}$. Now there is some ambiguity in the choice of suitable form of L, since for example, $f(x)L[x(t)] = 0$ will be equivalent to (1) for an $f(x)$. The scheme proposed will remove this ambiguity.

Consider a differential equation schematically represented by

$$L[x(t),\alpha] = 0. \tag{4}$$

where α is a parameter. As examples, take the following differential equations representing oscillations with linear and nonlinear dampings:

$$x'' + 2\alpha x' + x = 0 \tag{5}$$

and

$$x'' + \alpha(x')^3 + x = 0 \tag{6}$$

when $\alpha = 0$ a functional J can be found and a proper variational formulation exists. The $\Delta J = 0$ will lead to the following relation:

$$\int L[x(t);0]\Delta x \, dt = 0 \; . \tag{7}$$

This L is now unequivocally determined by the variational formulation. Then for α sufficiently small, we expect a similar relation

$$\int L[x(t);\alpha]\Delta x \, dt = 0 \tag{8}$$

is also valid. This is the essence of the general scheme of the proposed Galerkin method.

III. APPROXIMATE SOLUTIONS OF THE WAVE EQUATION INCLUDING AN ARBITRARY NONLINEAR SELF-ACTION EFFECT

We consider the following equation which can be transformed into a strongly nonlinear Schrödinger equation[2] with

$$E(\xi,t) = \frac{1}{\sqrt{2}} \psi(\xi)\exp(-i\omega_0 t) + c.c.$$

$$E_{tt} - V_g^2 E_{xx} + \omega_{p_0}^2 F(<E^2>)E = 0 \qquad (9)$$

where V_g is the group velocity of the wave, ω_{p_0} is the linear plasma frequency, $<>$ denotes time average, E is wave amplitude, and F characterizes the arbitrary nonlinearity. Strictly speaking, there is no equivalent proper variational formulation for the equation (9).

We use the Galerkin method treating the term with F as small, to obtain

$$\int_0^t \int_{-\infty}^{+\infty} (E_{tt} - V_g^2 E_{xx} + \omega_{p_0}^2 F(<E^2>)E)\Delta E \, dxdt = 0 \, . \qquad (10)$$

We take the trial solution of the form.

$$E = A(x,t)\phi(s(x,t)) \qquad (11)$$

where A and the derivative of S are both slowly varying functions of (x,t), and ϕ is assumed to satisfy

$$\begin{aligned}
\phi(s + 2\pi) &= \phi(s), \\
<\phi> &\equiv \frac{1}{2\pi}\int_0^{2\pi} \phi \, ds = 0, \\
<\phi^2> &= 1, \\
<\phi\dot\phi> &= 0, \\
<\dot\phi^2> &= \beta, \\
<\phi\ddot\phi> &= -<\dot\phi^2> = -\beta.
\end{aligned} \qquad (12)$$

Substituting (11) into (10), we obtain

$$\int_0^t \int_{-\infty}^{+\infty} dxdt \{\Delta A [A_{tt}\phi^2 + 2A_t S_t \phi\dot\phi + AS_{tt}\phi\dot\phi + AS_t^2 \phi\ddot\phi - V_g^2 A_{xx}\phi^2 - 2V_g^2 S_x\phi\dot\phi -$$
$$-V_g^2 AS_{xx}\phi\dot\phi - V_g^2 AS_x^2\phi\ddot\phi + \omega_{p_0}^2 F(A^2)A\phi^2] +$$
$$+\Delta S [AA_{tt}\phi\dot\phi + 2AA_t S_t \dot\phi^2 + A^2 S_{tt}\dot\phi^2 + A^2 S_t^2\dot\phi\ddot\phi - V_g^2 AA_{xx}\phi\dot\phi - 2V_g^2 AS_x\dot\phi^2 - \qquad (13)$$
$$-V_g^2 A^2 S_{xx}\dot\phi^2 - V_g^2 A^2 S_x^2\dot\phi\ddot\phi + \omega_{p_0}^2 F(A^2)A^2\phi\dot\phi]\} = 0 \, .$$

Using the average scheme as developed in the variational method[1] by retaining only the secular terms, and noting that ΔA and ΔB are independent, we obtain the approximate equations

$$\beta(S_t^2 - V_g^2 S_x^2) - \omega_{p_0}^2 F(A^2) = 0 \qquad (14)$$

and

$$(A^2 S_t)_t - V_g^2 (A^2 S_x)_x = 0 \qquad (15)$$

$1°$ $\beta = 1,$ $F(A^2) = 1/\omega_{p_0}^2 ,$

$$\phi(s) = \sqrt{2} \sin(S + \psi).$$

From (14) we obtain

$$S = \frac{k}{V_g} x - \sqrt{1 + k^2}\, t + m \qquad (16)$$

where k and m are two arbitrary constants. Substituting (16) into (15), we obtain

$$A = a(x - \frac{k}{V_g \sqrt{1 + k^2}}\, t)$$

where a is an arbitrary function.

$$E(x,t) = a(x - \frac{k}{V_g \sqrt{1 + k^2}}\, t) \sin\left[\frac{k}{V_g} x - \sqrt{1 + k^2}\, t + \psi\right]$$

$2°$ $\beta = 1,$ $F(A^2) = \frac{1}{\omega_{p_0}^2} A^2$

Then (14) and (15) become

$$S_t^2 - V_g^2 S_x^2 - A^2 = 0$$

and

$$(StA^2)_t - V_g^2(A^2Sx)_x = 0.$$

We obtain

$$E(x,t) = \frac{2}{3} a(t^2 - \frac{x^2}{V_g^2})^{-\frac{1}{6}} \phi\left[a(t^2 - \frac{x^2}{V_g^2})^{\frac{1}{3}} + b\right] \qquad (17)$$

where a and b are integration constants.

$3°$ $\beta = 1$, $\qquad F(A^2) = \frac{1}{\omega_{p_0}^2} A^n$

where n is even. Then (14) and (15) become

$$St^2 - V_g^2 Sx^2 - A^n = 0$$

and

$$(StA^2)_t - V_g^2(SxA^2)_x = 0.$$

We obtain

$$E(x,t) = a(t^2 - \frac{x^2}{V_g^2})^{-\frac{1}{n+4}} \phi\left[\frac{n+4}{4} a(t^2 - \frac{x^2}{V_g^2})^{\frac{2}{n+4}} + b\right] \qquad (18)$$

where a and b are integration constants.

$4°$ $\beta = 1$, $\qquad F(A^2) = \frac{1}{\omega_{p_0}^2} e^{-A^2}$.

We obtain

$$E(x,t) = A\phi\left[a(\ln A + \frac{1}{A^2} + b)\right] \qquad (19)$$

where A is a function satisfying

$$aA^{-2} e^{\frac{A^2}{2}} = \left(t^2 - \frac{x^2}{V_g^2}\right)^{\frac{1}{2}}. \qquad (20)$$

IV. APPROXIMATE SOLUTIONS OF STRONGLY NONLINEAR KdV EQUATIONS

We consider the equation[4]

$$u_t + u^3 u_x + u_{xxx} = 0. \qquad (21)$$

We use the Galerkin method as before

$$\int_0^t \int_{-\infty}^{+\infty} (u_t + u^3 u_x + u_{xxx}) \Delta u \, dx dt = 0. \qquad (22)$$

We take the trial solution of the form with single soliton

$$u = A(x,t)\phi(x - Ut). \qquad (23)$$

Substituting (23) into (22), we obtain

$$\dddot{\phi} + A^3 \phi^3 \dot{\phi} - U\dot{\phi} = 0 \qquad (24)$$

if A = const. Integrating twice, we obtain

$$\frac{\dot{\phi}^2}{2} + \frac{A^3 \phi^5}{20} - \frac{U\phi^2}{2} + c_1 \phi + c_2 = 0 \qquad (25)$$

In the special case when ϕ and its derivatives tend to zero at ∞, then $c_1 = c_2 = 0$.

$$\phi(x,t) = \frac{3\sqrt{10U}}{A} \operatorname{sech}^{2/3}\left(\frac{3}{2}\sqrt{U}\right)(x - Ut) \qquad (26)$$

$$u(x,t) = 3\sqrt{10U} \operatorname{sech}^{2/3}\left(\frac{3}{2}\sqrt{U}\right)(x - Ut)$$

Now we consider the KdV equation generalized by adding a time-dependent "forcing term",[4] if it is of some special form.

$$u_t + u^3 u_x + u_{xxx} = \psi(x - Ut). \qquad (27)$$

We use the Galerkin method as before and take the trial solution of the form with single soliton.

$$u = A(x,t)\phi(x - Ut) \qquad (28)$$

where $\dot{\phi} = \psi$. Substituting (28) into (27) and carrying the scheme as before, we obtain

$$\dddot{\phi} + A^3 \phi^3 \dot{\phi} - U\dot{\phi} - \frac{\dot{\phi}}{A} = 0 \qquad (29)$$

If $A = $ const. Then

$$\phi = \frac{\sqrt[3]{10(U + \frac{1}{A})}}{A} \operatorname{sech}^{2/3}(\frac{3}{2}\sqrt{U + \frac{1}{A}})(x - Ut) \qquad (30)$$

$$u(x,t) = \sqrt[3]{10(U + 1)} \operatorname{sech}^{2/3}(\frac{3}{2}\sqrt{U + 1})(x - Ut). \qquad (31)$$

Thus (31) is a solution of (27), only if the forcing term is of the form

$$\psi(X) = \left[\frac{\sqrt[3]{10(U + \frac{1}{A})}}{A} \operatorname{sech}^{2/3}(\frac{3}{2}\sqrt{U + \frac{1}{A}})X \right] .$$

V. DISCUSSION

By means of a Galerkin method, we have obtained approximate solutions of highly nonliner KdV equations and Schrödinger equations. It may be remembered, in contrast to many other asymptotic methods, the perturbation method is not easily adapted to solve highly nonlinear equations and the variational method is not intrinsically a perturbation method and therefore is well adapted to treat a certain class on nonlinear equations.

The method presented above is simple in concept and straightforward in application; yet it yields a great deal of information. It is clear that much work is still needed to answer the many questions raised by the proposed Galerkin method and to explore its wide ranging potentials.

ACKNOWLEDGEMENTS

The author wishes to thank Professor D. Y. Hsieh for many stimulating discussions and thank the Division of Applied Mathematics of Brown University for their hospitality.

REFERENCES

1. D. Y. Hsieh, J. Math. Phys. 16, 1630-36, (1975).
2. D. Anderson, M. Boundal and M. Lisak, Physics Fluids 22 (4), 788-9, (1979).
3. Fengsu Chen and D. Y. Hsiech, "A Galerkin Method and Nonlinear Oscilations and Waves". To appear.
4. R. M. Miura, SIAM Review 18, (3), 412-459, (1976).

MULTITIME ASYMPTOTIC METHODS FOR SLOWLY EVOLVING OSCILLATING SYSTEMS

Oded Regev and J. Robert Buchler
Physics Department, University of Florida, Gainesville, Fl 32611

ABSTRACT

This paper describes the application of asymptotic methods to dynamical oscillatory systems which undergo a slow evolution. Several cases, which are motivated by problems in the study of stellar pulsation, are discussed.

INTRODUCTION

We discuss the motion of a dynamical system governed by the equations:

$$\frac{d^2 y}{dt^2} + g(y,s) = 0 \qquad (1a)$$

$$\frac{ds}{dt} = \varepsilon h(y,s), \qquad (1b)$$

where $y = (y_1 \ldots y_n) \in D_y \subset R^n$, $s = (s_1 \ldots s_k) \in D_s \subset R^k$, $h = (h_1 \ldots h_k) \in D_h \subset R^k$ and the function $g = (g_1 \ldots g_n) = D_g \subset R^n$ is derivable (at fixed s) from a potential:

$$g(y,s) = \frac{\partial}{\partial y} V(y,s). \qquad (2)$$

D denote bounded domains and the functions g and h are sufficiently smooth. ε is a small parameter of the system.

The smallness of ε makes a straightforward numerical integration of system (1) very costly and often prohibitive. However, it allows for approximation methods. We describe here briefly such a method. The reader is referred to the cited publications for details.

GENERAL CASE

Assume that in the limit $\varepsilon=0$ (i.e. fixed s) the Hamiltonian system (1a) admits at least one stable, strictly periodic solution of period P with two rest points per period. Adapting a known two time asymptotic perturbation technique (see Cole 1968, Nayfeh 1980) to the problem at hand, we define a slow time $\tau = \varepsilon t$ and a fast time t_* (defined by $dt_*/dt \equiv \phi(\tau)$, with $\phi(\tau)$ undetermined at this point).

0094-243X/82/880327-05$3.00 Copyright 1982 American Institute of Physics

Changing the variables in system (1) from t to τ and t_* and introducing the asymptotic expansions:

$$y(t_*,\tau) = y_0(t_*,\tau) + \varepsilon y_1(t_*,\tau) + \ldots \qquad (3a)$$

$$s(t_*,\tau) = s_0(t_*,\tau) + \varepsilon s_1(t_*,\tau) + \ldots , \qquad (3b)$$

we obtain a hierarchy of PDE systems (the various orders in ε), the first two of which are:

$$\phi^2 \frac{\partial^2}{\partial t_*^2} y_0 + g(y_0, s_0) = 0 \qquad (4a)$$

$$\phi \frac{\partial s_0}{\partial t_*} = 0 \qquad (4b)$$

$$\phi^2 \frac{\partial^2}{\partial t_*^2} y_1 + \frac{\partial g}{\partial y_0} \cdot y_1 = - \frac{\partial g}{\partial s_0} \cdot s_1 - \left(\frac{d\phi}{d\tau} \frac{\partial y_0}{\partial t_*} + 2\phi \frac{\partial^2 y_0}{\partial t_* \partial \tau}\right)$$

$$(5a)$$

$$\phi \frac{\partial s_1}{\partial t_*} + \frac{\partial \overset{*}{s}_0}{\partial \tau} = h(y_0, s_0) \qquad (5b)$$

Let y_0 be a periodic solution of eq.(4a), whose existence is guaranteed by assumption. It is now possible to choose the function $\phi(\tau)$ in such a way as to make the period (in t_*) independent of τ. This, in turn, enables us to use a theorem by Kuzmak (1959). The theorem insures the uniform validity of expansions (3), for $0 < \tau < T/\varepsilon$, provided the terms in those expansions are also periodic in t_* and with the same (τ independent) period. Eq.(4b) implies that s_0 is a function of τ only. The function y_1 is periodic in t_* with the same period as y_0 provided that

$\oint x_i \cdot$ RHS (eq. 5a) $dt_* = 0$ for all independent solutions x_i of the adjoint homogeneous system corresponding to (5a). One of these solutions is $\partial y_0/\partial t_*$ and the condition gives

$$\frac{d}{d\tau} \{\phi \oint (\frac{\partial y_0}{\partial t_*})^2 dt_*\} = -\oint \frac{\partial y_0}{\partial t_*} \cdot \frac{\partial g}{\partial s_0} \cdot s_1 \, dt_* . \qquad (6a)$$

The other solution can be shown to set $s(0,\tau)$ in eq. 7 equal to zero.

The assumption that s_1 is also periodic gives:

$$\frac{ds_o}{d\tau} = \frac{1}{P} \oint h[y_o(t_*,\tau),s_o(\tau)]dt_* \qquad (6b)$$

(from (5b)). The function s_1 can be formally obtained by solving eq.(5b):

$$s_1(t_*,\tau) = s_1(0,\tau) + \frac{1}{\phi}\{\int_o^{t_*} h[y_o(t'_*,\tau),s_o(\tau)]dt'_* - t_*\frac{ds_o}{d\tau}\} \qquad (7)$$

and it can then be used in eq.(6a).

The system (6) (which we call the "reduced" system) is of order n+1 in the slow variable and advantageously replaces the original system (1) which is of order 3n and rapidly varying. When solved it furnishes the desired slow-time evolution of the periodic (in the fast-time) solution y_o. The problem is thus formally solved up to $O(\varepsilon)$. For details see Buchler and Perdang (1977), Buchler et.al. (1977) and Buchler (1978a).

In practice, it is often difficult to obtain the function y_o explicitly since in general the function g is nonlinear in y. In what follows we explore further the case in which the function $g(y,s)$ is linear in y and takes the form:

$$g(y,s) = A(s) \cdot [y - Y(s)], \qquad (8)$$

where A is a symmetric matrix and Y(s) denotes the "hydrostatic equilibrium position" of the system. Both are assumed to be known functions of s. We refer to this case as "linear oscillators" but remark that h can be still arbitrarily nonlinear in y and s.

If $g(y,s)$ is weakly nonlinear ($g(y,s) = A \cdot (y-Y) + \mu f(y,s,\mu)$ with μ small), the method described in the following section can be somewhat modified by introducing expansions in μ (Barranco et.al. 1981).

LINEAR OSCILLATORS

Let A(s) have the positive eigenvalues ω_i^2 and their corresponding eigenvectors e_i ($A(s) \cdot e_i(s) = \omega_i^2(s)e_i(s)$) for i=1...n. When eq.(8) is used for the function g, the solution of (4a) is trivial. It consists of a linear combination of oscillations with the eigenfrequencies ω_i. Assuming that all the ω_i are, and stay, incommensurate the number of excited modes will depend only on the initial conditions. In the case in which only one mode is initially present (say $\omega_1 \equiv \omega$) we speak of a monofrequency oscillation. In this case the solution of (4a) is:

$$y_o(t_*,\tau) = Y(s_o) + a(\tau)e(s_o)\cos[\frac{\omega(s_o)}{\phi} t_*] , \qquad (9)$$

where e is the eigenvector corresponding to ω. The choice $\phi(\tau) = \omega(s_o)$ will insure that the period (in t_*) of y_o is τ independent. Proceeding as in the general case one obtains the reduced system, which furnishes the slow evolution of s_o and the amplitude a:

$$\frac{da}{d\tau} = -\frac{1}{2} a \frac{d}{ds_o} (\ln \omega^2) \cdot (b_o - \frac{1}{2} b_2) - \frac{1}{2} e \cdot \frac{dY}{ds_o} \cdot b_1 \qquad (10a)$$

$$\frac{ds_o}{d\tau} = b_o(a,s_o), \qquad (10b)$$

where the b functions are the coefficients of the Fourier expansion:

$$h(y_o(t_*,\tau)s_o(\tau)) = \sum_{n=0}^{\infty} b_n(a,s_o)\cos nt_* \qquad (11)$$

A full exposition of the above procedure can be found in Buchler (1978b).

When several modes are present we speak of a multifrequency oscillations. This case can be treated in a formally similar manner. Generalizing the monomode case we introduce several short times (equal in number to the excited modes): $t_* = (t_*^1, t_*^2, \ldots t_*^P)$ and several functions $\phi(\tau) = (\phi_1(\tau), \phi_2(\tau) \ldots \phi_p(\tau))$. The requirement of periodicity is replaced by quasiperiodicity (i.e. periodicity in each of the t_*^i separately). The systems (4) and (5) are formally similar to those corresponding to the monomode case, but with $\phi \, \partial/\partial t_*$ replaced by the "scalar product" $(\phi, \frac{\partial}{\partial t_*}) \equiv \sum_{i=1}^{P} \phi_i \frac{\partial}{\partial t_*^i}$. The solution y_o is:

$$y_o(t_*,\tau) = Y(s_o) + \sum_{i=1}^{P} a_i e_i(s_o)\cos(t_*^i) \qquad (12)$$

where again we have set $\phi_i = \omega_i$.

A reduced sytem is obtained for s_o and all the amplitudes a_i and is similar to (10), but includes multiple Fourier components of h. The double mode case is worked out in detail in Regev and Buchler (1981) (see also Buchler and Regev (1981a)).

When two or more of the eigenfrequencies are commensurate a modification to the above procedures is needed. For example, if two frequencies satisfy $|n\omega_n - m\omega_m| < \epsilon$ a double mode expansion in these frequencies breaks down due to the appearance of small divisors. Any commensurate (resonant) mode has to be included in the solution in addition to the original ones, and the solution is then periodic with the common overall period.

Such cases can be treated in a systematic way by splitting the matrix A into a strictly resonant part A_0 and a "detuning" matrix A_1 (of order ϵ). For a full description of resonances see Regev et al. (1981).

APPLICATIONS

The problem discussed above arises in the study of stellar pulsations. The stellar structure and evolution equations, when discretized in a Lagrangian description, assume the form of eq.(1). The parameter y is here related to the radii of the mass shells and s denotes the specific entropy profile of the star. The parameter ϵ, which denotes the ratio of dynamic timescale to (thermal) evolutionary timescale, is in general very small. For some of the basic problems (e.g. existence of limit cycles, mode switching, multimode pulsation) the assumption of linear oscillators is often sufficient. The method has been applied to various problems in stellar pulsation (Buchler et.al. 1977; Buchler 1978b; Buchler and Regev 1981a,b) and has been shown to be a useful and promising analytical as well as computational tool.

REFERENCES

Kuzmak, G.E., 1959, J. Appl. Math. Mech., 23, 730.
Cole, J.D., 1968, Perturbation Methods in Applied Mathematics, Blaisdell.
Buchler, J.R. and Perdang, J., 1977, Int'l. J. Quantum Chem., Sympos. 13, 183.
Buchler, J.R., Yueh, W.R. and Perdang, J., 1977, Astrophys. J. 214, 510.
Buchler, J.R, 1978a, in A. Barut and F. Calogero, (eds.), Nonlinear Equations in Physics and Mathematics, D. Reidel.
Buchler, J.R., 1978b, Astrophys. J., 220, 629.
Nayfeh, A.H., 1980, Introduction to Perturbation Techniques, Wiley-Interscience.
Buchler, J.R. and Regev, O., 1981a, Astrophys. J. (Nov. 15).
Regev, O. and Buchler, J.R., 1981, Astrophys. J. (Nov. 15).
Buchler, J.R. and Regev, O., 1981b, Astrophys. J. (submitted).
Barranco, M., Buchler, J.R. and Regev, O., 1981, Astrophys. Space Sci. (in press).
Regev, O., Buchler, J.R. and Barranco, M., 1981, Astrophys. J. (submitted).

FLUCTUATION-DISSIPATION RELATIONS FOR SYSTEMS WITH INTERNAL MULTIPLICATIVE NOISE

Katja Lindenberg* & V. Seshadri
La Jolla Institute, P. O. Box 1434, La Jolla CA 92038

I. Introduction

Stochastic differential equations or "Langevin equations"[1] have been used to describe a large variety of systems interacting with a fluctuating environment. In principle, every differential equation of motion that describes a physical system on a macroscopic or "mesoscopic" level should be stochastic because of the inevitable interactions of the system with its surroundings. In practice, the importance of including stochastic effects is determined by the influence these effects have on the evolution of the system.

Most Langevin equations that are used in practice have been postulated phenomenologically. Among these is the famous stochastic equation that describes the Brownian motion of an oscillator of unit mass, momentum p and displacement x:[2]

$$\dot{x} = p \qquad (1a)$$
$$\dot{p} = -V'(x) - \lambda p + f(t) \qquad (1b)$$

Here $-V'(x)$ is the restoring force; $f(t)$ is a random, zero-centered, Gaussian-distributed force caused by the surrounding fluid and has the correlation function

$$\langle f(t)f(t')\rangle = 2D\delta(t-t') \qquad (2)$$

The brackets $\langle \; \rangle$ denote an average over an ensemble of realizations of $f(t)$. The dissipation parameter λ and the strength D of the fluctuations are related by the fluctuation-dissipation relation

$$\lambda = D/kT \qquad (3)$$

where T is the temperature of the fluid that surrounds the Brownian particle. The fluctuation-dissipation relation ensures that the energy of the oscillator is canonically distributed at equilibrium:

$$P_{eq}(p,x) \sim \exp -(\tfrac{p^2}{2} + V(x))/kT \qquad (4)$$

Another well known example of a stochastic differential equation is used to describe the evolution of a spin of magnetic moment $\underset{\sim}{M}$ in a solid:[3]

$$\dot{\underset{\sim}{M}} = -\gamma \, \underset{\sim}{H} \times \underset{\sim}{M} \qquad (5)$$

Here γ is the gyromagnetic ratio of the spin and $\underset{\sim}{H}$ is the magnetic field. This field is assumed to consist of an externally imposed portion $\underset{\sim}{H}_0(t)$ and a stochastic portion $\underset{\sim}{H}_1(t)$ caused by interactions of the spin M with the surrounding medium,

$$\underset{\sim}{H}(t) = \underset{\sim}{H}_0(t) + \underset{\sim}{H}_1(t) \qquad (6)$$

This system does not achieve a canonical distribution.

*Permanent address: Dept. of Chemistry, UCSD, La Jolla, CA 92093.

Recently there has been a great deal of interest in systems described by Langevin equations in which the fluctuations depend on the system variables. The spin equation (4) is an example of such a system, since the fluctuations $-\gamma \underset{\sim}{H}_1(t) \times \underset{\sim}{M}$ include the system variable M as a multiplicative factor. The Brownian motion example (1) is not of this type since the fluctuations in that case are independent of the system variables, i.e. f(t) is additive.

The derivation of Langevin equations for systems with additive fluctuations has been carried out rigorously in the "$\lambda^2 t$" limit starting from a full Hamiltonian description of the system coupled to a heat bath.[4] In these systems the dissipation is necessarily linear in the system variables. Until recently, very little was known about systems that include internally caused multiplicative noise, i.e. systems in which multiplicative fluctuations are brought about by the interaction of the system with a heat bath with which it ultimately equilibrates. In particular, such an equilibration would require appropriate dissipative contributions to be associated with the multiplicative fluctuations, and the nature of these dissipative contributions had not been investigated (not even phenomenologically).

To gain some physical insight into this problem we set out to study particular models in which one can derive exact system equations of motion analytically. This requires that one first be able to integrate the equations of motion of the heat bath, which in turn requires that these evolution equations be linear in the heat bath variables. The system can be nonlinear, and the coupling between the system and the heat bath can be nonlinear in the system variables (although linear in the heat bath variables.). We will now present two such models.

2. Mechanical Oscillator

Consider a mechanical oscillator in a heat bath, described by the following Hamiltonian:[5,6]

$$H = H_S + H_{HB} + H_{SB} \quad . \tag{7}$$

The system Hamiltonian H_S describes a classical mechanical oscillator of unit mass, momentum p, and displacement x,

$$H_S = \tfrac{1}{2} p^2 + V(x) \quad . \tag{8}$$

The potential energy $V(x)$ may contain a harmonic term $\omega_0^2 x^2 / 2$ as well as anharmonic contributions. The heat bath consists of N independent harmonic oscillators of mass m_ν, momentum p_ν, coordinate q_ν, and frequency ω_ν, with $\nu = 1, 2, \ldots, N$. The interaction of the oscillator and the heat bath is linear in q but arbitrary in x. It is convenient to write

$$H_B + H_{SB} = \tfrac{1}{2} \sum_\nu \left[\frac{p_\nu^2}{2 m_\nu} + \omega_\nu^2 (q_\nu - a_\nu(x))^2 \right] \quad . \tag{9}$$

If the contribution $\tfrac{1}{2} \sum_\nu \omega_\nu^2 a_\nu^2(x)$ is the "static" interaction of the oscillator and the heat bath, then V(x) in (8) is the potential energy of the isolated oscillator; otherwise V(x) must be interpreted as a suitably modified potential energy.

The equations of motion for the system and for the bath can be obtained from the Hamiltonian. The bath equations are linear, and they can be integrated and the result substituted in the system equations. This yields a generalized Langevin equation in which the bath enters only via initial conditions. To exhibit the important features of the resulting equations most clearly we go to the limit of short bath correlation times and take particular forms of the coupling function $a_\nu(x)$. If $a_\nu(x)$ is linear,

$$a_\nu(x) = \Gamma_\nu x , \qquad (10)$$

we recover Eq. (1), i.e. the classical Brownian motion with additive fluctuations and linear dissipation.[5]

If the coupling function $a_\nu(x)$ is quadratic,

$$a_\mu(x) = \tfrac{1}{2}\beta_\nu x^2 , \qquad (11)$$

then one obtains[5,6]

$$\boxed{\begin{aligned}\dot{x} &= p \\ \dot{p} &= -G(x) - (\omega_0^2 + \gamma(t))x - \mu x^2 p\end{aligned}} \qquad \begin{aligned}(12a)\\(12b)\end{aligned}$$

The term $G(x)$ is the arharmonic restoring force contained in $-V'(x)$. Microscopic expressions for the fluctuations $\gamma(t)$ and the dissipation μ are given in Refs. 5 and 6. Furthermore, $\gamma(t)$ and μ obey the fluctuation-dissipation relation.

$$\langle\gamma(t)\gamma(t')\rangle = 2\mu kT\delta(t-t') . \qquad (13)$$

Equation (12) is a stochastic differential equation with multiplicative fluctuations. We have studied this equation in some detail and have arrived at the following important conclusions.

a) The dissipative term $(-\mu x^2 p)$ that must be included to compensate for the multiplicative fluctuations is necessarily nonlinear. It is thus not possible to have a linear closed system with multiplicative fluctuations.

b) We have explicitly shown that the energy of the oscillator (12) is canonically distributed as $t\to\infty$, i.e.

$$P_{eq}(x,p) \sim \exp\left(-\frac{p^2}{2} + V(x)\right) \qquad (14)$$

c) The multiplicative fluctuations and corresponding nonlinear dissipation can greatly affect the evolution of the system towards equilibrium.

3. Spin System

A completely different system that can be considered by our procedure is a spin interacting with a surrounding heat bath. The simplest heat bath is one composed of other spins that interact with the spin of interest via a Heisenberg interaction. Let M be the magnetic moment of the spin of interest and let the heat bath be composed of spins with magnetic moments m_ν, $\nu = 1, 2, \ldots, N$. We assume that the bath spins are spatially fixed and interact with one another in a mean field sense. The differences in the local environment causes each spin to have a different precession frequency. Phenomenologically it is usually assumed that the average components of M relax to equilibrium exponentially according to the Bloch equations, with constant relaxation times T_1 for $<M_z>$ and T_2 for $<M_+> \equiv <M_x \pm i M_y>$.[3] We shall see that a Hamiltonian model leads to multiplicative fluctuations and does not predict such simple behavior.

The model Hamiltonian for the spin system is

$$H = H_S + H_B + H_{SB} \equiv -H_0 \cdot M - \sum_\nu (H_0 + H_\nu) \cdot m_\nu + \sum_\nu G_\nu M \cdot m_\nu \quad (15)$$

Here H_0 is the external field, $H_0 + H_\nu$ is the local field at spin m_ν, and the G_ν are coupling constants.

The feature that makes this problem different from (and more complicated than) the mechanical oscillator is the difference in the commutation relations (or Poisson bracket relations) for the variables in the two cases. While in the oscillator case the Poisson brackets $\{x,p\}$ are c-numbers, here one has the angular momentum relations $\{M_x, M_y\} \alpha M_z$ (and cyclic permutations thereof).

The system equation of motion obtained from the Hamiltonian (15) is

$$\dot{M} = \gamma M_0 \times M + \sum_\nu \gamma_\nu G_\nu m_\nu \times M \quad (16)$$

where γ and γ_ν denote the gyromagnetic ratios of M and m_ν respectively. The equations of motion of the bath spins are similar to (16) and, since they are formally linear in m_ν, can be integrated exactly in terms of the initial conditions $m_\nu(0)$.

Substitution of this solution in (16) gives an equation which although exact and dependent only on bath initial conditions, is highly nonlinear. In particular, it is difficult to clearly identify the fluctuating and dissipative contributions in this complicated result. Instead, we decided to work in the weak coupling approximation, expand the solution for the bath to linear order in the coupling constants G_ν, and substitute this approximate result in (16). In the system equations of motion we than retained fluctuating terms to $O(G_\nu)$ and dissipative terms to $O(G_\nu^2)$ (in order to establish a consistent fluctuation-dissipation relation). The resulting equation in the limit of a delta-correlated bath are [7]

$$\dot{M}_+ = i\Omega_0 M_+ - ih(t)M_z - \mu M_+ M_z \quad (17a)$$

$$\dot{M}_- = -i\Omega_0 M_- + ih^*(t)M_z - \mu M_- M_z \quad (17b)$$

$$\dot{M}_z = i/2[h(t)M_- - h^*(t)M_+] + \mu M_+ M_- \quad (17c)$$

Microscopic expressions for $h(t)$ and μ are given in Ref. (7) and are related by the fluctuation-dissipation relation

$$<h(t)h^*(t')> = \frac{2kT}{H_0} \mu \delta(t-t') \quad . \quad (18)$$

Several points are noteworthy about the result (17), which is again a set of stochastic differential equations with multiplicative fluctuations.

a) The linear portions correspond to the phenomenological model such as in Eq. (5). The "dissipative" terms $\mu M_i M_j$ are new and have never appeared in previous derivations. These terms are again necessary to compensate for the multiplicative fluctuations properly.

b) We have explicitly shown the equilibration of the spin as a consequence of Eqs. (17), i.e.

$$P_{eq}(\underline{M}) \sim \exp(\underline{H}_0 \cdot \underline{M}) \quad . \tag{19}$$

We also note that the equilibration does not occur in the absence of the nonlinear dissipative terms.

C) We have also shown that the multiplicative fluctuations and corresponding nonlinear dissipation play a crucial role in the rate of equilibration of the evolution which are different from that predicted by the simple Bloch equations.

References

1. N.G. van Kampen, Phys. Rep. 24, 171 (1976).
   J.D. Mason, ed., Stochastic Differential Equations and Applications (Academic Press, New York, 1977).
   A. Schenzle and H. Brand, Phys. Rev. A20, 1628 (1979).

2. G.E. Uhlenbeck and L.S. Ornstein, Phys. Rev. 36, 823 (1930).
   S. Chandrasekhar, Rev. Mod. Phys. 15, 1 (1943).
   M.C. Wang and G.E. Uhlenbeck, Rev. Mod. Phys. 17, 323 (1945).

3. R. Kubo, in Fluctuation, Relaxation and Resonance in Magnetic Systems, D. ter Haar, ed. (Oliver and Boyd, Edinburgh, 1962); in Stochastic Processes in Chemical Physics. K.E. Shuler, ed. (John Wiley, New York, 1969).
   P.W. Anderson and P.R. Weiss, Rev. Mod. Phys. 25, 269 (1953).
   M.Lax, Rev. Mod. Phys. 38, 359 (1966); ikid. 38, 541 (1966).

4. J.L. Lebowitz and E. Rubin, Phys. Rev. 131, 2381 (1963).
   P. Mazur and I. Oppenheim, Physica 50, 241 (1970).
   J.M. Deutch and I. Oppenheim, J. Chem. Phys. 54, 3547 (1971).
   E. Braun, Physica 103A, 325 (1980).
   G.W. Ford, M. Kac and P. Mazur, J. Math. Phys. 6, 504 (1965).
   P. Mazur and E. Braun, Physica 30, 1973 (1964).
   I.R. Senitzky, Phys. Rev. 119, 670 (1960).
   I.R. Senitzky, Phys. Rev. 124, 642 (1961).

5. R. Zwanzig, J. Stat. Phys. 9, 215 (1973).

6. K. Lindenberg and V. Seshadri, Physica 109A, 483 (1981).

7. V. Seshadri and K. Lindenberg, to appear in Physica.

ON THE SYMPLECTIC STRUCTURE OF INTEGRABLE SYSTEMS

Alice M. Roos
Center for Studies of Nonlinear Dynamics
La Jolla Institute
La Jolla, California 92037

INTRODUCTION

We show that the symplectic structure of many integrable systems is such that any of the infinite set of integrals can be considered the Hamiltonian of the system. This leads to a systematic method of determining Lax pairs. Using the symplectic structure, we find the sine-Gordon raising operator.

RAISING OPERATOR

For many completely integrable dynamical systems, a raising operator can be constructed that determines the infinite set of conserved integrals. In several cases (Korteweg - de Vries[1] (KdV), modified KdV (MKdV), nonlinear Schrodinger (NLS), Harry Dym[2], Toda Lattice[3] and as will be shown, sine-Gordon (s-G)) the powers of the raising operator L generate the gradients of the conserved integrals $H^{(n)}$.

$$L^n \frac{\delta H^{(0)}}{\delta u} = \frac{\delta H^{(n)}}{\delta u}, \text{ for integer } n \qquad (1)$$

For real infinite systems the gradient $\frac{\delta H}{\delta u} \equiv dH$ is defined by the Frechet derivative

$$\frac{dH(u+\varepsilon v)}{d\varepsilon} \bigg|_{\lim \varepsilon = 0} = (dH,v) \qquad (2)$$

where the bilinear form is

$$(F,G) = \int FG \, dx.$$

For the compact problem (periodic solutions) the integral is over the period. For non-compact boundary conditions (assuming solutions vanish at infinity) the limits of the integral are $\pm \infty$.

The raising operators for the integrable systems listed above are the product of two symplectic operators. An operator S is symplectic if the corresponding bracket [,] is Lie.

$$[F,G] \equiv (F,SG). \qquad (3)$$

When the recursion relation for the gradients can be expressed in terms of L in the form $L = S_1^{-1} S_0$ where S_0 and S_1 are symplectic, it is simple to prove that the infinite set of integrals are in involution. See for instance Lax's use of the Lenart recursion relation for KdV[4].

The raising operator leads one systematically to a Lax pair[5] for the evolution equation. The operator L determines the eigenvalue problem

$$\lambda \psi = L\psi \qquad (4)$$

if we define $\psi = \sum_n \dfrac{dH^{(n)}}{\lambda^n}$.

The operator B of the corresponding time dependent problem

$$\partial_t \psi = B\psi \qquad (5)$$

is found by linearizing the original evolution equation. The pairs L,B derived in this way[6] are different from but equivalent to the conventional pairs[7] used to integrate the evolution equations. The eigen-function ψ is a quadratic function of the conventional eigenfunctions.

HIERARCHY OF SYMPLECTIC OPERATORS

In this section it will be shown that the raising operator $L \equiv S_1^{-1} S_0$, where S_1 and S_0 are symplectic, determines an infinite set of symplectic operators of the form $S_1 F(L)$. Here F is any real-valued polynomial in L and L^{-1}. A consequence is that an appropriate Poisson bracket can be constructed so that any linear combination of the integrals $\tilde{H}$ with a gradient of the form $d\tilde{H} = F(L) dH^{(0)}$ can be considered the Hamiltonian of the integrable system.

We will prove by induction that the operators $S_1 L^n$ are symplectic. The steps necessary to show that any $S = \sum_n a_n S_1 L^n$, for real a_n, is also symplectic will then be indicated.

The antisymmetry of S_1 and S_1L implies

$$(F, S_1LG) = -(S_1LF, G) = (LF, S_1G). \tag{6}$$

Using (6) n-1 times one can show that S_1L^n for positive n is antisymmetric.

$$(F, S_1L^nG) = (L^{n-1}F, S_1LG),$$

and by the antisymmetry of LS_1

$$(L^{n-1}F, S_1LG) = -(S_1L^nF, G). \tag{7}$$

To establish that the brackets determined by S_1L^{+n} satisfy the Jacobi identity, we need an identity derived from the assumption that S_1 and S_1L do. Assume

$$(A, S_1L(B, S_1LC)) + (B, S_1L(C, S_1LA)) = -(C, S_1L(A, S_1LB)). \tag{8}$$

Using (6) twice we have that

$$-(C, S_1L(A, S_1LB)) = -(LC, S_1(LA, S_1B)). \tag{9}$$

But since the S_1-bracket also satisfies the Jacobi identity

$$-(LC, S_1(LA, S_1B)) = (LA, S_1(B, S_1LC)) + (B, S_1(LC, S_1LA)). \tag{10}$$

Comparing lines (8) and (10) and relabing (B→A, C→B, LA→C) yields the identity

$$(A, S_1L(B, S_1C)) = (A, S_1(LB, S_1C)). \tag{11}$$

With (11) it is simple to show by induction that S_1L^n-brackets satisfy the Jacobi identity. Assume that this is true for a particular S_1L^m. Then,

$$(LA, S_1L^m(LB, S_1L^mC)) = -(LB, S_1L^m(C, S_1L^mLA)) - (C, S_1L^m(LA, S_1L^mLB)). \tag{12}$$

On using (11) this equation becomes (13)

$$(A, S_1 L^{m+1}(B, S_1 L^{m+1} C)) = -(B, S_1 L^{m+1}(C, S_1 L^{m+1} A)) - (C, S_1 L^{m+1}(A, S_1 L^{m+1} B)).$$

We have shown for positive n that the operators $S_1 L^n$ are symplectic. From this proof we automatically have that the $S_1 L^{-n}$ are also symplectic since nothing in the proof distinguishes between the operators S_0 and S_1. That is, we can interchange S_0 and S_1 so that the raising operator is $M \equiv S_0^{-1} S_1$ and the proof we have gives us that $S_0 M^{n+1}$ is symplectic, but $S_0 M^{n+1} = S_1 M^n = S_1 L^{-n}$.

Therefore, for the integrable systems with a raising operator of the form $L \equiv S_1^{-1} S_0$, the two symplectic operators S_1 and S_0 determine an infinite set of symplectic operators $S_1 L^n$ for $n = 0, \pm 1, \ldots$. Each of the $S_1 L^n$ defines a Poisson bracket in terms of which a different conserved integral can be said to generate the equations of motion. In fact it is straight forward to show that any operator $S = \sum_n a_n S_1 L^n$ determines a Lie bracket and is therefore symplectic. The proof of this is given in the Appendix.

SINE-GORDON RAISING OPERATOR

Here we will construct the raising operator for the s-G equation in light-cone coordinates. In doing so a connection is made between the s-G equation

$$u_{xt} = \sin u \tag{14}$$

and the integrated MKdV (IMKdV) equation

$$u_t + \frac{u_x^3}{2} + u_{xxx} = 0. \tag{15}$$

In Part A we show that these apparently very different equations conserve the same infinite set of integrals which are determined by the positive and negative powers of L. In Part B we give an algorithm for calculating the integrals determined by L^{-n}.

A. The s-G and IMKdV equations conserve an infinite set of polynomial integrals. The first several (non-compact) integrals are

$$H^{(0)} = \int_{-\infty}^{+\infty} \left(-\frac{u_x^2}{2} \right) dx. \qquad (16)$$

$$H^{(1)} = \int_{-\infty}^{+\infty} \left(-\frac{u_{xx}^2}{2} + \frac{u_x^4}{8} \right) dx. \qquad (17)$$

$$H^{(2)} = \int_{-\infty}^{+\infty} \left(-\frac{u_{xxx}^2}{2} - \frac{u_x^6}{16} + \frac{5}{4} u_x^2 u_{xx}^2 \right) dx. \qquad (18)$$

The integral $H^{(1)}$ is the IMKdV Hamiltonian with the Poisson bracket

$$\{F,G\} = (dF, S_1 dG) \qquad (19)$$

determined by S_1,

$$S_1 F \equiv 1/2 \left(\int_{-\infty}^{x} F dx - \int_{x}^{+\infty} F dx \right). \qquad (20)$$

A raising operator $L \equiv S_1^{-1} S_0$ for the polynomial integrals is readily constructed from the conditions that

i) $S_0 dH^{(0)} = S_1 dH^{(1)}$

ii) $\{F,G\} \equiv (dF, S_0 dG)$ is a Lie bracket.

One finds

$$S_0 \phi = \partial_x \phi + u_x \partial_x^{-1} (u_x \phi). \qquad (21)$$

Thus, we have the raising operator $L = \partial_x S_0$ for the polynomial integrals of the s-G equation.

The question we ask (and answer) is what is L for the s-G trigonometric integrals[8]? The simplest such integral is

$$H_{s-G} = \int_{-\infty}^{+\infty} (1-\cos u) \, dx. \qquad (22)$$

In terms of the Poisson bracket given by (19) and (20), this integral is the s-G Hamiltonian.

One is lead to consider the quantities generated by the inverse powers of L on finding that

$$LdH_{s-G} = dH^{(0)}. \tag{23}$$

Using the antisymmetry of S_1 and S_0 it is simple to prove that the $H^{(\pm n)}$ determined by $L^{\pm n}$ are in involution. Therefore, the $H^{(\pm n)}$ are conserved by both equations. If it is true that the H^{-n} are the s-G trigonometric integrals we will have determined that their raising operator is L^{-1}, and surprisingly, we will have shown that the IMKdV equation also conserves the trigonometric integrals.

B. The integrals formally determined by the inverse powers of L are constructed by an algorithm using the space-time symmetry of s-G equation. Each of the polynomial integral $H^{(n)}$ defines a conserved current

$$\partial_t J_0^{(n)} + \partial_x J_1^{(n)} = 0 \tag{24}$$

where
$$H^{(n)} \equiv \int_{-\infty}^{+\infty} J_0^{(n)} \, dx .$$

Because of the symmetry in (x,t) of the equation of motion (14) another set of conserved currents can be constructed from (24)

$$\partial_t J_1^{(n)\prime} + \partial_x J_0^{(n)\prime} = 0 . \tag{25}$$

The prime denotes the operation of interchanging the derivatives with respect to x and t, and then eliminating the time derivatives using $u_t = \partial_x^{-1} \sin u$.

We find that the integrals

$$H^{-(n+1)} \equiv \int J_1^{(n)\prime} \, dx \tag{26}$$

are indeed the s-G trigonometric integrals satisfying

$$L^{-n} dH^{(0)} = dH^{(-n)} \quad n=1,2,\ldots \; . \tag{27}$$

For example, we construct $H^{(-2)}$. From (17) we have

$$J_0^{(1)} = -\frac{u_{xx}^2}{2} + \frac{u_x^4}{8} .$$

Since

$$\partial_t J_0^{(1)} = -u_{xx}u_{xxt} + \frac{u_x^3}{2}u_{xt} = -\partial_x\left(\frac{u_x^2}{2}\cos u\right),$$

$$J_1^{(1)'} = \frac{(\partial_x^{-1}\sin u)^2 \cos u}{2}. \tag{28}$$

Therefore

$$H^{(-2)} = -\frac{(\partial_x^{-1}\sin u)^2 \cos u}{2}. \tag{29}$$

One can easily check that the sinh-Gordon equation and the IMKdV equation of the form

$$u_t - \frac{u_x^3}{2} + u_{xxx} = 0 \tag{30}$$

also conserve the integrals generated by the positive and negative powers of a raising operator L. In this case

$$L\phi = \partial_x(\partial_x\phi - u_x\partial_x^{-1}(u_x\phi)). \tag{31}$$

REFERENCES

1. C. S. Gardner, J. M. Greene, M. D. Kruskal and R. M. Miura, Phys. Rev. Lett. 19 (1967), 1095-1097.
2. F. Magri, J. Math. Phys. 19 (1978), 1156-1162.
3. H. Flaschka, Theoretical Phys. 51 (1974), 703.
4. P.D. Lax, Siam Review 18, 3 (1976), 351-375.
5. P.D. Lax, Comm. Pure Appl. Math. 21 (1968), 467-490.
6. K. M. Case, A. M. Roos, J. Math. Phys. 23 (1982), 392-395.
7. M. J. Ablowitz, D. J. Kaup, A. C. Newell, and H. Segur, Stud. Appl. Math. 53 (1974), 249-315.
8. B. Yoon, Phys. Rev. D. 13 (1976), 3440.

APPENDIX

We show here that the operators $S = \sum_n a_n S_1 L^n$, a_n real, determine Lie brackets.

It is clear that the brackets (F,SG) are antisymmetric since (,) is bilinear and each $a_m S_1 L^m$ determines an antisymmetric bracket (See 7).

To show that the brackets (F,SG) satisfy the Jacobi identity we use (6) and (11).

$$(A, \sum_m a_m S_1 L^m (B, \sum_n a_n S_1 L^n C)) = \sum_m a_m (\sum_n b_n (A, S_1 L^m (B, S_1 L^n C))) \quad \text{A1}$$

For each m there are two cases to consider: $n \geq m$ or $n = m-p$ for p a non-negative integer, and $n < m$ or $n = m+r$ for r a positive integer. Then

$$\sum_m a_m (\sum_n b_n (A, S_1 L^m (B, S_1 L^n C)))$$

$$= \sum_m a_m \sum_p b_{m-p} (A, S_1 L^m (B, S_1 L^{m-p} C)) + \sum_m a_m \sum_r b_{m+r} (A, S_1 L^m (B, S_1 L^{m+r} C)), \quad \text{A2}$$

which using (11) p times and (6) r times can be written as

$$= \sum_m a_m \sum_p b_{m-p} (A, S_1 L^{m-p} (L^p B, S_1 L^{m-p} C)) + \sum_m a_m \sum_r b_{m+r} (A, S_1 L^m (L^r B, S_1 L^m C)).$$

Since each $S_1 L^n$ is a symplectic operator from A3 we then have

$$\sum_m a_m \left(\sum_n b_n (A, S_1 L^m (B, S_1 L^n C)) \right)$$

$$= \sum_m a_m \Big\{ \sum_p b_{m-p} \big((L^p B, S_1 L^{m-p} (C, S_1 L^{m-p} A)) + (C, S_1 L^{m-p} (A, S_1 L^m B)) \big)$$

$$+ \sum_r b_{m+r} \big((L^r B, S_1 L^m (C, S_1 L A)) + (C, S_1 L^m (A, S_1 L^{m+r} B)) \big) \Big\} \quad \text{A4}$$

$$= -\sum_m a_m \sum_n b_n \big((B, S_1 L^m (C, S_1 L^m A)) + (C, S_1 L^m (A, S_1 L^n B)) \big) . \quad \text{A5}$$

Thus, the operators $\sum_n a_n S_1 L^n$ are symplectic.

NON-GAUSSIAN WATER-WAVE FIELDS

Bruce J. West
Center for Studies of Nonlinear Dynamics
La Jolla Institute
POB 1434, La Jolla, CA 92038

The generation of waves both on and in the deep ocean and their evolution toward an asymptotic statistical steady state have long been topics of interest to Ken Watson.[1-4] In this note we consider a linear dynamic model of the ocean surface in which deep-water waves relax to a steady state with non-Gaussian statistics. This model provides a direct interpretation of the concept of wave "groupiness" recently introduced into the oceanographic literature.[5] Due to limitations of space we will discuss a generic linear model which lacks certain of the features of the sea surface but maintains those properties thought to be essential. The model is motivated by the exact dynamic equations and provides a possible explanation for the tendency of water waves on the ocean to appear in groups.

We assume the vertical displacement of the sea surface $\zeta(\underline{x},t)$ and the velocity potential on this surface $\phi_s(\underline{x},t)$ to be expressible as a superposition of N linear modes $\{\zeta_n(t), \phi_n(t)\}$ with wave vector $\underline{k}_n$ and frequency ω_n, n=1, 2, ... N. The equations of motion are given by

$$[\frac{\partial}{\partial t} + \lambda_n - \gamma_n(t)]\phi_n(t) + g\zeta_n(t) = f_n(t), \tag{1}$$

$$\frac{\partial}{\partial t} \zeta_n(t) = k_n \phi_n. \tag{2}$$

In general, the influx of energy to the spectral interval of interest is produced by: 1) the incoherent fluctuations in the airflow at the sea surface; 2) the average coupling between the air and sea surface; 3) the coherent fluctuations in air flow at the sea surface; and 4) the transfer of energy from one spectral interval to another by wave-wave interactions. The linear air-sea coupling model of Miles and Phillips describes the generation and initial growth of the individual waves in the wave field.[6] This model provides a fluctuating contribution to $f_n(t)$ and a negative definite contribution to λ_n. The generalization of this model by West and Seshadri[7] incorporates the modulation of the wind fluctuations into the equations of motion through the fluctuating parameter $\gamma_n(t)$. The nonlinear wave-wave interactions provide two additional contributions to the linearized model. Firstly they give rise to a second source of additive fluctuations to drive the waves of interest. Secondly, and more

00940243X/82/880347-06$3.00 Copyright 1982 American Institute of Physics importantly, they provide a mechanism to saturate the growth of energy in a given spectral interval, i.e., the wave-wave interactions transfer energy out of a spectral interval after the energy content within the interval exceeds some critical value, thereby providing a positive contribution to λ_n. Observationally the existence of an asymptotic statistical steady state of the surface wave field implies $\lambda_n > 0$ for all n in the linear model (1) and (2) at late times.

Equations (1) and (2), as they stand, are a fairly good first-order model of our present level of understanding of the evolution of wind-generated waves, provided the proper values of λ_n are used initially and asymptotically. The same "constant" λ_n cannot be used at both early and late times. Initially when the waves have small amplitudes the test wave responds primarily to the wind field, i.e. $\lambda_n < 0$. Asymptotically when the wave is near its steady state level the "growth" parameter must include the average wave-wave interactions as well as the coupling to the wind so that $\lambda_n > 0$. West,[8] using the method of statistical linearization, has described the possible development of a statistical steady state for gravity-capillary waves using a model quite close to that described by (1) and (2). In this model $\lim_{t \to 0} \lambda_n < 0$ and $\lim_{t \to \infty} \lambda_n > 0$ for all n. The exact time at which λ_n changes sign depends on the growth of the nonlinear interactions. Henceforth we assume $\lambda_n > 0$, i.e. the water waves are near their steady state level.

In the remainder of our discussion we assume that both $f_n(t)$ and $\gamma_n(t)$ are zero-centered Gaussian processes that are delta correlated in time. The incoherent $f_n(t)$-fluctuations have a spectral strength $2D_n (\equiv \int <f_n(t)f_n(t')> dt)$, and the coherent $\gamma_n(t)$-fluctuations have a spectral strength $2\Lambda_n (\equiv \int <\gamma_n(t)\gamma_n(t')> dt)$. We also find it convenient to replace the velocity potential and the surface displacement by the action-angle variable $(J_n(t), \theta_n(t))$. In this representation we have

$$\frac{k_n}{\omega_n} \phi_n(t) \mp i\zeta_n(t) = \sqrt{\frac{2\omega_n}{g\rho_0}} J_n(t) e^{\mp i\theta_n(t)} \qquad (3)$$

so that the equations of motion (1) and (2) are replaced with

$$\dot{J}_n(t) + 2\lambda_n J_n(t) = 2\gamma_n(t) J_n(t) + \text{Real}\left[e^{-i\theta_n(t)} f_n(t)\right] [2\rho_0 k_n J_n(t)/\omega_n]^{\frac{1}{2}} \qquad (4)$$

along with an equation for $\dot{\theta}_n(t)$. We are not interested in this latter equation, as will become apparent, and so we do not write it down. The statistical properties of the surface waves can be determined by constructing an equation of evolution for the probability density $P(J_n, \theta_n, t)$.[9,10]

For the assumed statistics of the fluctuations we can construct a Fokker-Planck equation for $P(J_n,\theta_n,t)$. A condensed description is obtained in terms of the reduced distribution $P(J_n,t)$ defined by averaging $P(J_n,\theta_n,t)$ over the angle variable, i.e.

$$P(J_n,t) \equiv \frac{1}{2\pi}\int_0^{2\pi} d\theta_n P(J_n,\theta_n,t) \qquad (5)$$

The Fokker-Planck description of the evolution of $P(J_n,t)$ is given by

$$\frac{\partial}{\partial t}P(J_n,t)=\frac{\partial}{\partial J_n}\left[2\lambda_n J_n+2\Lambda_n J_n\frac{\partial}{\partial J_n}J_n + \bar{D}_n J_n^{\frac{1}{2}}\frac{\partial}{\partial J_n}J_n^{\frac{1}{2}}\right]P(J_n,t), \qquad (6)$$

where $\bar{D}_n \equiv 2\rho_0 D_n/V_n = 2\rho_0\omega_n D_n/g$. The solution to (6) gives the full statistics of the action of the surface wave as a function of time. The steady state (asymptotic) solution to (6) is given by

$$\frac{\partial}{\partial t}P_{ss}(J_n) = \lim_{t\to\infty}\frac{\partial P}{\partial t}(J_n,t) = 0, \qquad (7)$$

yielding a distribution which is independent of time and the initial conditions.

Using (7) we have, from (6),

$$[\bar{D}_n + (2\lambda_n + \Lambda_n)J_n]P_{ss}(J_n) + 2(\bar{D}_n J_n + \Lambda_n J_n^2)\frac{\partial}{\partial J_n}P_{ss}(J_n) = 0, \qquad (8)$$

which has the normalized solution

$$P_{ss}(J_n) = \left(\frac{\bar{D}_n}{\Lambda_n}\right)^{\lambda_n/\Lambda_n} \frac{\Gamma\left(\frac{\lambda_n}{\Lambda_n}+\frac{1}{2}\right)}{\Gamma\left(\frac{\lambda_n}{\Lambda_n}\right)\Gamma(\frac{1}{2})} \frac{1}{J_n^{\frac{1}{2}}\left[\frac{\bar{D}_n}{\Lambda_n}+J_n\right]^{\lambda_n/\Lambda_n+\frac{1}{2}}} \qquad (9)$$

This distribution has a number of interesting features. Firstly if we consider the case when the coherent fluctuations vanish, i.e. $\Lambda_n \to 0$, we obtain

$$\lim_{\Lambda_n\to 0} P_{ss}(J_n) \cong \left(\frac{\lambda_n}{\pi\bar{D}_n}\right)^{\frac{1}{2}} \exp[-\lambda_n J_n/\bar{D}_n]/J_n^{\frac{1}{2}} \qquad (10)$$

where we have used the asymptotic forms of the gamma functions in (9). Note that (10) is a Gaussian distribution in the square root of the action for the test wave, i.e. $x_n = J_n^{\frac{1}{2}}$

$$\lim_{n\Lambda_n \to 0} dJ_n P_{ss}(J_n) = \left(\frac{\lambda_n}{\pi \bar{D}_n}\right)^{\frac{1}{2}} \exp[-\lambda_n x_n^2/\bar{D}_n]dx_n. \qquad (11)$$

This distribution is consistent with our past intuition about the statistics of water waves.

The distribution (9) is clearly non-Gaussian. However, if we evaluate the average action using (9) we obtain

$$\langle J_n \rangle_{ss} = \frac{\bar{D}_n}{2\lambda_n(1 - \Lambda_n/\lambda_n)} \qquad (12)$$

which gives a correction $(1 - \Lambda_n/\lambda_n)^{-1}$ to the average action calculated using the Gaussian distribution (10). For $\Lambda_n/\lambda_n \ll 1$ the two results are identical. This is the region where the action transfer due to nonlinear interaction, i.e. the effective dissipation, dominates the action influx from the coherent fluctuations in the wind. If $\Lambda_n \geq \lambda_n$ then, since the average action is positive definite, (12) tells us that the action spectral density diverges, i.e. the surface wave becomes unstable. Such a situation arises when the action flux to this spectral region by the coherent fluctuations is too great to be transferred away by the wave-wave interaction terms. There is then a "pileup" of energy in this spectral interval and an energetic instability results, i.e., the wave breaks.

Let us consider the ν-th moment of the action

$$\langle J_n^\nu \rangle_{ss} = \left(\frac{\bar{D}_n}{\Lambda_n}\right)^{\lambda_n/\Lambda_n} \frac{\Gamma\left(\frac{\lambda_n}{\Lambda_n} + \frac{1}{2}\right)}{\Gamma\left(\frac{\lambda_n}{\Lambda_n}\right)\Gamma(\frac{1}{2})} \int_0^\infty \frac{J_n^{\nu - \frac{1}{2}} dJ_n}{[\bar{D}_n/\Lambda_n + J_n]^{\lambda_n/\Lambda_n + \frac{1}{2}}} \qquad (13)$$

This integral is finite for $-\frac{1}{2} < \nu < \frac{\lambda_n}{\Lambda_n}$ but diverges for $\nu > \lambda_n/\Lambda_n$.

Thus no matter how small the spectral strength of the coherent fluctuations there is some moment of the action which will diverge. This is a generic property of linear dissipative systems with state-dependent fluctuations. These systems are quite different from those with Gaussian fluctuations, which have finite moments to all orders. The divergence is actually quenched by the nonlinear terms in physical systems which are treated more exactly.[10] One expects to see the vestige of this effect in the physical system, however.

Consider now the ocean surface represented by the linear superposition of N such modes

$$\zeta(\underline{x},t) = -\sum_{n=1}^{N} \left[\frac{2\omega_n J_n(t)}{\rho_0 g}\right]^{1/2} \sin[\underline{k}_n \cdot \underline{x} - \theta_n(t)], \quad (14)$$

where $J_n(t)$ is a random variable described by the distribution function (9) and $\theta_n(t)$ is uniformly distributed in the interval $(0, 2\pi)$. The mean surface elevation is zero due to the phase average, and the variance is given by

$$\overline{\langle\zeta^2(\underline{x},t)\rangle_{ss}}^{\theta} = \sum_{n=1}^{N} \frac{4D_n}{V_n^2 \lambda_n (1 - \Lambda_n/\lambda_n)}, \quad (15)$$

where the overbar denotes a uniform phase average. This moment is finite for $\lambda_n > \Lambda_n$ but has the instability already noted. The correlation between points on the surface separated by a distance $\underline{L}$ is given by

$$\overline{\langle\zeta(\underline{x})\zeta(\underline{x}+\underline{L})\rangle_{ss}}^{\theta} = \sum_{n=1}^{N} \frac{4D_n}{V^2(\lambda_n - \Lambda_n)} \cos(\underline{k}_n \cdot \underline{L}). \quad (16)$$

When $\lambda_n \sim \Lambda_n$ the Fourier transform of the correlation function, i.e. the spectrum, has a peak at $\underline{k}_n$. For $\lambda_n \gg \Lambda_n$ the spectrum is given by $D_n/\lambda_n V_n^2$ and can be quite broad in wave number space. West[8,11] has calculated this quantity for gravity-capillary waves and found that $D_n/\lambda_n V_n^2 \sim 1/\omega_n^4$. The observed spectrum in this region is closer to $1/\omega_n^3$,[12] but then one would not expect such a simple model to be more than semiquantative.

Even when $\lambda_n \gg \Lambda_n$, where the energy instability noted above does not occur, other moments can be unstable. The moments $\overline{\langle\zeta^{2\nu}\rangle_{ss}}^{\theta}$ always have a component $\overline{\langle J_n^\nu\rangle_{ss}}$ which diverges for $\nu > \lambda_n/\Lambda_n$. This divergence can be viewed as a "bunching up" of the wave field on the sea surface, i.e. the action is not distributed as smoothly as it is for a Gaussian process. This behavior is a consequence of the asymptotic form of the probability density (9), i.e.

$$P_{ss}(J_n) \sim \text{constant}/J_n^{\lambda_n/\Lambda_n + 1}. \quad (17)$$

Such distributions are known to exhibit clustering in the variate;

see e.g. Mandelbrot[13] or Montroll and West.[14] The extremes of such clustering are the so-called fractals for which there are no inner or outer scales in the process. Such scales for wind-generated wave fields on the ocean are of course determined by viscosity on the one end and the average wind speed on the other. Thus the statistical self-similarity of the wave spectrum is confined to well defined limits in this model.

In summary we observe that this linearized model of the surface wave dynamics describes the asymptotic statistical steady state as a non-Gaussian wave field. The coherent pressure fluctuations $[\gamma_n(t)]$ give rise to the hyperbolic probability density in the action, [cf. (9) and (13)], resulting in a clustering of the waves on the sea surface, i.e. the waves appear in groups. In addition the model predicts a steady state frequency spectrum which is not unreasonable.

ACKNOWLEDGMENTS

I thank V. Seshadri and K. Lindenberg for many discussions on which this work is based. This work was supported in part by the Office of Naval Research.

REFERENCES

1. K. M. Watson and B. J. West, J. Fluid Mech. $\underline{70}$, 815-826 (1975).
2. K. M. Watson, B. J. West, and B. I. Cohen, J. Fluid Mech., $\underline{77}$, 185-208 (1976).
3. J. D. Meiss, N. Pomphrey, and K. M. Watson, Proc. Nat. Acad. Sci., USA, $\underline{76}$, 2109 (1979).
4. See additional references in Nonlinear Properties of Internal Waves, ed. B. J. West, AIP. Proceed, $\underline{76}$, 203 (1981).
5. See e.g. J. Hamilton and W. H. Hui, J. Geophys. Res. $\underline{84}$, C8, 4875 (1979).
6. See e.g. the discussion in O. M. Phillips, Dynamics of the Upper Ocean, 2nd Ed., Cambridge Univ. Press (1977).
7. B. J. West and V. Seshadri, J. Geophys. Res. $\underline{86}$, C5, 4293 (1981).
8. B. J. West, J. Geophys. Res. $\underline{86}$, C11, 11073 (1981).
9. R. L. Stratonovich, Topics in the Theory of Random Noise, Vol. 1, Gordon and Breach, N.Y., 115 (1967).
10. V. Seshadri, B. J. West and K. Lindenberg, Physica $\underline{107A}$, 219 (1981).
11. B. J. West, to appear in J. Fluid Mech. (1982).
12. G. T. Lleonart and D. R. Blackman, J. Fluid Mech. $\underline{87}$, 455 (1980).
13. B. B. Mandelbrot, Fractals, Form, Chance and Dimension, W. H. Freeman and Co., San Francisco (1977).
14. E. W. Montroll and B. J. West, in Fluctuation Phenomena, eds. E. W. Montroll and J. L. Lebowitz, North-Holland, Amsterdam (1979).

AIP Conference Proceedings

		L.C. Number	ISBN
No. 1	Feedback and Dynamic Control of Plasmas	70-141596	0-88318-100-2
No. 2	Particles and Fields - 1971 (Rochester)	71-184662	0-88318-101-0
No. 3	Thermal Expansion - 1971 (Corning)	72-76970	0-88318-102-9
No. 4	Superconductivity in d-and f-Band Metals (Rochester, 1971)	74-18879	0-88318-103-7
No. 5	Magnetism and Magnetic Materials - 1971 (2 parts) (Chicago)	59-2468	0-88318-104-5
No. 6	Particle Physics (Irvine, 1971)	72-81239	0-88318-105-3
No. 7	Exploring the History of Nuclear Physics	72-81883	0-88318-106-1
No. 8	Experimental Meson Spectroscopy - 1972	72-88226	0-88318-107-X
No. 9	Cyclotrons - 1972 (Vancouver)	72-92798	0-88318-108-8
No. 10	Magnetism and Magnetic Materials - 1972	72-623469	0-88318-109-6
No. 11	Transport Phenomena - 1973 (Brown University Conference)	73-80682	0-88318-110-X
No. 12	Experiments on High Energy Particle Collisions - 1973 (Vanderbilt Conference)	73-81705	0-88318-111-8
No. 13	π-π Scattering - 1973 (Tallahassee Conference)	73-81704	0-88318-112-6
No. 14	Particles and Fields - 1973 (APS/DPF Berkeley)	73-91923	0-88318-113-4
No. 15	High Energy Collisions - 1973 (Stony Brook)	73-92324	0-88318-114-2
No. 16	Causality and Physical Theories (Wayne State University, 1973)	73-93420	0-88318-115-0
No. 17	Thermal Expansion - 1973 (lake of the Ozarks)	73-94415	0-88318-116-9
No. 18	Magnetism and Magnetic Materials - 1973 (2 parts) (Boston)	59-2468	0-88318-117-7
No. 19	Physics and the Energy Problem - 1974 (APS Chicago)	73-94416	0-88318-118-5
No. 20	Tetrahedrally Bonded Amorphous Semiconductors (Yorktown Heights, 1974)	74-80145	0-88318-119-3
No. 21	Experimental Meson Spectroscopy - 1974 (Boston)	74-82628	0-88318-120-7
No. 22	Neutrinos - 1974 (Philadelphia)	74-82413	0-88318-121-5
No. 23	Particles and Fields - 1974 (APS/DPF Williamsburg)	74-27575	0-88318-122-3
No. 24	Magnetism and Magnetic Materials - 1974 (20th Annual Conference, San Francisco)	75-2647	0-88318-123-1
No. 25	Efficient Use of Energy (The APS Studies on the Technical Aspects of the More Efficient Use of Energy)	75-18227	0-88318-124-X

No.26	High-Energy Physics and Nuclear Structure - 1975 (Santa Fe and Los Alamos)	75-26411	0-88318-125-8
No.27	Topics in Statistical Mechanics and Biophysics: A Memorial to Julius L. Jackson (Wayne State University, 1975)	75-36309	0-88318-126-6
No.28	Physics and Our World: A Symposium in Honor of Victor F. Weisskopf (M.I.T., 1974)	76-7207	0-88318-127-4
No.29	Magnetism and Magnetic Materials - 1975 (21st Annual Conference, Philadelphia)	76-10931	0-88318-128-2
No.30	Particle Searches and Discoveries - 1976 (Vanderbilt Conference)	76-19949	0-88318-129-0
No.31	Structure and Excitations of Amorphous Solids (Williamsburg, VA., 1976)	76-22279	0-88318-130-4
No.32	Materials Technology - 1976 (APS New York Meeting)	76-27967	0-88318-131-2
No.33	Meson-Nuclear Physics - 1976 (Carnegie-Mellon Conference)	76-26811	0-88318-132-0
No.34	Magnetism and Magnetic Materials - 1976 (Joint MMM-Intermag Conference, Pittsburgh)	76-47106	0-88318-133-9
No.35	High Energy Physics with Polarized Beams and Targets (Argonne, 1976)	76-50181	0-88318-134-7
No.36	Momentum Wave Functions - 1976 (Indiana University)	77-82145	0-88318-135-5
No.37	Weak Interaction Physics - 1977 (Indiana University)	77-83344	0-88318-136-3
No.38	Workshop on New Directions in Mossbauer Spectroscopy (Argonne, 1977)	77-90635	0-88318-137-1
No.39	Physics Careers, Employment and Education (Penn State, 1977)	77-94053	0-88318-138-X
No.40	Electrical Transport and Optical Properties of Inhomogeneous Media (Ohio State University, 1977)	78-54319	0-88318-139-8
No.41	Nucleon-Nucleon Interactions - 1977 (Vancouver)	78-54249	0-88318-140-1
No.42	Higher Energy Polarized Proton Beams (Ann Arbor, 1977)	78-55682	0-88318-141-X
No.43	Particles and Fields - 1977 (APS/DPF, Argonne)	78-55683	0-88318-142-8
No.44	Future Trends in Superconductive Electronics (Charlottesville, 1978)	77-9240	0-88318-143-6
No.45	New Results in High Energy Physics - 1978 (Vanderbilt Conference)	78-67196	0-88318-144-4
No.46	Topics in Nonlinear Dynamics (La Jolla Institute)	78-057870	0-88318-145-2
No.47	Clustering Aspects of Nuclear Structure and Nuclear Reactions (Winnepeg, 1978)	78-64942	0-88318-146-0
No.48	Current Trends in the Theory of Fields (Tallahassee, 1978)	78-72948	0-88318-147-9
No.49	Cosmic Rays and Particle Physics - 1978 (Bartol Conference)	79-50489	0-88318-148-7

No.	Title		
No. 50	Laser-Solid Interactions and Laser Processing - 1978 (Boston)	79-51564	0-88318-149-5
No. 51	High Energy Physics with Polarized Beams and Polarized Targets (Argonne, 1978)	79-64565	0-88318-150-9
No. 52	Long-Distance Neutrino Detection - 1978 (C.L. Cowan Memorial Symposium)	79-52078	0-88318-151-7
No. 53	Modulated Structures - 1979 (Kailua Kona, Hawaii)	79-53846	0-88318-152-5
No. 54	Meson-Nuclear Physics - 1979 (Houston)	79-53978	0-88318-153-3
No. 55	Quantum Chromodynamics (La Jolla, 1978)	79-54969	0-88318-154-1
No. 56	Particle Acceleration Mechanisms in Astrophysics (La Jolla, 1979)	79-55844	0-88318-155-X
No. 57	Nonlinear Dynamics and the Beam-Beam Interaction (Brookhaven, 1979)	79-57341	0-88318-156-8
No. 58	Inhomogeneous Superconductors - 1979 (Berkeley Springs, W.V.)	79-57620	0-88318-157-6
No. 59	Particles and Fields - 1979 (APS/DPF Montreal)	80-66631	0-88318-158-4
No. 60	History of the ZGS (Argonne, 1979)	80-67694	0-88318-159-2
No. 61	Aspects of the Kinetics and Dynamics of Surface Reactions (La Jolla Institute, 1979)	80-68004	0-88318-160-6
No. 62	High Energy e^+e^- Interactions (Vanderbilt, 1980)	80-53377	0-88318-161-4
No. 63	Supernovae Spectra (La Jolla, 1980)	80-70019	0-88318-162-2
No. 64	Laboratory EXAFS Facilities - 1980 (Univ. of Washington)	80-70579	0-88318-163-0
No. 65	Optics in Four Dimensions - 1980 (ICO, Ensenada)	80-70771	0-88318-164-9
No. 66	Physics in the Automotive Industry - 1980 (APS/AAPT Topical Conference)	80-70987	0-88318-165-7
No. 67	Experimental Meson Spectroscopy - 1980 (Sixth International Conference, Brookhaven)	80-71123	0-88318-166-5
No. 68	High Energy Physics - 1980 (XX International Conference, Madison)	81-65032	0-88318-167-3
No. 69	Polarization Phenomena in Nuclear Physics - 1980 (Fifth International Symposium, Santa Fe)	81-65107	0-88318-168-1
No. 70	Chemistry and Physics of Coal Utilization - 1980 (APS, Morgantown)	81-65106	0-88318-169-X
No. 71	Group Theory and its Applications in Physics - 1980 (Latin American School of Physics, Mexico City)	81-66132	0-88318-170-3
No. 72	Weak Interactions as a Probe of Unification (Virginia Polytechnic Institute - 1980)	81-67184	0-88318-171-1
No. 73	Tetrahedrally Bonded Amorphous Semiconductors (Carefree, Arizona, 1981)	81-67419	0-88318-172-X
No. 74	Perturbative Quantum Chromodynamics (Tallahassee, 1981)	81-70372	0-88318-173-8

No. 75	Low Energy X-ray Diagnostics-1981 (Monterey)	81-69841	0-88318-174-6
No. 76	Nonlinear Properties of Internal Waves (La Jolla Institute, 1981)	81-71062	0-88318-175-4
No. 77	Gamma Ray Transients and Related Astrophysical Phenomena (La Jolla Institute, 1981)	81-71543	0-88318-176-2
No. 78	Shock Waves in Condensed Matter - 1981 (Menlo Park)	82-70014	0-88318-177-0
No. 79	Pion Production and Absorption in Nuclei - 1981 (Indiana University Cyclotron Facility)	82-70678	0-88318-178-9
No. 80	Polarized Proton Ion Sources (Ann Arbor, 1981)	82-71025	0-88318-179-7
No. 81	Particles and Fields - 1981: Testing the Standard Model (APS/DPF, Santa Cruz)	82-71156	0-88318-180-0
No. 82	Interpretation of Climate and Photochemical Models, Ozone and Temperature Measurements (La Jolla Institute, 1981)	82-071345	0-88318-181-9
No. 83	The Galactic Center (Cal. Inst. of Tech., 1982)	82-071635	0-88318-182-7
No. 84	Physics in the Steel Industry (APS.AISI, Lehigh University, 1981)	82-072033	0-88318-183-5
No. 85	Proton-Antiproton Collider Physics - 1981 (Madison, Wisconsin)	82-072141	0-88318-184-3
No. 86	Momentum Wave Functions - 1982 (Adelaide, Australia)	82-072375	0-88318-185-1
No. 87	Physics of High Energy Particle Accelerators (Fermilab Summer School, 1981)	82-072421	0-88318-186-X
No. 88	Mathematical Methods in Hydrodynamics and Integrability in Dynamical Systems (La Jolla Institute, 1981)	82-072462	0-88318-187-8